Molecular foundations of drug–receptor interactio

Molecular foundations of

drug–receptor interaction

P.M.DEAN

Department of Pharmacology, University of Cambridge

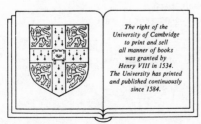

The right of the
University of Cambridge
to print and sell
all manner of books
was granted by
Henry VIII in 1534.
The University has printed
and published continuously
since 1584.

CAMBRIDGE UNIVERSITY PRESS

Cambridge

New York   Port Chester

Melbourne   Sydney

Published by the Press Syndicate of the University of Cambridge
The Pitt Building, Trumpington Street, Cambridge CB2 1RP
40 West 20th Street, New York, NY 10011, USA
10 Stamford Road, Oakleigh, Melbourne 3166, Australia

First published 1987
Reprinted 1989

Printed in Great Britain at the University Press, Cambridge

British Library Cataloguing in Publication Data
Dean, P. M.
Molecular foundations of drug-receptor
interaction.
1. Drugs – Structure-activity relationships
I. Title
615′.7   RM301.42

Library of Congress Cataloguing in Publication Data
Dean, P. M. (Philip Michael), 1944–
Molecular foundations of drug-receptor interaction.
Bibliography:
Includes indexes.
1. Molecular pharmacology. 2. Drug receptors.
I. Title
RM301.65.D43   1987   615′.7   86-28381

ISBN 0 521 30255 2

MP

CONTENTS

Preface   xi

1   The development of theories about drug-receptor interaction   1
1.1   The concept of receptors   2
    1.1.1. J. N. Langley   2
    1.1.2. P. Ehrlich   3
    1.1.3. Langley again   6
1.2   Early chemical theories of drug–receptor interaction   7
    1.2.1. A. V. Hill   7
    1.2.2. A. J. Clark and occupation theory   8
1.3   Antagonism   11
1.4   Biological activity   13
    1.4.1. Intrinsic activity   13
    1.4.2. Efficacy and stimulus   15
    1.4.3. Spare receptors   18
    1.4.4. Intrinsic efficacy   19
1.5   Stimulus–response relationships   20
    1.5.1. Stimulus chains and functional interactions   20
    1.5.2. Transducer models for receptor–effector coupling   22
    1.5.3. Transducer coupling between receptors and ion channels 25
    1.5.4. Transducer coupling and functional antagonism   27
    1.5.5. Ion gating   27
1.6   Cooperativity   29
    1.6.1. Negative cooperativity   30
    1.6.2. Positive cooperativity at the nicotinic acetylcholine
    receptor (nAchR)   30
1.7   Molecular mechanisms in drug–receptor interaction   31
    1.7.1. Molecular factors that determine the formation of the
    drug–receptor complex   32
1.8   Drug–receptor interaction and computational chemistry   33

2   Structural geometry of molecules   35
2.1   Why do we need precise geometrical information?   35
2.2   Analytic geometry and matrix transformation   36
    2.2.1. Molecular coordinate systems   36
    2.2.2. Elementary relationships in analytic geometry   38
    2.2.3. Elementary matrix transformation   40
2.3   Crystallographic terms   42
    2.3.1. The unit cell   42
    2.3.2. Crystal and orthogonal axes   45
    2.3.3. Structure determination   47
    2.3.4. Neutron diffraction   51
2.4   The use of crystal data   51
2.5   Databases of molecular structures   57
    2.5.1. Cambridge Structural Database (CSD)   58
    2.5.2. Computer analysis of data files of molecular geometry 59
    2.5.3. The Brookhaven Protein Data Bank   64
2.6   Protein structures   66
    2.6.1. Chemical composition   66
    2.6.2. Protein crystals for x-ray diffraction   67
    2.6.3. Electron density maps   67
    2.6.4. Structure refinement   69

3   Intra- and intermolecular forces   70
3.1   Methods to calculate different forces   73
    3.1.1. The electron distribution   73
    3.1.2. Electrostatic energy   76
    3.1.3. Induction energy   78
    3.1.4. Dispersion energy   79
    3.1.5. Short-range energies   80
    3.1.6. Interaction energy calculations   81
    3.1.7. Atom–atom pair-potentials   81
3.2   Molecular dynamics   83
3.3   Intramolecular forces   87
3.4   Hydrogen bonds in drug–receptor interaction   91
    3.4.1. Geometries of hydrogen bonds   92
3.5   Molecular electrostatic potentials   96
    3.5.1. Calculation of the electrostatic potential   96
    3.5.2. Display of molecular electrostatic potentials   97
    3.5.3. The use of molecular electrostatic potentials   99
3.6   Hydrophobic interactions   103
    3.6.1. Evaluation of hydrophobicity   104
    3.6.2. Molecular shape and hydrophobic interactions   106

4   Characterization of molecular shape   109
4.1   Electron density distribution and shape   109

4.2	Distance matrices (DMs) and data manipulation	111
	4.2.1. Animated DMs in molecular dynamics	112
	4.2.2. Molecular superposition	115
4.3	Molecular surfaces	118
	4.3.1. Geometrical complementarity between molecular surfaces	121
4.4	Intramolecular flexibility and conformation	122
	4.4.1. Conformational energy computations	123
	4.4.2. Conformation of small peptide units	126
	4.4.3. Graphical representation of multiple torsion angles	128
4.5	Macromolecular receptor structures	130
	4.5.1. Classification of protein structure	130
	4.5.2. Protein structure correlations with function	134
	4.5.3. Nucleic acid structure and conformation	134
4.6	Prediction of polypeptide structure from amino acid sequence	139
4.7	Dynamic conformational changes in proteins	142
	4.7.1. Atom mobility in proteins	145
4.8	Software systems for shape display	147
	4.8.1. Molecular structure	147
	4.8.2. Accessible surface representation	148
	4.8.3. Graphical display of quantum-mechanical properties on molecular surfaces	149
5	**Ligand-binding sites**	150
5.1	Identification of receptors	150
5.2	Identification of binding sites for drug molecules	153
	5.2.1. DHFR	153
	5.2.2. The active site	157
	5.2.3. Site-points for binding NADPH	158
	5.2.4. Site-points for methotrexate	161
	5.2.5. Substrate binding	162
	5.2.6. A structural mechanism for catalysis	162
5.3	Geometrical searching for binding sites	165
5.4	Identification of latent binding sites	169
5.5	Subsets of site-points	176
	5.5.1. Coenzyme binding to DHFR	177
	5.5.2. Inhibitor binding to DHFR	179
5.6	Topographical mapping of ligand-binding sites by inference methods	185
6	**The importance of solvent in drug–receptor interactions**	188
6.1	The structure of water	189
	6.1.1. Liquid water	189
	6.1.2. Water structure at a macromolecular surface	191

6.2   Organized water in crystal structures   193
       6.2.1. Water surrounding small molecules   194
       6.2.2. Water at macromolecular surfaces   197
       6.2.3. Water surrounding drug–receptor complexes   201
6.3   Simulation of the solvent environment   205
       6.3.1. The supermolecule approach to hydration   206
       6.3.2. Water isoenergy contour maps   208
       6.3.3. Monte Carlo simulations of hydration   211
       6.3.4. Simulation of solvation by molecular dynamics   216
6.4   Hydrophilicity and hydrophobicity   221
       6.4.1. Hydropathy   221
       6.4.2. Hydropathy and tertiary structure of proteins   223
       6.4.3. Hydrophobic moments   225
6.5   The effect of water on intermolecular interactions   227
       6.5.1. Dielectric theory   228
       6.5.2. Water as a dielectric medium   230
       6.5.3. Ion interactions in water   231

7   **Ligand docking at a binding site**   236
7.1   Mechanical flexibility and the formation of intercalative sites   237
7.2   Molecular energy changes in the unstacking of adjacent base
       pairs   241
       7.2.1. Interactions between the backbone and base pairs   241
       7.2.2. Base-pair interactions in an electric field   243
7.3   Molecular electrostatic potentials   247
       7.3.1. Calculation of electronic charge distributions   247
       7.3.2. Molecular electrostatic potential surrounding (dC-dG)·
       (dC-dG)   251
       7.3.3. Molecular electrostatic potentials surrounding
       polynucleotides   252
       7.3.4. Electrostatic complementarity   254
       7.3.5. Perturbation of the receptor electrostatic potential by a
       drug molecule   258
7.4   Receptor-induced ligand orientation   258
       7.4.1. Roll rotation   260
       7.4.2. Yaw rotation   261
       7.4.3. Pitch rotation   262
7.5   Electric fields and ligand orientation   263
       7.5.1. The electric field   263
       7.5.2. Direction fields surrounding base pairs of nucleic acids   264
       7.5.3. Direction fields along the helix axis of nucleic acid   266
       7.5.4. Ligand orientation in the receptor's electric field   270
7.6   Stability of the drug–receptor complex   275
       7.6.1. Interaction energies for drug–dinucleotide complexes   276

7.6.2. Drug-induced conformational changes in the dinucleotide
receptor 278
7.7 Intramolecular changes during docking 279
7.7.1. Modification of the electron distribution in the drug
molecule 280
7.7.2. Changes in orbital energies in the drug molecule 281
7.7.3. Intramolecular changes in substrate docking at an active
site 283
7.8 Software for docking interactions 285

8 Ligand design techniques 287
8.1 The problem of design 287
8.1.1. The biochemical lead 288
8.1.2. Combinatorial modifications to a lead structure 289
8.1.3. Combinatorial reduction by feature selection 290
8.1.4. Emerging pathways for rational design of ligands 291
8.2 Ligand design to fit a known structural site 292
8.2.1. The diphosphoglycerate binding site on haemoglobin 292
8.2.2. Construction of a ligand to fit the DPG site 294
8.2.3. Testing the putative ligands for biological activity 295
8.2.4. Ligand binding at genetically modified sites 298
8.3 Ligand design in the absence of information about site structure 300
8.3.1. Chemometrics 300
8.3.2. Multivariate analysis 301
8.3.3. Multivariate regression analysis 303
8.3.4. Principal component analysis 303
8.3.5. Descriptors of molecular structure 306
8.3.6. Molecular descriptors in linear free enthalpy relationships 308
8.3.7. Optimization of molecular design from chemometric data 313
8.4 Computer-aided ligand design 316
8.4.1. Representation of receptor structure 317
8.4.2. Construction of a hypothetical ligand 318
8.4.3. Molecular graphical displays 321
8.4.4. Computer-aided ligand construction 323
8.4.5. Computer-aided QSAR 323
8.4.6. Inhibitors of angiotensin converting enzyme (ACE) 324

9 Future studies in drug–receptor interaction 328
9.1 Protein engineering 328
9.1.1. Site-directed mutagenesis 328
9.1.2. Protein engineering and macromolecular recognition 331
9.1.3. Modifications to the acetylcholine receptor by site-directed
mutagenesis 333
9.1.4. Redesigning binding sites for structure-function studies 334

9.2 Knowledge engineering   337
    9.2.1. Drug design using computer-aided synthesis   340
    9.2.2. Synthesis planning   341
    9.2.3. Logic and heuristics applied to synthetic analysis (LHASA)   341
    9.2.4. Artificial intelligence in QSAR   344
    9.2.5. Representation of molecular structure by computer codes   345
    9.2.6. Automatic selection of structural descriptors for QSAR   346
    9.2.7. Automatic allocation of hydrogen-bonding site-points to macromolecular surfaces   348
9.3 Computations for drug–receptor interaction by supercomputers   348
    9.3.1. Parallel processing   349
    9.3.2. Distributed array processors (DAP)   350
    9.3.3. Molecular dynamics calculations on DAPs   351

Postscript   353

References   355

Subject index   372

Author index   377

PREFACE

In the beginning, before *Molecular Foundations* was formed, a meeting with the Biological Sciences Editor of Cambridge University Press was held to discuss possible approaches to the subject. The editor explained that she did not require either another standard text on molecular pharmacology or a catalogue of facts and chemical formulae compiled into a massive scientific 'Bible'.

'We are in the computer age now and it is time there was a book dealing with molecular pharmacology to reflect this', she finally stated.

The principal objective of this book, therefore, has been to draw together those strands of research, carried out since 1970, that widen our understanding of the molecular details of drug–receptor interaction. Hopefully, the book is more substantial than an introduction, but it is not an exhaustive treatment of the subject. Only what I consider to be primary material is surveyed here; much detailed background literature has not been handled. Technical items can be found by returning to the original papers cited in the references. Part of this material has been used in Part 2 of the Natural Sciences Tripos (Pharmacology) and so could be useful as an advanced undergraduate textbook. However, most of the book is of more specialist interest and should appeal to research scientists working in the wider field of biomolecular interactions.

One of the major difficulties in writing a book to the brief that it must be modern and up to date, lies in the limitation that the research one has to consider is of necessity novel, speculative and largely untested. The writer has to run the risk, if he includes contemporary material, of drawing attention to work which may, at a later date, prove to be unreliable. Nevertheless, research is a restless occupation and there is much to be said for propagating the stimulus of new ideas to provoke further thought, so that out of the melting pot of controversy these ideas can be refined to stand the test of time.

The main thrust of this book is to shift the emphasis away from phenomenological experiments of pharmacology to molecular theory, that is, to move to the place where we can ask questions that are too difficult, or impossible, to answer by conventional experiments. At this level of questioning we can only construct hypotheses based on molecular theory and provide computational simulations to test our ideas. For example, we cannot measure the molecular electrostatic potential in the cleft of a drug binding site but, given the atomic charges obtained from a quantum chemistry calculation, we can compute the potential and make some deductions from those computations about drug action and specificity. How could we test our prediction of the value of the potential? In this brave new world of computational chemistry there are many pitfalls to trip the unwary. Rash generalizations from limited data can be very tempting when most calculations deal only with simulating the *in vacuo* state and neglect the solvent environment. It is easy to get drunk on enthusiasm about new developments, to be blind to mistakes and to be hypnotized by clever speculation.

The underlying assumption of this book is that it is possible to treat drug–receptor interaction purely as a chemical problem. This problem lies at the heart of pharmacology, but many pharmacologists seem unaware that it is surreptitiously being stolen from them by scientists from other specialist disciplines. A careful perusal of the references at the end of this book reveals that only a small number of pharmacological journals are quoted. Many of the important recent advances are to be found in literature that few pharmacologists would regularly scan. This is due, in part, to the fact that the overwhelming mass of research in pharmacology is, naturally, concerned with how particular drugs act on the body's physiological systems; there seems to be little commitment to understanding what is happening at the molecular level. Current awareness of recent developments in drug–receptor interaction is essential if the molecular nature of the problem is to remain accessible to us as the theoretical basis of pharmacology.

This book is limited to one small step in the chain of events of drug action. Only the very first stage is considered, namely the initial interaction between the drug molecule and its receptor site to form a molecular complex. Biochemical and biophysical consequences that follow from this step are not described since very little is, as yet, clearly understood about them. The initial events can loosely be described as molecular recognition. If recognition can be understood in the precise language of intermolecular forces, there will undoubtedly be important applications of molecular theory to the design of new drug molecules. This theme of recognition forms a constant undercurrent throughout the book and its link to the rational design of novel ligands is frequently alluded to.

One of the great advantages that an author has when commissioned to write a book is that he is forced to sit down and read through the earlier literature to trace the historical development of his subject. It seems to me that one paper is pre-eminent in laying the molecular foundations of drug–receptor interaction. Sir Arnold Burgen's (1966) concise essay on 'The drug–receptor complex' is almost a manifesto for subsequent investigations, the results of which are now outlined here. What is surprising is that his paper was written before the application of computational chemistry to biomolecular interactions. These new techniques have, since then, simply filled in some of the details of the overall plan.

My own research interests are heavily invested in three chapters written here and so reflect the importance that I personally attach to those areas. Close involvement with one part of a topic may have led me to overemphasize it with respect to other related viewpoints. It is extremely difficult, when one is intimately involved with a subject, to stand at some vantage point and view the whole field impartially; thus I am aware of my own particular perspectives.

Chapter 1 outlines the historical development of theories about drug–receptor interaction beginning with the birth of the receptor concept by Langley and Ehrlich. Major steps in understanding drug action are highlighted over a period of nearly 100 years. This chapter will be familiar to classically trained pharmacologists, but I believe there is little mileage left in this approach which is essentially one of empirically fitting equations to gross measurements of tissue responses. The rest of the book concentrates on molecular details. There is a strong emphasis throughout the text on the need to understand and manipulate three-dimensional molecular structure. Chapter 2 provides some basic mathematical and crystallographic methods for handling molecular architectures. Work in the early 1980s on molecular databases is also outlined to provide the reader with ready access to these new and important facilities. Structural geometry gives rise to different spatial dispositions of intermolecular forces. Chapter 3 examines these forces and deals in detail with some of those that are believed to be crucial in drug–receptor interaction. Quantum chemistry is not dealt with in this book; an excellent treatment of it, and its application to molecular pharmacology, can be found elsewhere (Richards, 1983). If molecular shapes are to be handled by computer programs, methods must be developed to characterize and handle shape in any form (chapter 4). Only certain portions of the molecular receptor are concerned with ligand binding. Different ways to examine these sites are explored in chapter 5. Most computational models for drug–receptor interaction ignore the solvent environment. This neglect is a substantial defect in current models. The behaviour of solvent molecules over small intermolecular distances is not well understood. What inroads have been made into this complex problem in the physical sciences (chapter 6)? The docking manoeuvre of a drug molecule into its receptor site is illustrated in chapter 7 where emphasis is placed on electrostatic interactions. Practical applications of computational chemistry to the rational design of new drugs are outlined in chapter 8. The final chapter attempts to predict the major avenues along which drug–receptor interaction will progress. Three important developments are already surfacing: protein engineering offers the pharmacologist the ability to re-design the receptor; knowledge engineering draws together key strands from artificial intelligence that are making it possible for new comprehensive and automated techniques to emerge for drug design; new computer hardware will, if the fifth generation project is successful, provide sufficient computing power to make molecular simulations significantly more complex and bring the research on drug–receptor interaction nearer to completeness.

I would like to acknowledge with gratitude the encouragement and financial support given to me over many years by the Wellcome Trust. Without the Trust's generosity it would not have been possible for me to pursue a research career largely unfettered by a heavy academic teaching load. Thanks go to colleagues in the department, and elsewhere in Cambridge, for providing stimulus and a lively environment for the exchange of new ideas. Lastly, but not least, I am indebted to Diana for patiently deciphering my writing for input to the word-processor and then nit-picking through the manuscript to remove many grammatical obscurities.

Cambridge, May 1986.   P. M. Dean

ACKNOWLEDGEMENTS

I would like to thank the following for permission to publish copyright material: The American Chemical Society and the authors for permission to reproduce tables 1 and 2 from *Biochemistry* (1980) **19**, 3732; tables 2 and 3 from *Biochemistry* (1980) **19**, 3723; table 3 from *Biochemistry* **23**, 4733; figures 2b, 3b and tables 4, 5 and 6 from *Journal of the American Chemical Society* (1979) **101**, 825; Macmillan Press and the authors for permission to reproduce tables 1 and 2 from *The British Journal of Pharmacology* (1979) **65**, 535; table 1 from *Nature* (1985) **313**, 364; Oxford University Press and authors for permission to reproduce table 1 from *X-ray Crystallography and Drug Action* (1984) edited by A. Horn and C. de Ranter, p. 169: International Union of Crystallography and the authors for permission to reproduce tables 3, 4b and 4c from *Acta Crystallographica* (1978) **B34**, 2534; table 2 from *Acta Crystallographica* (1982) **B38**, 2516: Pharmaceutical Society of Japan and the authors for permission to reproduce figures 2a, 2b, 3a and 3b from the *Chemical and Pharmaceutical Bulletin* (1984) **32**, 3313: Cold Spring Harbor Symposia and the author for permission to reproduce table 1 from *Cold Spring Harbor Symposium* (1983) **47**, 251: American Society of Biological Chemists and authors for permission to reproduce figure 6 and table 3 from the *Journal of Biological Chemistry* (1982) **257**, 13650; figure 2 and tables 2, 3 and 4 from the *Journal of Biological Chemistry* (1982) **257**, 13663: Academic Press and authors for permission to reproduce table 3 from the *Journal of Molecular Biology* (1983) **168**, 621; table 2 from the *Journal of Molecular Biology* (1982) **157**, 105.

1

The development of theories about drug–receptor interaction

The reading of seminal papers in any scientific discipline, in the cool silence of a good library, is akin to the experience of visiting the cellars of a well-established wine-merchant. The pleasure of tasting the classed growths of Bordeaux in sequence through the years is punctuated at definite and widely recognized intervals by the great vintages. At these points in history the wines are remarkable for their intensity, complexity and finish; they manage to stand the test of time without deterioration. So with drug–receptor interaction, there are extraordinary periods of intellectual growth within the discipline, landmarks in time where developments have held together despite the changing tides of fashion.

For the pharmacologist, drug–receptor interaction is the very heart of his subject; yet despite the centrality of this theory to the practical development of new drugs, surprisingly little is known about the process. Only in the last decade or so, have we been able to characterize the molecular structures of a few receptors. Naturally, the pharmacologist's understanding has lagged behind the evolution of chemical theories of molecular reactions. Only with the advent of high-speed computers has it been possible to apply the rigours of computational chemistry to molecular biological problems. Recent progress in these different research areas has led to a flurry of new work; chemical theories inaccessible to experiment can be evaluated by careful computational simulation. This novel work forms the subject matter of the book. However, we must begin by placing these new ideas within an historical perspective. Ideas do not just happen in isolation, they emerge from a fertile milieu through a process of accretion. Ehrlich described this process as 'a host of individual facts which, being stored at a subconscious level, lead one to take involuntarily the right direction'. (Dale, 1956).

Inevitably, whenever a review of drug–receptor interaction is contemplated the discussion revolves around the axis of Paul Ehrlich. His astonishing abilities, sheer hard work, and prophetic vision have contributed more to this subject than that of any other worker. In a key paper on structure–activity relationships which Ehrlich published in 1902 he expressed his hopes, his fears and his rationale in these words: 'Hence the expectation to be able to construct new drugs of predetermined action on the basis of theoretical conceptions will probably have to be deferred for a long time. To the initiate, the lack of sufficient positive knowledge is revealed by the inactivity which now characterizes a field once entered upon with so much promise. The innumerable drugs which have overwhelmed medicine in the past few years, of which only a few are of any value and thus denote any real progress, have sufficed speedily to allay the original enthusiasm. A feeling of indifference has thus been engendered, which is constantly being increased by the advertisements which are daily becoming more and more evident. Apart from these evils, however, this line of study is at present suffering especially from two other evils:

(1) The habit, when a drug has been partially accepted, of immediately following it with a dozen rivals of similar composition, and,

(2) The exclusive preference given to drugs acting purely symptomatically, which are not true curative agents.

A change for the better will occur only when purely biological points of view are adopted, ie if the initiative is transferred from the chemical to the biological laboratory. As physicians we must cease to be content with the auxiliary role of advisers in these important questions. In this subject, our very own since time immemorial, we must insist on taking first place. Now is the time that we must turn to more general, biological conceptions, and it is therefore the duty of everyone to contribute his brick to the construction of this new theory.' The cornerstone was destined to be laid by Ehrlich with his notion of receptors.

1.1   The concept of receptors
1.1.1   J. N. Langley

The notion of specific receptors for drug molecules or natural neurotransmitters developed slowly over the course of about 20 years. Two strands in the research can be detected. J. N. Langley, working in Cambridge under the influence of Michael Foster, began his studies of the nervous system linked initially to the mechanism of secretion. His early studies were on the action of the recently discovered cholinomimetic compound jaborandi. The drug stimulated salivary secretion and could be antagonized by atropine (Langley, 1873, 1878). In the discussion of his

1878 paper Langley speculated 'Until some definite conclusion as to the point of action of the poisons is arrived at it is not worth while to theorize much on their mode of action: but we may, I think, without much rashness, assume that there is some substance or substances in the nerve endings or gland cells with which both atropine and pilocarpin are capable of forming compounds. On this assumption then the atropine or pilocarpin compounds are formed according to some law of which their relative mass and chemical affinity for the substance are factors. In the analogous case with inorganic substances, other things being equal, these are the sole factors. To take the simplest case, if a and b are both able to form, with y, the compounds ay and by, then ay and by are both formed, the quantity of ay and by depending on the relative masses of a and b present and their relative chemical affinity to y.' Receptors, antagonism, chemical affinity, drug–receptor complex, all the ingredients for receptor theory are here, but they lay dormant for many years. Meanwhile, the results of these initial findings led to a number of papers on antagonism of various poisons by physiologically active alkaloids (Langley, 1880; Langley & Dickinson, 1890a, b). This research pathway focused on the nerve endings in ganglia, the neuromuscular junction and neural control of secretion.

1.1.2 P. Ehrlich

Ehrlich's contribution forms the second strand and was independent of Langley's work; it approached the notion of receptors from a different direction. An early training in the chemistry of dyestuffs provided Ehrlich with a great facility for research in staining biological tissues and he made numerous important contributions to histology. One observation in histology that fascinated Ehrlich was the fact that staining could be specific for a particular tissue. Furthermore, he noted that substances which stain neuronal tissues are lipid soluble. If the lipid solubility of the chromophore group is changed, for example by making a sulphonic acid derivative, although the colour remains the dye is not taken up by the tissue and the biological material is not stained. This simple observation led to the idea that different chemical structures could show different biological distributions and the distribution of a chromophore could be changed by modifying the molecular structure. The possibility that dyestuffs could combine with specific parts of biological tissue was not appreciated for many years because there were many competing theories about the staining process.

Ehrlich's work with dyestuffs had bearing on another idea that was popular at the end of the nineteenth century; this was the emerging notion of affinity. It had been observed that many of the biologically active alkaloids combine very tightly with tissues. In fact there was a theory that the molecules were synthetically linked to the tissue. High affinity and high toxicity often appeared hand in hand. What was not understood was why these neuroactive alkaloids, that Langley had worked with, had such dramatically and specifically different actions. Ehrlich (1902) pointed out that the action of the drugs was often transitory; furthermore, some of the specific dyestuffs could be washed out of the tissues. These observations provided evidence that these molecules were not synthetically combined to the tissues but were bound reversibly with high affinity.

A distinction began to be drawn between molecules that could be incorporated in tissues and those which could exert a specific biological effect but did not form part of the cell protoplasm. For example, nutrients were observed to be metabolized in cells and assimilated in a converted form into cellular protoplasm. Other substances, like histological stains, which did not act as nutrients even though they could enter the cell, were not necessarily metabolized and could be washed out. Thus an idea was forming that substances which act on cells fall into two groups: nutrients which are incorporated into the protoplasm and non-nutrients such as alkaloids, drugs and dyes which, although they may possess activity, can be released from the cell unchanged. These notions may to us seem self-evident, but at the end of the nineteenth century they were not understood and for some years were matters of great controversy.

Another avenue of research to which Ehrlich had made substantial contributions was immunology. These studies began in 1891 and progressed steadily to yield a theory of immune reactions. This work received a wide publication in Ehrlich's Croonian lecture (1900). From a series of experiments on diphtheria toxin, Ehrlich postulated that toxin molecules possessed two different combining groups: a haptophore that could bind to the antitoxin and a toxophore that caused the toxic action. These groups on the toxin were physically separated since the toxophore could be destroyed by special treatment without affecting the haptophore. Cells possessed what were termed toxophile groups which were capable of combining with the toxin. These toxophile groups were not believed to be present specifically for binding to toxins but to have other biological functions. It was thought to be simply a matter of probability that toxophiles possessed molecular structures capable of binding to a toxophore. Thus we can see that Ehrlich was beginning to formulate the concept of specific surface groups on cells that could interact with toxin molecules showing specific binding properties. These binding groups at the surface could be visualized as molecular side-chains (figure 1.1). Each haptophore side-chain has a structure that is related to that of the toxin rather like that of a male and female screw, or as a lock and key. The toxic properties of the toxin are manifest when the side-chain of the toxophile is occupied by the toxin and can thus no longer carry out its normal function.

The side-chain theory of Ehrlich was renamed a little later by Ehrlich & Morgenroth (1900) in this quotation from their famous paper: 'For the sake of brevity, that combining group of the protoplasmic molecule to which the introduced group is anchored will hereafter be termed *receptor*. The side-chain, for example, which combined with the tetanus toxin in the organism is such a receptor.' Six brilliant papers on haemolysins by Ehrlich & Morgenroth proved the value of the receptor concept in understanding the mechanism of haemolysis in the immune reaction; different types of receptor for lysis were discovered. The fifth paper of this series (Ehrlich & Morgenroth, 1901) compared the association of an immune body with its receptor as a purely chemical phenomenon, in many respects similar to the reaction of diazobenzaldehyde and its congeners with cells. For the first time, rudimentary algebra was used to link the number of receptors occupied with the lytic response of the haemolysin. This algebraic notation although proposed by Ehrlich & Morgenroth, was not followed up by them; pharmacology had to wait another 25 years before it resurfaced again.

All this histochemical and immunological data was stacked in Ehrlich's subconscious mind. An interest in chemotherapy was beginning to take

Figure 1.1. Ehrlich's side-chain theory. (*a*) Cell receptor protruding from the cell surface as a side-chain; (*b*) toxin molecule with its haptophore group (shaded) and the toxophore group; (*c*) toxin molecule adsorbed onto the molecular side-chain.

(*a*)   (*b*)   (*c*)

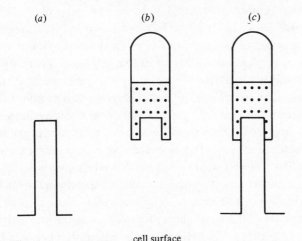

cell surface

a hold on him. All that he now had to do was make the right connections. His paper in 1902 linked the ideas together and the concept of drug–receptor interaction was born. Furthermore, the observation of the pharmacological properties of a series of cocaine derivatives indicated to Ehrlich that congeneric drug molecules might contain a similar skeletal molecular structure on which toxophore groups could be hung to modify the properties of the drug molecule. His hopes are conveyed in the quotation: 'If only we had a deeper understanding of this function we might hope, by means of substitutive action on the toxophore groups, to modify the action of the alkaloids to suit our purpose.' This was a new concept for the rational design of novel drugs. From that moment Ehrlich spent the rest of his life putting the concept – or, as it was popularly known, the magic bullet hypothesis – into chemotherapeutic practice. An outline of the developments in his own research that led to the concept of receptors is given in his Nobel lecture (Ehrlich, 1909a); but this was not the zenith of Ehrlich's endeavours. The greatest achievement of his life was the synthesis of the organic arsenical salvarsan (Ehrlich, 1909b). This was the first effective chemotherapeutic agent against syphilis. His whole research effort had led him inexorably along this pathway. He was able to put into practice all that had been discovered about receptors; nevertheless the search for salvarsan was tedious and the appropriate molecular structure was discovered only at the 606th attempt.

1.1.3 Langley again

Ehrlich laid the foundations of drug–receptor theory. However, we need to return back to Cambridge and to John Langley to see how the concept of receptors began to be incorporated into the pharmacology of the nervous system. Undoubtedly Langley would have read of Ehrlich's receptor work in the Croonian Lecture; but in his own lecture (Langley, 1906) there is no mention of Ehrlich. Langley put forward the view that at the nerve ending there were substances (receptive substances) capable of receiving or transmitting stimuli. These receptive substances at the terminal region could be stimulated or paralyzed by various poisons. The receptive substances were located in ganglia and at the nerve endings on muscle. Langley then went on to produce a series of papers providing experimental evidence in support of specific receptive substances on muscle (Langley, 1907, 1908a, 1909). Curiously, in the 1908a paper Langley suggested that the contraction of frog skeletal muscle depends on the rate of combination of nicotine with the receptive substance; this concept was to emerge 50 years later in Paton's rate theory for drug–receptor interaction. Only in the paper of 1908b did Langley make the connection between his own research on 'receptive' substances on muscles with Ehrlich's theory of immunity. Thus Ehrlich's side-chain theory became linked with drug action and normal function of the nervous system. The alkaloid nicotine could form a dissociable complex with the skeletal muscle receptor (Langley, 1909); this finding echoes Ehrlich's belief that drug combination with receptors could be non-synthetic and rather like an ion-dissociation reaction. Langley believed, however, that there were differences between his theory and Ehrlich's side-chain hypothesis in that, on muscle the receptive substances had a clearly defined physiological function whereas Ehrlich's side-chains responsible for binding arsenic compounds did not appear to be related to specific functional receptive substances.

1.2 Early chemical theories of drug–receptor interaction

The pioneering work of Ehrlich and Langley gave birth to the concept of receptors. However, the concept was only a tentative one; what was needed was stronger evidence that drug molecules could interact chemically, but not synthetically, with these receptor substances. Obviously this whole question of the molecular properties of the receptors was at the back of Langley's mind: how could their features be deduced through experiment?

1.2.1 A. V. Hill

A. V. Hill, a gifted student of Langley, took up the challenge. Mathematical and physicochemical methods were employed to analyse the time course of the contractile response of frog *rectus abdominus* muscle to nicotine and curare (Hill, 1909). The contraction of the muscle to a single dose of nicotine could be described by the equation

$$y = k\{1 - \exp(-\lambda t)\} \tag{1.1}$$

where y is the height of contraction, t is the time measured from the start of the contraction and λ is a constant. On washout of nicotine, the relaxation of the muscle obeyed the equation

$$y = k \exp(-\lambda t) \tag{1.2}$$

Two explanations for the equations were possible: (*a*) the diffusion of nicotine, (*b*) the combination of nicotine with the receptor. These two possibilities could be distinguished by determining the temperature coefficients of the contractile response. The equation to calculate the temperature coefficient μ is

$$V_2 = V_1 \exp\left\{\frac{\mu}{2}\left(\frac{T_2 - T_1}{T_2 T_1}\right)\right\} \tag{1.3}$$

where V_2 and V_1 are the velocities of contraction at temperatures T_2 and T_1. Experimental determinations of μ gave a value of 17 340 for nicotine interacting with frog muscle. On the other hand, the temperature coefficient for diffusion is only 289. Thus it would appear that the kinetics of the contraction of muscle by nicotine are not described by diffusion; the temperature coefficient is too large. Similarly, Hill found that curare antagonized the action of nicotine with a large temperature coefficient and so provided evidence that antagonism of drug action also occurs by chemical interactions with receptors.

1.2.2   *A. J. Clark and occupation theory*

Significant developments in the theory of drug–receptor interaction were sparse until A. J. Clark (1926a) made a quantitative study of the action of acetylcholine on isolated cardiac ventricle muscle and the frog *rectus abdominus*. The time course of drug action confirmed the observations of Hill (1909). A new development was a study of the dose–response relationship. If the response at a particular dose of drug was plotted as a percentage of the maximum response against the $\log_{10}$ molar concentration of acetylcholine, then the relationship was sigmoidal (figure 1.2). The relationship can be expressed as

$$Kx = y/(100 - y) \tag{1.4}$$

where x is the molar concentration of drug, y is the response as a percentage

Figure 1.2. The relationship between the isotonic contractile response of frog *rectus abdominus* muscle to acetylcholine. Response is plotted as a percentage of the maximum response attainable against $\log_{10}$ concentration of acetylcholine. (Re-drawn from Clark 1926a.)

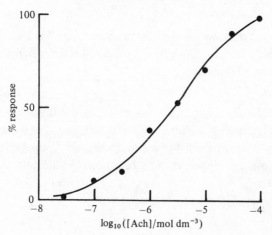

of the maximum possible response, K is a constant. If $y = 50\%$, $x = 1/K$. If $\log_{10} x$ is plotted against $\log_{10}\{y/(100-y)\}$ a linear relationship can be produced (figure 1.3). Thus the formula

$$\log_{10} K + n\log_{10} x = \log_{10}\{y/(100-y)\} \tag{1.5}$$

can be obtained and the value of K determined experimentally; values for the constant n were found to be about unity.

The quantity of drug entering muscle cells could be calculated as 14×10^6 molecules per cell; this value correlated with a 50% response. Only about $20\,000$ molecules per cell are associated with a just detectable response. The surface area of $20\,000$ acetylcholine molecules is about $10^{-10}\,cm^2$ compared with a cell surface area of $2 \times 10^{-5}\,cm^2$. Therefore, the acetylcholine molecules would appear to act on a small portion of the cell surface. Further calculations on the amount of drug entering cells, and the response produced, showed little correlation. Clark therefore assumed that penetration of the cell and the pharmacological response of contractile muscle to acetylcholine are independent of each other.

Atropine antagonizes the action of acetylcholine on frog ventricular and *rectus abdominus* muscle. Clark (1926b) showed that the relationship between $\log_{10}\{y/(100-y)\}$ vs $\log_{10} x$ was shifted in a parallel way by atropine (figure 1.4). The antagonism between atropine and acetylcholine can be re-expressed by finding the concentrations of atropine and acetylcholine required to reduce the response of the heart by 50%. A $\log_{10}$–$\log_{10}$ plot is then a straight line above a threshold concentration of acetylcholine (figure 1.5). In the region where the acetylcholine and

Figure 1.3. Contraction of frog *rectus abdominus* muscle. A plot of $\log_{10}$ dose against $\log_{10}$ response. (Re-drawn from Clark 1926a.)

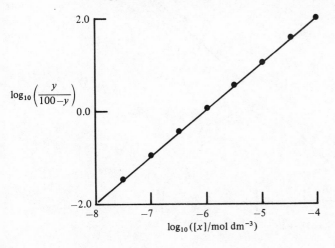

atropine concentrations are above their threshold values, the relationship can be expressed as

$$[Ach]/[Atr] = \text{constant} \tag{1.6}$$

Thus the dependence between response and drug concentration is given by

$$K[Ach]/[Atr] = y/(100 - y) \tag{1.7}$$

Figure 1.4. The antagonism of acetylcholine by atropine on heart muscle. Line (*a*) is the control response in the absence of atropine, lines (*b*) and (*c*) are plotted for increasing concentrations of atropine. (Re-drawn from Clark 1926b.)

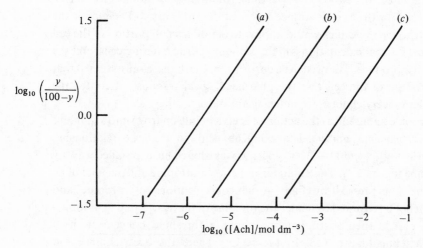

Figure 1.5. Concentrations of acetylcholine and atropine which cause a 50 % reduction in the inotropic contraction of frog ventricles. (Re-drawn from Clark 1926b.)

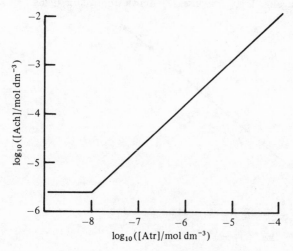

Clark (1933) pointed out that equation (1.4) can be derived from Langmuir's adsorption isotherm. If x is the concentration of drug, A is the total number of receptors and y is the number of receptors occupied, then the rate of combination k_c is

$$k_c = K_1 x(A - y) \tag{1.8}$$

and the rate of dissociation k_d is

$$k_d = K_2 y \tag{1.9}$$

therefore

$$Kx = y/(A - y) \tag{1.10}$$

where $K = K_1/K_2$. Clark assumed that the response of the tissue to an agonist was proportional to the number of receptors occupied by the agonist drug.

This method of studying drug–receptor interaction by measuring a drug response at equilibrium has the advantage that kinetic parameters affecting the response, such as the rate of diffusion onto the receptors and the rate of reaction, are minimized. Thus in an equilibrium state the drug response can be considered to arise from a reaction of drug A with the receptors R that follows the law of mass action to form a complex RA. This reaction is reversible and can be expressed by the simple equations

$$R + A \rightleftharpoons RA; \quad K_A = [RA]/[R][A] \tag{1.11}$$

where K_A is the affinity constant.

One difficulty in relating equation (1.11) to a true mass action equation is the lack of knowledge about the actual concentrations [RA] and [R]; these are unknown, but they are believed to be in a fixed ratio and their units cancel anyway. Nevertheless, despite this difficulty the law of mass action has been assumed to apply to drug–receptor interaction ever since Clark's initial application.

1.3 Antagonism

Gaddum (1936) extended Clark's theory of antagonism by proposing that antagonist drugs act by competing with agonist drugs for receptors. The antagonist simply occupies the receptors without any effect. Suppose that $y\%$ of receptors are occupied by agonist and $z\%$ are occupied by antagonist, then the percentage of free receptors is $(100 - y - z)$. At equilibrium the rate of combination of drug with the free receptors will be equal to the rate of dissociation. Therefore if C_1 and C_2 are the concentrations of agonist and antagonist,

$$K_1 C_1 (100 - y - z) = y$$

and

$$K_2 C_2 (100 - y - z) = z$$

then elimination of z gives

$$K_1 C_1 = (1 + K_2 C_2) y / (100 - y) \qquad (1.12)$$

Since y is directly proportional to response, equation (1.12) can be used to express the relationship between agonists and antagonists in generating the pharmacological response. The solution for the general case is given by

$$K_1 C_1 = (1 + K_2 C_2^n) y / (100 - y) \qquad (1.13)$$

where n is a parameter of value close to unity.

Comparative scales for drug antagonism were developed by Schild (1947, 1949), but these have their origin in a paper by Clark & Raventos (1937). These scales can be formulated for the general case as pA_x defined as the negative logarithm of the molar concentration of antagonist which reduces the effect of a multiple dose, x, of an agonist drug to that of a single dose of agonist in the absence of antagonist. For example, if x is 2 then the pA_2 value is $-\log_{10}$ [antagonist] which reduces the effect of a double agonist dose to that of a single agonist dose.

Following Gaddum, Schild derived the expression for relating the percentage response to drug concentrations as

$$y / (100 - y) = K_1 A \qquad (1.14)$$

therefore

$$y / (100 - y) = K_1 x A / (K_2 B_x + 1) \qquad (1.15)$$

and

$$K_2 B_x = x - 1 \qquad (1.16)$$

where B_x is the concentration of antagonist required to reduce the effect of a concentration of xA to that of the agonist concentration A. If the law of mass action is accepted as a basis for reversible drug combination with the receptors then K_1 and K_2 can be identified as the dissociation constants of the agonist and antagonist. The utility of Schild's pA_x parameter lies in the ability of it to indicate whether antagonists are acting competitively at the same receptor. By the definition of pA_x as

$$pA_x = -\log_{10} B \qquad (1.17)$$

therefore

$$\log_{10} (x - 1) = \log_{10} K_2 - npA_x \qquad (1.18)$$

and a plot of $\log_{10} (x - 1)$ vs pA_x gives a straight line of slope $-n$ and an intercept of $\log_{10} K_2$. Therefore if $n = 1$

$$pA_2 - pA_{10} = \log_{10} 9 \qquad (1.19)$$

and this difference in pA_x value can be used as a test for competitive antagonism (Arunlakshana & Schild, 1959). The method has been useful in determining the affinity constant of the antagonist for the receptor; no assumptions are made about the relationship between response and fraction of receptors occupied.

1.4   Biological activity
        The intriguing problem of what gives a drug its activity when it combines with its receptor was investigated by Raventos (1937, 1938) and Clark & Raventos (1937). These workers studied the cholinergic activity of a number of quaternary ammonium salts; the biological activity varied significantly with molecular structure. Some members of a homologous series showed agonist actions whereas others were found to be antagonists. Moreover, the cholinergic actions varied for different tissues. These tissues could be divided into three groups depending on the pharmacological actions: (*a*) paralysis of motor nerve endings, (*b*) contracture of frog *rectus abdominus*, (*c*) contracture of the gut and inhibition of cardiac tissue. For the series $Me_3N^+ - R$, where R is a linear alkyl chain containing 1–16 carbon atoms, it was noticed that at parasympathetic junctions contractile activity increased up to C_5 and beyond this level the increase in chain length rendered the molecules antagonistic. Atropine, at parasympathetic junctions, inhibited the actions of acetylcholine and Me_3N^+—R derivatives.

1.4.1   Intrinsic activity
        Ariens (1954) noticed that some choline ester derivatives and bisquaternary ammonium compounds showed dual action; they could, depending on the tissue being treated, show agonist or antagonist activity (table 1.1). These observations were similar to those of Raventos (1937) on the alkyl trimethylammonium salts. The simple ideas of competitive inhibition developed by Gaddum (1936) were not sufficient to cope with these phenomena. The concept of intrinsic activity of a drug was developed by Ariens to explain these dual actions; a complete antagonist had an intrinsic activity of 0 and a full agonist had an intrinsic activity of 1; intermediate values could be used to describe partial agonists.
        The Michaelis–Menten theory for enzyme action was applied by Ariens to drug–receptor interaction since the equations can equally be derived from a Langmuir adsorption isotherm. The theory is only applicable to the case of one receptor type and one response. Consider a reversible reaction of a drug A with a receptor R

$$A + R \rightleftharpoons AR \qquad (1.20)$$

Let the total concentration of receptors r_0 be $[R] + [RA]$, then if only a small fraction of A is combined with receptors

$$K_A = (r_0 - [RA])[A]/[RA]; \quad [RA] = r_0[A]/(K_A + [A]) \tag{1.21}$$

where K_A is the dissociation constant. Affinity is the reciprocal of the dissociation constant and determines how much of the drug–receptor complex is formed. The equation for $[RA]$ can be rearranged

$$[RA] = r_0/\{1 + (K_A/[A])\} \tag{1.22}$$

In the Michaelis–Menten equation the velocity, E_A, of the enzyme reaction is proportional to $[RA]$

$$E_A = \alpha[RA] = \alpha r_0/\{1 + (K_A/[A])\} \tag{1.23}$$

where α is a proportionality constant. By analogy with the enzyme reaction, α is the intrinsic activity of the drug for the receptor system. Therefore, the intrinsic activity, α, is a constant, specific to the drug molecule, that determines the biological activity. Two parameters, K_A and α are needed to describe drug action.

The equations can be extended to consider two drugs A and B acting on the same receptor system

$$2R + A + B \rightleftharpoons RA + RB \tag{1.24}$$

The effect E_{AB} is given by

$$E_{AB} = \alpha[RA] + \beta[RB] \tag{1.25}$$

$$= r_0\left(\frac{\alpha K_B[A] + \beta K_A[B]}{K_A[B] + K_B[A] + K_A K_B}\right) \tag{1.26}$$

Experimental evidence to support the theory of drug action based on affinity and intrinsic activity was provided by Ariens & Simonis (1954) and Ariens & de Groot (1954). Particular types of dose–response curves could

Table 1.1. *Different agonist and antagonist effects of bisquaternary-ammonium compounds on skeletal muscle*

$$R_2 \overset{\displaystyle R_1}{\underset{\displaystyle R_3}{\diagdown\!\!\!-N^+-(CH_2)_{10}-N^+-}}\overset{\displaystyle R_1}{\underset{\displaystyle R_3}{\diagup\!\!\!-R_2}}$$

R_1	R_2	R_3	Agonist	Antagonist
CH_3	CH_3	CH_3	+	−
CH_3	CH_3	C_2H_5	+	+
CH_3	C_2H_5	C_2H_5	−	+

Data taken from Ariens (1954).

be predicted from theory and were found by experiment. For example, if the intrinsic activity of two drugs was 1 and they have different K_A values, separate dose–response curves are simply displaced but have the same maximum. On the other hand, drugs having the same K_A but different α values have different maxima (figure 1.6). Similarly, the administration of two drugs with different intrinsic activities follows a predicted dose–response curve. An example is shown in figure 1.7 where a full agonist, ie with $\alpha = 1$, is given with a partial agonist. The dose–response curves are shifted to the right indicating that the partial agonist competes inhibitively with the full agonist.

1.4.2 Efficacy and stimulus

Stephenson (1956) introduced a different explanation for partial agonism by reducing the restrictions of Clark's receptor theory. He proposed three modifications: (a) a maximum response may be produced when only a small proportion of receptors is occupied, (b) the relationship between response and the fraction of receptors occupied is not necessarily linear, (c) drugs may have very different capacities to induce a response, this property he called efficacy. The response, R, of the tissue was thought of as a function of what was termed the stimulus, S,

$$R = f(S) \tag{1.27}$$

Figure 1.6. Dose–response curves for drugs with different affinity and intrinsic activity. Lines (a) and (b) are for full agonists with $\alpha = 1$ but different K_A values. Lines (c) and (d) are for partial agonists with $\alpha < 1$ but with the same K_A values. E_A is the fractional response of the tissue to drug concentration [A].

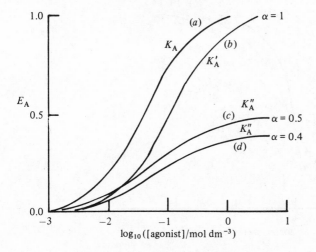

This value of the stimulus is defined as the product of the fraction of receptors occupied, y, and the efficacy e.

$$S = ey \qquad (1.28)$$

Thus S can be expressed as

$$S = eKA/(1 + KA) \qquad (1.29)$$

The parameter e can have any positive value.

Suppose we have a full agonist in concentrations A_1, A_2 and A_3 with an efficacy e_a and a partial agonist in a concentration P with an efficacy e_p. Let P produce the same response as A_1 and let A_2 produce the same response as $P + A_3$. There are therefore two different responses $R_1 = f(S_1)$ and $R_2 = f(S_2)$. Let x be the proportion of receptors occupied by the partial agonist, and if the agonist is a full agonist occupying only a small proportion of the receptors to produce its effect then

$$S_1 = e_p x; \quad S_1 = e_a K_a A_1 \qquad (1.30)$$

thus

$$e_p x = e_a K_a A_1 \qquad (1.31)$$

Similarly

$$S_2 = e_a K_a A_2 \qquad (1.32)$$

Figure 1.7. Dose–response curves for a full agonist ($\alpha = 1$) constructed in the presence of a partial agonist ($\alpha = 0.4$). Line (a) is for the agonist alone, lines (b) and (c) are for the full agonist in increasing concentrations of partial agonist.

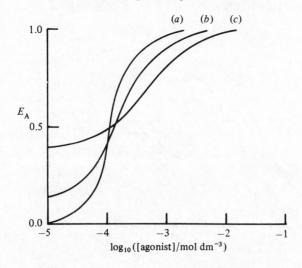

and if the values of S are assumed to be additive

$$S_2 = e_p x + e_a K_a A_3 (1 - x) \tag{1.33}$$

thus we obtain

$$1 - x = (A_2 - A_1)/A_3 \tag{1.34}$$

so that the proportion of receptors occupied by the partial agonist can be determined readily.

The conceptual development put forward by Stephenson meant that the old idea of the response being proportional to the fraction of receptors occupied had to be abandoned. Experimental evidence in support of Stephenson's theory was available from systematic studies of alkyl trimethylammonium compounds on the contraction of guinea-pig ileum. Efficacy is not the same as intrinsic activity since the latter parameter is directly proportional to the size of the maximal response. A maximum response may be produced by compounds with very different efficacies (figure 1.8). Nevertheless, although the precise definitions of intrinsic activity and efficacy are not the same, the essence in both concepts is very similar. Each molecule has a property, distinct from the affinity constant, which characterizes the biological activity of the molecule. The differences in definition between intrinsic activity and efficacy can lead to the production of dissimilar affinity constants for the same drug molecule determined experimentally by each method. Therefore great caution must be exercised in accepting values for affinity taken from the early literature.

Figure 1.8. The agonist action of a series of pharmacologically similar compounds with the same K_A but with different efficacies e.

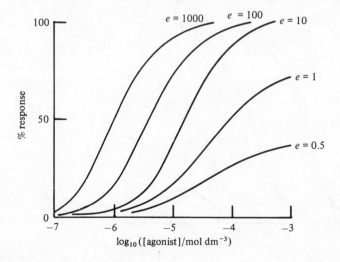

1.4.3   Spare receptors

Further strong evidence in support of the view that agonist action was not directly proportional to the number of receptors occupied was provided by Nickerson (1956). Isolated strips of guinea-pig ileum treated with GD-121 (a histamine antagonist) shifted the dose–response curve in parallel to the right; the blockade remaining after ten minutes of washout was irreversible. The dose–response curve to histamine was shifted 2 $\log_{10}$ units without a significant change in slope or maximum response. Larger concentrations of GD-121 reduced the maximum response to histamine (figure 1.9). GD-121 appeared to behave as an irreversible competitive antagonist decreasing the amount of free receptors available for interaction with histamine. The experimental evidence suggested that only about 1% of receptors need be stimulated by histamine molecules to gain a maximum response. Thus in this tissue most receptors are spare or held in reserve. However, the concept of 'spare' receptors does not exist for a compound whose efficacy is such that only in large doses can a maximal response be reached. In other words, the number of 'spare' receptors is dependent on the efficacy of the drug; there are only 'spare' receptors when the agonist efficacy is greater than necessary for a full agonist response to be produced at high doses (Ariens, Van Rossum & Koopman, 1960).

Figure 1.9. Dose–response curves for the agonist histamine acting on isolated strips of guinea-pig ileum. Line (*a*) is the control response to histamine; lines (*b*), (*c*) and (*d*) are constructed after treatment with increasing concentrations of GD-121, an irreversible competitive antagonist. (Re-drawn from Nickerson, 1956.)

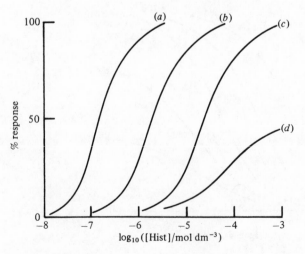

1.4.4   Intrinsic efficacy

The terms efficacy and stimulus as defined by Stephenson (1956) yield dimensionless parameters. Efficacy has no chemical identity and is a parameter that describes a relative effect. Stimulus is also a dimensionless parameter set in the scale 0–1. Furchgott (1966) introduced the term, intrinsic efficacy, ε, to describe the stimulus; ε has the dimensions of the reciprocal of the concentration of receptors. Thus

$$E_A/E_{max} = f(s) = f(\varepsilon[RA]) \tag{1.35}$$

$$= f(\varepsilon[R_T][RA]/[R_T]) \tag{1.36}$$

where $[R_T]$ is the initial concentration of active receptors, E_A is the measured response to an agonist concentration $[A]$ and E_{max} is the maximum response when the stimulus approaches a high value, and in terms of Stephenson's efficacy e

$$E_A/E_{max} = f(e[RA]/[R_T]) \tag{1.37}$$

e is equivalent to $\varepsilon[R_T]$. At equilibrium

$$[RA]/[R_T] = [A]/(K_A + [A]) \tag{1.38}$$

where K_A is the dissociation constant of the drug–receptor complex RA. Therefore we obtain

$$E_A/E_{max} = f\{\varepsilon[R_T][A]/(K_A + [A])\} \tag{1.39}$$

Suppose that an irreversible inactivating agent, specific for the receptors, is used to decrease the number of receptors to a fraction, q, of the initial concentration of $[R_T]$. Equation (1.39) is then modified to

$$E_A/E_{max} = f(s') = f\{\varepsilon q[R_T][A']/(K_A + [A'])\} \tag{1.40}$$

where $[A']$ is the concentration of agonist after receptor reduction. If a dose–response curve to A has been determined before inactivation of receptors and the value of K_A is known, then it is possible to determine the fractional occupation necessary for any response.

So far no assumptions about the relationship between the stimulus and response have been made. If it is assumed that an equal stimulus before inactivation gives the same response as an equal stimulus after inactivation then if equal pairs of responses are obtained the following relationship holds

$$1/[A] = (1/q[A']) + (1-q)/qK_A \tag{1.41}$$

Therefore a plot of $1/[A]$ against $1/[A']$ would be a straight line with a slope of $1/q$ and an intercept $(1-q)/qK_A$. Hence q and K_A can be determined.

1.5 Stimulus–response relationships

Clark (1933) made a careful study of the mathematical nature of the dose–response curve. He showed that three relationships could describe the curve to within $\pm 5\%$ of the maximum value of the response when the response varied in the range 20–60 % of the maximum possible. If x is the dose and y is the response the curves are

$$kx = y/(100-y) \tag{1.42}$$

a hyperbola;

$$ky = \log_{10}(ax+1) \tag{1.43}$$

an exponential curve;

$$kx^n = y \tag{1.44}$$

a parabola. The hyperbola has the same mathematical relationship as the Langmuir adsorption isotherm; the exponential curve follows the Weber–Fechner law and the parabola approximates to a Freundlich adsorption formula.

1.5.1 Stimulus chains and functional interactions

Furchgott (1966) drew attention to different functions of the stimulus and the corresponding model dose–response curves that they could produce. Mackay (1981) has analysed the stimulus producing the response as a chain of individual stimulus steps that can originate from different receptors. This is a new development for drug–receptor theory and attempts to account for functional antagonism or synergism. For example, figure 1.10 shows the chain of events when different receptors R_1 and R_2 combine with the agonists A_1 and A_2; the stimulus chain stretches between α and Ω. Each stimulus in the chain S_{x+1} is related to S_x by a step function

$$1/S_{x+1} = a_x + b_x/S_x \tag{1.45}$$

where a_x and b_x are the step constants. For the overall chain an

Figure 1.10. The chain of stimulus steps $S_{1\alpha}\ldots S_{1\Omega}$ and $S_{2\alpha}\ldots S_{2\Omega}$ leading to a control stimulus S_N which triggers a response. Two different receptor types, R_1 and R_2, are shown together with their different agonists A_1 and A_2. (Re-drawn from Mackay, 1981.)

$$A_1 + R_1 \rightleftharpoons A_1R_1 \rightarrow S_{1\alpha}\cdots \rightarrow S_{1\Omega}$$
$$\searrow$$
$$S_N \rightarrow \cdots \text{response}$$
$$\nearrow$$
$$A_2 + R_2 \rightleftharpoons A_2R_2 \rightarrow S_{2\alpha}\cdots \rightarrow S_{2\Omega}$$

approximation can be made

$$1/S_\Omega = a + b/S_\alpha \tag{1.46}$$

At the confluence of stimulus pathways it is assumed that there exists a relationship between a control stimulus S_N, and $S_{1\Omega}$ and $S_{2\Omega}$ and that equal amounts of S_N produce the same response. Furthermore, Furchgott's relationship for the primary stimulus S_α is assumed

$$S_\alpha = \varepsilon R_T / (1 + 1/K_A[A]) \tag{1.47}$$

where ε is the intrinsic efficacy, K_A is the affinity constant of the drug A for the receptors and R_T is the total concentration of receptors in arbitrary units.

Now consider the response to two agonists A_1 and A_2 given simultaneously each acting on their separate receptors R_1 and R_2, then if the doses of each are varied, equivalent states can be represented by

$$(S_{1\Omega})_1 + (S_{2\Omega})_1 = (S_{1\Omega})_2 + (S_{2\Omega})_2 \tag{1.48}$$

where the bracket subscripts 1 and 2 show that the stimulus is derived from different agonist concentrations. The equation can be rearranged so that

$$(S_{1\Omega})_2 - (S_{1\Omega})_1 = (S_{2\Omega})_1 - (S_{2\Omega})_2$$
$$= \Delta S_{2\Omega} \tag{1.49}$$

This difference in stimulus $\Delta S_{2\Omega}$ is also related to the step functions of equation (1.45).

$$1/\{a_1 + b_1/(S_{1\alpha})_2\} - 1/\{a_1 - b_1/(S_{1\alpha})_1\} = \Delta S_{2\Omega} \tag{1.50}$$

where a_1 and b_1 are chain constants for the stimuli generated by agonist A_1 to step $S_{1\Omega}$. Mackay (1981) was able to show that equation (1.50) could be rearranged to

$$[A_1]_2 / [A_1]_1 = \alpha + \beta[A_1]_2 + \gamma/[A_1]_1 \tag{1.51}$$

where $[A_1]_1$ and $[A_1]_2$ are concentrations of A_1 that produce an equivalent state in the presence of agonist A_2 concentrations $[A_2]_1$ and $[A_2]_2$. The constants are given by the equations

$$\alpha = (1+v)/(1-v)$$
$$\beta = u/(1-v)$$
$$\gamma = w/(1-v)$$

where

$$u = \Delta S_{2\Omega}(\varepsilon_A K_A R_1/b_1)(a_1 + b_1/\varepsilon_A R_1)^2$$
$$v = \Delta S_{2\Omega}(a_1 + b_1/\varepsilon_A R_1)$$
$$w = \Delta S_{2\Omega}(b_1/\varepsilon_A K_A R_1)$$

where R_1 is the concentration of receptors of type 1 (arbitrary units).
Equation (1.51) has been tested by Emmerson & Mackay (1981) for

functional antagonism. The tissue employed was the isolated guinea-pig atria preparation to determine the inotropic effects of $(-)$ isoprenaline and antagonism by muscarinic agents. These drug types act on separate receptors. Propylbenzilylcholine mustard was used as a selective irreversible cholinergic antagonist; this compound made it possible to determine the affinity constants of the muscarinic agonists. The general form of the null equation fitted the experimental data satisfactorily; the parameters α, β and γ were calculated and used to estimate accurately the affinity constants of the antagonists. Identification of the stimulus-chain steps will be discussed in section 1.5.4.

1.5.2 *Transducer models for receptor–effector coupling*
The scheme

$$A + R = AR \rightarrow ? \rightarrow \text{pharmacological effect}$$
$$[AR] = [R_T][A]/(K_A + [A]) \tag{1.52}$$

is commonly used to describe the events from drug–receptor interaction to the response (K_A = dissociation constant). In the previous subsection the question mark was modelled as a function of a stimulus chain such that

$$E = f([AR]) \tag{1.53}$$

or in terms of the mass-action equation

$$E = f\{[R_T][A]/(K_A + [A])\} \tag{1.54}$$

where E is the effect. However, another way of formulating the problem of linking the drug–receptor complex to the effect is to regard the function as a transducer function (Black & Leff, 1983). The concentration of the drug–receptor complex $[AR]$ is a hyperbolic function of $[A]$. Therefore, if the relationship between E and $[AR]$ is hyperbolic, the transducer function, f, must either be linear or a hyperbolic function itself. Receptor reserve rules out the linear alternative so that the transducer function has to be hyperbolic and can be expressed as

$$E/E_{\max} = [AR]/(K_E + [AR]) \tag{1.55}$$

where $E_{\max}$ is the maximum effect and K_E is the value of $[AR]$ at $E/E_{\max} = 0.5$. Equation (1.55) can be re-expressed in terms of equation (1.52) so that

$$E/E_{\max} = [R_T][A]/(K_A K_E + [A]([R_T] + K_E)) \tag{1.56}$$

This relationship can be expressed graphically in figure 1.11.

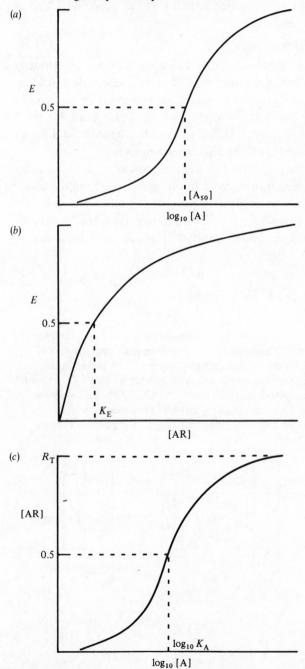

Figure 1.11. Graphical relationships for the transducer model of agonist action. (a) The log dose–response curve; it is generated by two components expressed in (b) and (c). (b) The relationship of the response E to the concentration of drug–receptor complex [AR]; it is a rectangular hyperbolic relationship; the parameter K_E is the value of [AR] that gives the half maximal effect. (c) The mass-action relationship between drug concentration and concentration of drug–receptor complex formed.

(a)

E

0.5

$[A_{50}]$

$\log_{10} [A]$

(b)

E

0.5

K_E

[AR]

(c) R_T

[AR]

0.5

$\log_{10} K_A$

$\log_{10} [A]$

A transducer ratio, τ, can be defined as

$$\tau = [R_T]/K_E \tag{1.57}$$

and it is a measure of the efficiency by which the occupation of receptors is converted into the response; it describes agonism. Furthermore, the concentration of agonist producing a half-maximal response $[A_{50}]$ is related to τ

$$[A_{50}] = K_A/(1+\tau) \tag{1.58}$$

and the asymptote is revealed at $\tau/(\tau + 1)$. The transducer ratio, τ, is loosely a measure of the efficacy of the agonist. If the $[A_{50}]$ value is located at a lower value of $[A]$ than K_A then there is a receptor reserve. For example, if τ is large, $\log_{10} \tau = 99$ then $[A_{50}] = K_A/100$; on the other hand, for small values of τ, $\tau = 0.01$, $[A_{50}] \approx K_A$. Differences between agonism in different tissues can be accounted for by statistical variations in $[R_T]$.

In drug–receptor interaction the first step is the formation of the complex AR; how is this complex transduced into the signal triggering the effect? One possible transduction step is to postulate another molecule, the transducer T, to combine with AR to form a ternary complex ART (figure 1.12). The effect is then related to $[ART]$ which may be linear or hyperbolic.

If the effect is linearly related to $[ART]$ then

$$[ART] = [T_T][AR]/(K_{AR} + [AR]) \tag{1.59}$$

Figure 1.12. A scheme for the combination of agonist (A), receptor (R) and transducer (T) molecules. An agonist receptor complex (AR) is formed first by a mass action relationship with a dissociation constant K_A. This complex then reacts with a transducer molecule (T) to form an agonist–receptor–transducer complex (ART) with a dissociation constant K_{AR}. This ternary complex (ART) leads directly to a biophysical or biochemical event that triggers the pharmacological response.

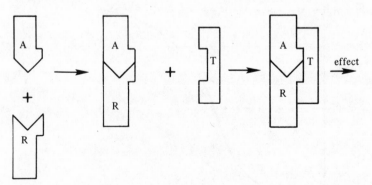

where $[T_T]$ is the total concentration of transducer molecules available and K_{AR} is the dissociation constant of the complex ART. This equation can be re-expressed as

$$[ART] = [T_T][R_T][A]/\{K_A K_{AR} + [A]([R_T] + K_{AR})\} \qquad (1.60)$$

Now if the effect E is proportional to $[ART]$ we have

$$E/E_{max} = [R_T][A]/\{K_A K_{AR} + [A]([R_T] + K_{AR})\} \qquad (1.61)$$

which is similar to equation (1.56) so that $K_E = K_{AR}$ and the transducer ratio $\tau = [R_T]/K_{AR}$.

If we now consider E to be a hyperbolic function of $[ART]$ we have

$$E/E_{max} = [ART]/(\beta + [ART]) \qquad (1.62)$$

where β is a coupling parameter between stimulus and effect, and therefore

$$E/E_{max} = [T_T][AR]/\{\beta K_{AR} + [AR]([T_T] + \beta)\} \qquad (1.63)$$

Thus two forms of transduction, linear and hyperbolic, can be modelled by Black & Leff's (1983) approach. Their model assumes that agonist binding to the receptor is non-cooperative and postulates that a transducer molecule may be responsible for transmitting the effect of an agonist–receptor complex to the events regulating the response. They propose that the transducer may act either by a biochemical mechanism involving second messengers, or by a conformational change in a channel protein thus opening ion channels.

1.5.3 *Transducer coupling between receptors and ion channels*

The adenylate–cyclase system has long been recognized to function as a second messenger relay between the reception of stimuli from numerous hormones, and some neurotransmitters, and internal biochemical events of the cell. Other second messengers may also be implicated in agonist action. However, the adenylate–cyclase system is very complex and consists of numerous proteins functionally linked with each other in the membrane. Receptors, in many cases, appear to be coupled to a pair of guanine-nucleotide-binding proteins, one involved in stimulation G_s and the other is concerned with inhibition G_i (Gilman, 1984). Both G-proteins appear to share a common subunit. Two subunits have been characterized; α subunits are different in G_s and G_i but the β subunits are identical. Guanine nucleotides ($GTP_\gamma S$) activate the proteins to perform either their stimulatory or inhibitory functions. Activation of the G-proteins causes dissociation of the β subunits of the stimulatory system.

$$G_{s\alpha} \cdot \beta + GTP_\gamma S \rightleftharpoons GTP_\gamma S \cdot G_{s\alpha} \cdot \beta \overset{Mg^{2+}}{\rightleftharpoons} GTP_\gamma S \cdot G_{s\alpha} + \beta$$

An analogous reaction occurs with the inhibitor G_i-protein.

The G-proteins do not necessarily act through the adenylate–cyclase system. In cardiac muscle, acetylcholine binds to the muscarinic acetylcholine receptor (mAchR), which in turn eventually leads to the opening of a potassium channel; the result is that the atrial pacemaker slows down. A G-protein acts as a transducer but without stimulating a second messenger as a link between receptor and ion channel. A schematic arrangement of receptor and proteins with the K channel is shown in figure 1.13. An islet activating protein (IAP), which is known to ADP-ribosylate G-proteins, uncouples mAchR from the K^+ channel. Treatment with IAP abolishes the action of Ach on K^+ channel currents (Pfaffinger *et al.*, 1985). The G-proteins can be activated independently of the mAchR link by the GTP analogue 5'-guanylylimododiphosphate (GppNHp). This analogue acts intracellularly and stimulates the K^+ channel even in the presence of atropine. Thus GppNHp would appear to uncouple the channel from the muscarinic receptor (Breitwieser & Szabo, 1985). Transduction of stimulus from receptor to ion channel in cardiac muscle should provide an ideal test system for the hypothesis of operational modes of agonism proposed by Black & Leff (1983).

Figure 1.13. A scheme for the coupling between the muscarinic acetylcholine receptor (mAchR) and the inwardly rectifying K^+ channel in atrial muscle. Acetylcholine (Ach) combines with mAchR and is competitively antagonized by atropine. The drug–receptor complex is coupled via the G-proteins to the K^+ channel. Islet activating protein (IAP) can uncouple the transduction. Guanosine triphosphate (GTP) is required for the coupling but it can be mimicked by the analogue GppNHp.

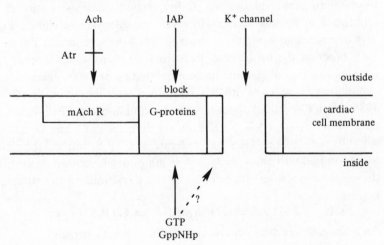

1.5.4   Transducer coupling and functional antagonism

Two receptors on the same cell may have functionally antagonistic actions; an example of this has been alluded to in the work of Mackay (1981) and Emmerson & Mackay (1981) with their study of muscarinic and adrenergic agents on the contractility of heart muscle. What are the relationships between transducer coupling in each event?

Isoprenaline stimulates the cardiac β-adrenergic receptor to induce a slow inward Ca^{2+} current. This electrical stimulation is accompanied by the production of cyclic-AMP. It appears that adenylate cyclase is regulated by G_s-protein (Sperelakis, 1984). GTP and GppNHp can activate the G_s-protein. The β-adrenergic blocking agent propranolol antagonizes the action of isoprenaline but not GppNHp. Thus the regulation of the slow inward Ca^{2+} current is in many respects similar to that of Ach-induced K^+ current at the mAchR; there is one important difference, however, the mAchR activation of the K^+ channel is not dependent on adenylate cyclase activation.

Voltage clamp studies of single atrial cells show that the slow inward Ca^{2+} channel linked to the β-adrenergic receptor can be modified by acetylcholine acting on mAchR (Breitwieser & Szabo, 1985). Stimulation of the receptor produces a G_i-protein. G_i can then compete with G_s for the regulation of adenylate cyclase. The two G-proteins can only bind GppNHp if their attached receptor is stimulated. This mechanism provides a very precise control of adenylate cyclase activity and hence intracellular levels of cyclic-AMP. The second messenger in turn regulates the opening of the Ca^{2+} channel (figure 1.14). Possible kinetic mechanisms for transducer coupling are discussed extensively by Levitzki (1984); five mathematical models for the coupling process are derived, but will not be developed here.

1.5.5   Ion gating

At the motor end-plate of the skeletal neuromuscular junction, spontaneously occurring miniature end-plate potentials can be recorded by microelectrodes placed in the cell. These small potential changes are caused by the release of quanta of the transmitter acetylcholine. Iontophoretic applications of acetylcholine can increase the electrical noise across the end-plate (Katz & Miledi, 1970, 1972). This noise is of low amplitude and appears to be derived from statistical fluctuations of elementary current pulses, or shot effects, due to the bombardment of nicotinic acetylcholine receptors (nAchR) by acetylcholine molecules. The time scale for this discrete shot event is about 1 ms and the depolarization is about $0.3 \mu V$; charge transfer of $10^{-14} C$ is equivalent to 5×10^4 cations moving down their channel. Experiments with carbachol, a powerful nicotinic agonist, showed that the drug had different noise characteristics from acetylcholine; carbachol channels opened only for 0.3–0.4 ms. The antagonist (+)tubocurarine did not alter the time course or size of the acetylcholine shot effect.

A short series of nicotinic agonists was studied by Colquhoun et al. (1975); they showed that small structural variations in the molecules appeared to affect significantly single-channel opening times and conductances. Colquhoun & Hawkes (1977) then provided a mathematical foundation for drug-induced membrane currents which could be used to test different theories of drug action. However, even in simulation experiments, distinctions between different theories would be difficult for those involving all but a few rate constants. This work is still in its infancy and more recent experiments by Gardner, Ogden & Colquhoun (1984) have shown, using single-channel recording techniques by the patch-clamp method, that conductance changes induced by ten different nicotinic agonists are indistinguishable. Differences in agonist efficacy are not related to single-channel conductances. These results are a salient warning

Figure 1.14. A scheme for transducer coupling and functional antagonism in cardiac cells by stimulation of adrenergic receptors (βAdrR) with isoprenaline (ISO) and inhibition by interaction of acetylcholine (Ach) with the muscarinic receptors (mAchR); the receptors are competitively inhibited by propranolol and atropine (Atr). G-proteins are coupled to adenylate cyclase which controls the breakdown of adenosine triphosphate (ATP) to cyclic AMP; in turn this second messenger regulates the Ca^{2+} channel.

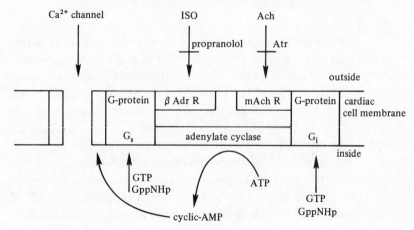

to workers in this field not to overinterpret experimental data into elaborate mathematical theories before more detailed analysis has been carried out.

1.6 Cooperativity

So far in our discussion of drug–receptor interaction it has been assumed that the binding of a drug to its receptor follows a kinetic pattern that can be represented at equilibrium by the law of mass action applied to a biomolecular interaction. In other words, drug binding has been assumed to follow a Langmuir adsorption isotherm. No account has been taken of dose–response curves that depart from a hyperbolic function. However, if the dose–response curve is not hyperbolic and instead is S-shaped, then the simple mathematical analysis used previously to describe drug–receptor interaction cannot really be applied. S-shaped dose–response curves are characteristic of cooperative effects in agonist–receptor binding. But cooperativity, where it is suspected, needs to be examined extensively to differentiate it from other effects on non-cooperative binding (Levitzki, 1981).

Hill (1913) was the first to investigate cooperative phenomena in the binding of oxygen to haemoglobin. Consider the case where a number of agonist molecules bind to the receptor.

$$nA + R \rightleftharpoons A_nR \tag{1.64}$$

then the dissociation constant, K, is given by

$$K = [R][A]^n/[A_nR] \tag{1.65}$$

If it is assumed that there are no intermediate species between A_1R and A_nR then the total number of receptors, $[R_T]$, is

$$[R_T] = [R] + [A_nR] \tag{1.66}$$

Thus the fractional occupancy y is

$$y = [R]/[R_T] \tag{1.67}$$

and

$$1 - y = [A_nR]/[R_T] \tag{1.68}$$

and

$$\log_{10}\{y/(1-y)\} = \log_{10}(1/K) + n\log_{10}[A] \tag{1.69}$$

Therefore a plot of $\log_{10}\{y/(1-y)\}$ against $\log_{10}[A]$ should be a straight line of slope n, where n is the number of binding sites on the receptor. In general, values of $n > 1$ indicate positive cooperativity, whereas if $n < 1$ negative cooperativity or site heterogeneity is probable. Although these plots have extensive usage they only represent an approximation because intermediate species are usually formed.

1.6.1   Negative cooperativity

In contrast to positive cooperativity, negative cooperativity is a commonly observed phenomenon of drug binding to receptors. There are two reasons why negative cooperativity is common. Firstly, if drug molecules bind to a set of site points in the binding site region, then it is possible that two molecules may bind only partially to a different subset of site points and neither molecule binds to the full set which is needed for activity (see section 5.5). Secondly, if drug binding causes a conformational transition in the receptor, then the likelihood of the ligand inducing a structural change which would be highly specific for a further ligand is much less than the drug inducing a change that is structurally non-specific.

Genuine negative cooperativity can be distinguished from non-interacting binding at heterogeneous sites by using irreversible affinity labels. It is assumed that this affinity label is specific for one site and leaves the other site(s) unaffected. Ligand binding to the receptor is then studied in the absence of the affinity label and after progressive treatment of the receptor with the label; the effect on the Hill slope indicates whether true cooperativity is present. If the Hill slope shifts progressively from a value < 1 towards 1 with increased receptor blockade then the drug binding shows pure negative cooperativity. On the other hand, if the Hill slope is unchanged by treatment with an irreversible affinity label then it would appear that there is a heterogeneous population of drug-binding sites (Levitzki, 1984).

1.6.2   Positive cooperativity at the nicotinic acetylcholine receptor (nAchR)

When activated by acetylcholine the nAchR shows an extremely fast coupling to ion gating; channel opening occurs within microseconds and remains open for 1 ms or so. This rapid coupling is unlikely to have a synthetic biochemical intermediary; coupling of receptor to ion channel appears to be direct. Early experiments by Podelski & Changeux (1970) on the excitable electroplax membrane of *Electrophorus electricus* showed that the dose–response curve to acetylcholine was S-shaped. Other nicotinic agonists such as carbachol, phenyltrimethylammonium and deca-methonium all produce S-shaped curves with Hill slopes greater than 1 and close to 2. This preliminary evidence indicated positive cooperativity: further checks by studying the voltage–current relationships in the presence of caesium chloride showed that the cooperativity was not ion-dependent. The competitive antagonist, flaxedil, shifts the dose–response curve of carbachol to the right but does not affect the Hill slope. The effect of another competitive agonist decamethonium on the response to carbachol is not simply additive but cooperative.

The origin of the cooperativity in nAchR appears to lie in its structure; the receptor and ion channel are composed of five subunits $\alpha_1 \, \alpha_2 \, \beta \, \gamma \, \delta$. The α-subunits appear to bind cholinergic agonists, but they do not lie in the same environments within the oligomeric structure (see section 9.1.3 for experiments on gene cloning of the receptor subunits). This difference in local environment for the two α-subunits could affect the affinity of agonists for each subunit because of non-identical conformations of each binding site (Levitzki, 1984). Speculation about the binding constants and state constants for the interconversion of different conformations would, in the mid-1980s, appear premature.

1.7   Molecular mechanisms in drug–receptor interaction

The historical development of theories about drug–receptor interaction has been dominated by the phenomenological approach. Although of great practical use to stimulate further research, this method is, in many ways, unsatisfactory. Theories are developed in response to experimental findings; rarely have theories been developed from sound *a priori* principles with the objective of stimulating further research to test the theory. The reason for this historical and philosophical weakness probably lies in the reluctance of many pharmacologists to come to terms with modern theoretical chemistry. Chinks in the armour of the phenomenological approach first surfaced with Stephenson's (1956) notion of efficacy where a dimensionless parameter, without chemical identity, was necessary to link the proportion of receptors occupied to the stimulus produced. A further shift in thinking was prompted by Paton's (1961) rate theory of drug action. He postulated a departure from occupation theory; stimulus was proposed to be proportional to the rate of drug–receptor combination and not to the proportion of receptors occupied by the drug. The act of combination was thought to produce the stimulus as a quantal event; persistence of a drug–receptor complex did not lead to stimulus. The stimulant or antagonistic properties of the drug were specified by two rate constants, k_1, the association rate constant and k_2 the dissociation rate constant

$$D + R \underset{k_2}{\overset{k_1}{\rightleftharpoons}} DR \qquad (1.70)$$

The ratio $k_2/k_1 = k_e$ the equilibrium constant and is then the reciprocal of the affinity; k_2 determines whether the drug molecule is a powerful stimulant; D and R represent molecules of drug and receptor.

The proportion, p, of receptors occupied at equilibrium was given by

$$p = k_1 x/(k_2 + k_1 x) \tag{1.71}$$

where x is the drug concentration. The rate of association A of drug with receptors is

$$A = k_1 x (1 - p) = k_2 p$$
$$= k_2 x/(x + k_2/k_1) \tag{1.72}$$

The response y is then equal to ΦA, where Φ is a factor relating intensity of stimulus to the response. Under equilibrium conditions

$$dA/dx = k_2^2/\{k_1(x + k_2/k_1)^2\} \tag{1.73}$$

and at low doses, ie when $x \to 0$

$$dA/dx = k_1 \tag{1.74}$$

Paton's rate theory drew attention to the possible importance of kinetics in drug–receptor interaction and provided a large stimulus to make experimental measurements of the rate constants.

A succinct paper by Burgen (1966) signalled a new approach to the study of drug–receptor interactions which shifted the investigations away from the phenomenological theories and channelled the research effort along the serious track of sound chemical theory. In many respects this paper is the forerunner for much of the important research in molecular pharmacology that has been undertaken since that time.

1.7.1   Molecular factors that determine the formation of the drug–receptor complex

It is worth considering the skeletal outline for drug–receptor interaction proposed by Burgen (1966) before an attempt is made in the rest of this book to add the flesh of new research to the early ideas. The distance over which intermolecular forces may lead to a reaction is small; 1 or 2 molecular diameters, or about 12 Å, is the approximate distance. A reaction may occur when the reactants approach each other to their collision radius. In solution the collision rate is controlled by the relative diffusion rates of the reactants. All intermolecular forces are electromagnetic in origin and show a marked distance dependence. The forces between two reactants can be considered as a summation of individual atom-pair interactions.

Translational and rotational diffusion, by Brownian motion, control the movement of drug molecules assuming that the receptors are fixed macromolecular structures. The probability of a drug molecule colliding with a receptor in an orientation suitable for complex formation can be calculated from the Einstein equations for Brownian motion (Einstein,

1956) and from the Boltzmann distribution. If the drug molecule does not have a rigid structure but is free to assume different conformations, the probability of an effective collision is reduced further. However, the conformational restrictions to drug binding may be modified by a stepwise binding process. Another important restriction to drug binding to a receptor may be access to the site. If the site points on the receptor surface are held in a crevice, the pathway to the site points may present energetic hurdles that have to be overcome by the approaching ligand.

The fact that all receptors with a physiological function are immersed in solution would indicate that solvent and hydration effects should have an important role in drug–receptor interaction. Solvent will introduce dielectric shielding to reduce the magnitude of some intermolecular forces. Reorganization of solvent molecules can be expected in the microscopic environment of the drug's, or receptor's, molecular surface. Temporary binding of the water of hydration may obscure, or impede, recognition and binding of an approaching molecule to the receptor site.

Once the drug–receptor complex has formed, and this is a reversible complex, the strength of the association is determined by the energy difference at the equilibrium position compared with that at infinite separation. Different types of intermolecular forces hold the molecules together, for example, ionic forces, dispersion forces, and hydrogen bonds. Hydrophobic interactions arise through the presence of solvent interactions. The average lifetime of the complex is determined by the equilibrium value of the binding energy and the Boltzmann distribution.

1.8   Drug–receptor interaction and computational chemistry

The development of very large and fast mainframe computers in the mid-1960s transformed theoretical chemistry from the pursuit of pure theory into a practical and applicable science. Large-scale calculations that were only dreamed of in 1960 could be put into effect five years later. This availability of computing resources has increased progressively ever since. Huge calculations even by 1970s' aspirations are now commonplace. Two areas of theoretical chemistry that can be applied to biological interactions are pre-eminent; structural chemistry – by which it is meant the elucidation of molecular structure by x-ray crystallography – and molecular energy calculations. These developments form the cornerstone of model simulations for understanding drug–receptor interaction.

This book deals only with the first stage in drug action, that is, the approach of a drug molecule towards its binding site which leads subsequently to a molecular association and formation of a drug–receptor complex. Each of the conceptual steps in this process is examined through the use of structural and computational chemistry. No account is taken of events that occur after the initial formation of a complex. Stimulus–response relationships, briefly touched on in this chapter, are not developed any further; the basic biochemistry and molecular physics is missing. No single coherent model system for a drug–receptor interaction exists; therefore different examples, where they have been studied, are used to illustrate general aspects of the problem. Although this is unsatisfactory in terms of model completeness, nevertheless related research can be used to piece together the main concepts in drug–receptor interaction. As this work deals only with the steps leading to formation of a drug–receptor complex, much of the thrust of the book is directed towards the problem of molecular recognition. Recognition is treated as a problem of identifying, displaying and examining the forces surrounding the reactants.

2

Structural geometry of molecules

2.1   Why do we need precise geometrical information?

From the early work of Langley and Ehrlich at the turn of the century it rapidly became obvious to pharmacologists that small modifications to chemical structure could alter the efficacy of some drug molecules in a remarkable way. This work gave rise to the notion of two groups of compounds, one group acting at specific binding sites or receptors and a second group having non-specific actions by a general physicochemical effect on cell constituents. The idea of specific chemoreceptors, or targets for drug action, proposed by Ehrlich, rapidly gained acceptance with the increasing synthesis and pharmacological testing of variant drug structures. Many congeneric series of drug compounds were developed which, along with quantitative structure–activity relationships, indicated that there exist separate subtypes of receptor. These different receptor subtypes may have different cellular physiological functions. All the pharmacological evidence points towards precisely defined structural binding sites located on each receptor. Recognition between this three-dimensional array of putative interaction points is the mechanism that selects whether a particular complementary drug structure will bind and, possibly as a second action, exert its pharmacological effect. In many respects the process of drug–receptor recognition is analogous to Fischer's 'lock and key' hypothesis for enzyme–substrate interaction.

Molecular recognition and fitting of a ligand to the binding site are governed by two major constraints. Firstly, steric accessibility is the deciding factor in recognition, irrespective of the pattern of intermolecular forces surrounding the ligand. If a bulky group is present that does not fit into the three-dimensional jigsaw presented by the receptor then no effective recognition will be possible. Methods for defining the steric constraints have become a major research area within medicinal chemistry. Secondly, if a recognition event, in which the atoms of the ligand fit the receptor without steric repulsion, is possible, then for the transient molecular complex to exist for an effective time there must be a stabilization energy generated by that complex. In many examples of x-ray crystallographic determinations of the structure of macromolecules cocrystallized with ligands, hydrogen bonding appears to play a major role in the relative positioning of moieties. Since hydrogen bonding is a directed interaction with a narrow spatial geometrical distribution, the positioning of acceptor and donor atoms in the complex is critical for the summation of hydrogen-bond energies that lead to the stability of the complex.

A study of three-dimensional molecular geometry is therefore an essential foundation for our understanding of drug–receptor interaction. The objectives of this chapter are three-fold. Firstly, to outline the mathematics of analytic geometry so that the reader can become competent at quantitative manipulations of molecular structures. Secondly, to sketch out how molecular structures are determined crystallographically and how crystal data can be used. Thirdly, since crystal data is appearing in the literature at an exponential growth rate, a brief introduction to molecular structure databases and how to access them has been included. It is hoped that this will enable the reader to make a foray into this fertile field of structural data.

2.2   Analytic geometry and matrix transformation

2.2.1   *Molecular coordinate systems*

The analytical approach to geometry began to be applied systematically in the seventeenth century. Points in space were ascribed numbers (coordinates) and these numbers could be related algebraically. Rene Descartes was the first to develop the subject and the Cartesian coordinate system was named after him. The great importance of analytic geometry to the study of molecular structures is that it enables atomic positions to be handled conveniently by simple algebra. There are two approaches to applying coordinate systems to molecular structure. Firstly, by imposing on the molecule an external framework of axes, each point is represented by a number related to the axis; in this case the coordinate system might be rectangular, cylindrical or spherical. Secondly, an internal coordinate system may be established and defined by local atom positions in a non-axial system; here spatial points are related to each other by a length, an angle and a dihedral angle.

In a Cartesian coordinate system the axes form a rectangular frame with identical axis scaling and, unless stated otherwise, the convention is for a right-handed frame with axis labels x, y and z. A single point is described by the triple x, y, z (figure 2.1(a)). Cylindrical coordinates are described by the triple r, θ, z where r is the radius of the cylinder, z is the length along the cylindrical axis and θ is the angle (figure 2.1(b)). The cylindrical coordinate system is related to the Cartesian system by the relationship

$$\left.\begin{array}{l} x = r\cos\theta; \quad y = r\sin\theta; \quad z = z \\ r = (x^2 + y^2)^{1/2}; \quad \theta = \arctan{(y/x)} \end{array}\right\} \tag{2.1}$$

Cylindrical coordinates are useful in describing helical molecular structures. In a spherical coordinate system a point is described by the triple r, θ, ϕ (figure 2.1(c)). The relationship to the Cartesian system is given by:

$$\begin{aligned} & x = r\sin\phi\cos\theta; \quad y = r\sin\phi\sin\theta, \\ & z = r\cos\phi, \quad r^2 = x^2 + y^2 + z^2 \\ & \cos\phi = z/r; \quad \sin\theta = y/(x^2 + y^2)^{1/2} \\ & \cos\theta = x/(x^2 + y^2)^{1/2} \end{aligned} \tag{2.2}$$

Spherical coordinates have been used to describe the structure of icosahedral viruses.

An internal coordinate system for a molecule can be developed progressively, describing the structure by a distance between linked atoms, a bond angle linking a third atom, and a torsion angle to a fourth atom. Thus the four-atom molecular system is described by three distances, two bond angles and one torsion angle. For N atoms there are $3N - 6$ internal coordinates.

Figure 2.1. Commonly used coordinate systems: (a) right-handed Cartesian coordinates; (b) cylindrical coordinates; (c) spherical coordinates.

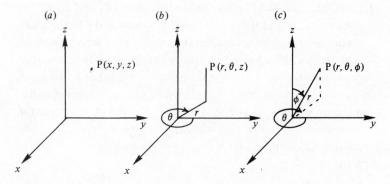

2.2.2   Elementary relationships in analytic geometry

The simplest example is provided by a four-atom molecule with separated atoms labelled i, j, k, l. The interatomic distance d_{ij} can be calculated from Cartesian coordinates of the positions $i(x_i, y_i, z_i)$ and $j(x_j, y_j, z_j)$

$$d_{ij} = \{(x_i - x_j)^2 + (y_i - y_j)^2 + (z_i - z_j)^2\}^{1/2} \tag{2.3}$$

Bond angle θ_{ijk} can be obtained from the cosine rule

$$\cos \theta_{ijk} = (d_{ij}^2 + d_{jk}^2 + d_{ik}^2)/(2d_{ij}d_{jk}) \tag{2.4}$$

where each distance can be expressed in Cartesian coordinates using equation (2.3). The torsion angle ϕ_{ijkl} can be obtained from the internal parameters

$$\cos \phi_{ijkl} = \frac{d_{ik}^2 - d_{il}^2 + d_{jl}^2 - d_{jk}^2 + 2d_{ij}d_{kl} \cos \theta_{ijk} \cos \theta_{jkl}}{2d_{ij}d_{kl} \sin \theta_{ijk} \sin \theta_{jkl}} \tag{2.5}$$

again, each internal coordinate can be expressed in terms of the Cartesian system. The calculated angle ϕ_{ijkl} is the angle between the planes ijk and jkl. The sign of this angle needs to be determined since l can lie on either side of the plane ijk. If the Cartesian coordinate system is right-handed then the sign of the torsion angle ϕ_{ijkl} is opposite to the sign of the volume of the tetrahedron with vertices i, j, k, l, the volume of the tetrahedron is obtained by the determinant

$$V = \frac{1}{6} \begin{vmatrix} x_i & y_i & z_i & 1 \\ x_j & y_j & z_j & 1 \\ x_k & y_k & z_k & 1 \\ x_l & y_l & z_l & 1 \end{vmatrix} \tag{2.6}$$

If we have a line defined by two points $P_1(x_1, y_1, z_1)$ and $P_2(x_2, y_2, z_2)$ then the equation for any point $P(xyz)$ on the line is given by

$$(x - x_1)/(x_2 - x_1) = (y - y_1)/(y_2 - y_1) = (z - z_1)/(z_2 - z_1) \tag{2.7}$$

The direction angles made by a ray emerging from the origin of the Cartesian coordinate parallel to a line defined by points P_1 and P_2 are the angles α between the ray and the x-axis, β to the y-axis, and γ to the z-axis (figure 2.2). The direction cosines of the ray are λ, μ and ν where $\lambda = \cos \alpha$, $\mu = \cos \beta$ and $\nu = \cos \gamma$. If the line is a vector in the opposite direction, then the direction angles are α', β', γ'. The relationship between the two direction angles is $\alpha + \alpha' = \pi$, $\beta + \beta' = \pi$, $\gamma + \gamma' = \pi$; therefore $\cos \alpha = -\cos \alpha'$, $\cos \beta = -\cos \beta'$ and $\cos \gamma = -\cos \gamma'$. The sum of the squares of the direction cosines λ, μ, ν is 1. For the line determined by $P_1(x_1, y_1, z_1)$ and $P_2(x_2, y_2, z_2)$ the direction numbers l, m, n are given by $l = x_2 - x_1$, $m = y_2 - y_1$, and $n = z_2 - z_1$.

The equation of a plane in three-dimensional space is given by

$$lx + my + nz + P = 0 \tag{2.8}$$

where the coefficients l, m, n are direction numbers of the normals to the plane; P = perpendicular distance from the origin to plane. The line of intersection between two planes can be derived by solving the equations of the two planes

$$l_1 x + m_1 y + n_1 z + P_1 = 0$$
$$l_2 x + m_2 y + n_2 z + P_2 = 0 \tag{2.9}$$

The perpendicular distance, h, between a point $Q_1(x_1 y_1 z_1)$ and the plane $lx + my + nz + P = 0$ is given by

$$h = |lx + my + nz + P| / (l^2 + m^2 + n^2)^{1/2} \tag{2.10}$$

The equation of the plane defined by three points $P_1(x_1 y_1 z_1)$, $P_2(x_2 y_2 z_2)$, and $P_3(x_3 y_3 z_3)$ is of the form

$$lx + my + nz + P = 0$$

and can be obtained by substituting for x, y and z into this equation to solve the linear equations:

$$lx_1 + my_1 + nz_1 + P = 0$$
$$lx_2 + my_2 + nz_2 + P = 0$$
$$lx_3 + my_3 + nz_3 + P = 0 \tag{2.11}$$

Figure 2.2. Representation of the line $P_1 P_2$ in Cartesian coordinates. Direction angles are labelled α, β and γ; direction numbers are represented by the lengths l, m and n.

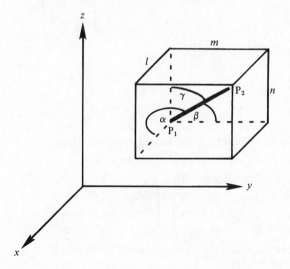

The best plane through N points is given by the least squares method using the equations

$$l\sum x^2 + m\sum xy + n\sum xz - P\sum x = 0$$
$$l\sum xy + m\sum y^2 + n\sum yz - P\sum y = 0 \qquad (2.12)$$
$$l\sum xz + m\sum xy + n\sum z^2 - P\sum z = 0$$

which can be solved to obtain l, m, n and p.

2.2.3   Elementary matrix transformation

In considering the pathway of a molecular reaction it is necessary to move the reactants; the coordinates may be changed by a translation and a rotation. If we want to view a molecule on a visual display unit we may wish to incorporate scaling as well. The computation of these steps is achieved easily by matrix manipulation using a square 4×4 matrix and the coordinate triad (x, y, z) represented by a column vector.

The translation matrix to shift the origin to the point (T_x, T_y, T_z) in the old system is

$$\begin{pmatrix} 1 & 0 & 0 & -T_x \\ 0 & 1 & 0 & -T_y \\ 0 & 0 & 1 & -T_z \\ 0 & 0 & 0 & 1 \end{pmatrix} \begin{pmatrix} x \\ y \\ z \\ 1 \end{pmatrix}$$

Thus a point (x, y, z) in the old system of coordinates becomes a point (x', y', z') in the new system

$$\left. \begin{aligned} x' &= x - T_x \\ y' &= y - T_y \\ z' &= z - T_z \end{aligned} \right\} \qquad (2.13)$$

For a change of scale S_x, S_y, S_z the matrix is

$$\begin{pmatrix} S_x & 0 & 0 & 0 \\ 0 & S_y & 0 & 0 \\ 0 & 0 & S_z & 0 \\ 0 & 0 & 0 & 1 \end{pmatrix}$$

For a rotation of the coordinate system by θ (measured anticlockwise) the matrix transformation is

$$\begin{pmatrix} \cos\theta & \sin\theta & 0 & 0 \\ -\sin\theta & \cos\theta & 0 & 0 \\ 0 & 0 & 1 & 0 \\ 0 & 0 & 0 & 1 \end{pmatrix}, \quad \begin{pmatrix} \cos\theta & 0 & -\sin\theta & 0 \\ 0 & 1 & 0 & 0 \\ \sin\theta & 0 & \cos\theta & 0 \\ 0 & 0 & 0 & 1 \end{pmatrix}, \quad \begin{pmatrix} 1 & 0 & 0 & 0 \\ 0 & \cos\theta & \sin\theta & 0 \\ 0 & -\sin\theta & \cos\theta & 0 \\ 0 & 0 & 0 & 1 \end{pmatrix}$$

round the z-axis   round the y-axis   round the x-axis

If a transformation T is given by a sequence of elementary transformations

$$A_1 \ldots A_n \quad (\text{ie } T = A_n \times A_{n-1} \times \cdots \times A_1),$$

then the inverse of the transformation to return to the original system is

$$T^{-1} = A_1^{-1} \times A_2^{-1} \times \cdots \times A_n^{-1} \tag{2.14}$$

The formula for multiplying two equal matrices M and N each with i rows and j columns is

$$(MN)_{ij} = \sum_n (M_{in} + N_{nj}) \tag{2.15}$$

In molecular interactions it is often necessary to study the tumbling motion of the molecule round its centre of mass, for example, if we were to calculate the energy of rotation of a molecule in an external field. If the origin of the coordinate system is placed at the centre of gyration then the rotation of the molecule can be expressed as rotation round the Euler angles. If the point initially has Cartesian components (x, y, z) then an anticlockwise rotation ϕ round the z-axis is achieved by the transformation

$$A_1 = \begin{bmatrix} \cos \phi & -\sin \phi & 0 \\ \sin \phi & \cos \phi & 0 \\ 0 & 0 & 1 \end{bmatrix} \tag{2.16}$$

so that the point now has components $(x_1, y_1, z_1(=z))$.

A second rotation round the new x-axis by θ is obtained from the matrix

$$A_2 = \begin{bmatrix} 1 & 0 & 0 \\ 0 & \cos \theta & -\sin \theta \\ 0 & \sin \theta & \cos \theta \end{bmatrix} \tag{2.17}$$

so that the point has components (x_2, y_2, z_2).

A final rotation round the new z-axis by ψ is given by

$$A_3 = \begin{bmatrix} \cos \psi & -\sin \psi & 0 \\ \sin \psi & \cos \psi & 0 \\ 0 & 0 & 1 \end{bmatrix} \tag{2.18}$$

The point finally has components (x_3, y_3, z_3). Therefore

$$\begin{pmatrix} x_3 \\ y_3 \\ z_3 \end{pmatrix} = A_3 A_2 A_1 \begin{pmatrix} x \\ y \\ z \end{pmatrix} \tag{2.19}$$

Cartesian coordinates can be constructed from $3N - 6$ internal coordinates by assigning coordinates to three selected sequential atoms. The first atom is placed at the origin, the second atom is placed on the y-axis at a y value of d_{12}, the third atom lies in the y–z plane with positive y- and z-coordinates. The fourth atom is then positioned according to the value of its torsion angle; if the torsion angle is positive then the *x*-coordinate is positive. Thus the sequence of atoms can be built up by a sequence of a translation and three orthogonal rotations. The convention for the direction or sign of a torsion angle in a four-atom sequence is shown in figure 2.3.

2.3   Crystallographic terms

2.3.1   *The unit cell*

Crystals are built up from a geometrical arrangement of molecular subunits. The reason for this orderly pattern of subunits is that the molecular components are at, or near to, the minimum energy position for each molecule. A subunit is formed when the energy of packing the molecules is also at a minimum. Thus we have a case where the sum of the bond energies within a molecule together with the influence of a neighbouring molecule affect the relative orientations and interaction of the two assemblies. The arrangement of successive molecules held at the minimum energy position eventually forms a repeating structure which is termed the unit cell. The arrangement of the unit cells enables the crystal to grow with an ordered geometry and a characteristic shape. Two transformations govern the relationship between adjacent unit cells; these transformations are translation and rotation steps.

The unit cell is characterized by three vectors in different planes. These vectors form the edges of a parallelepiped, and are given the symbols **a, b, c**; the magnitudes of the respective vectors are a, b, c (figure 2.4). The

Figure 2.3. The convention for the sign of the torsion angle ϕ in a four-atom sequence.

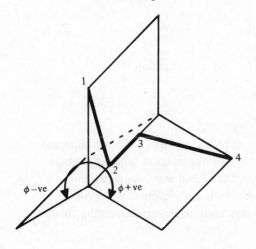

directions of the sides of the unit cell are given as the coordinate axes x, y and z. The interaxial angles yz, zx and xy are denoted by α, β, γ. The coordinate axes x, y, z are internal coordinates and are known as the notional axes and are not necessarily orthogonal to each other. Thus the unit cell can be specified completely by the lengths a, b, c and the interaxial angles α, β, γ. Special situations arise where edges are of equal length and where the interaxial angles are equivalent, 90° or 120°. These unit cell geometries form seven crystal systems, see table 2.1. Thus all crystals fall into one of these systems: triclinic, monoclinic, orthorhombic, tetragonal, trigonal, hexagonal and cubic.

The arrangement of adjacent unit cells is governed by translational and rotational operations. Only five arrangements of unit cell are able to fill a plane and exclude empty space. These shapes have rotational symmetry of 1-, 2-, 3-, 4-, 6-fold axes; there are no 5-fold axes because regular pentagons cannot be arranged in an ordered array (figure 2.5). In a three-dimensional array only 14 different types of unit cell can be stacked properly; these are known as the Bravais lattices. The Bravais lattice symbols for each crystal system and point group class are given in table 2.1. Lattice symbols represent the following relationships:

A, B or C centred on one pair of opposite faces of the unit cell as well as having points at the corners.

F all faces centred, in addition to points at the corners.

I body centred, in addition to points at the corners.

Figure 2.4. The unit cell with three non-orthogonal axes x, y, z; the angles between the axes are α, β, γ and the edge lengths of the cell are a, b, c.

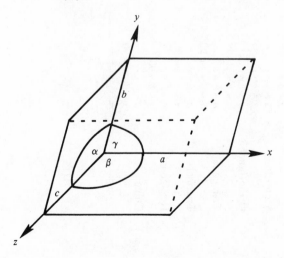

Table 2.1. *The seven crystal systems, their axial relationships and the 32 point groups*

Crystal system	Axes	Essential symmetry	Point group classification	Symbols of lattice types
Triclinic	$a \neq b \neq c$ $\alpha \neq \beta \neq \gamma$	None	$1, \bar{1}$	P
Monoclinic	$a \neq b \neq c$ $\gamma \neq \alpha \sim \beta = 90°$	Diad axis or inverse diad axis	$2, m, 2/m$	PB, PC
Orthorhombic	$a \neq b \neq c$ $\alpha \sim \beta \sim \gamma = 90°$	Three orthogonal diad or inverse diad axes	$222, 2mm, 2/m2/m2/m$	PC IF
Tetragonal	$a \sim b \neq c$ $\alpha \sim \beta \sim \gamma = 90°$	One tetrad or inverse tetrad axis	$4, \bar{4}, 4/m, 422, 4mm,$ $\bar{4}2m, 4/m2/m2/m$	PI
Trigonal	$a \sim b \sim c$ $\alpha \sim \beta \sim \gamma \neq 90°$	One triad or inverse triad axis	$3, \bar{3}, 32, 3m, \bar{3}2/m$	R
Hexagonal	$a \sim b \neq c$ $\alpha \sim \beta = 90°,\ \gamma = 120°$	One hexad or inverse hexad axis	$6, \bar{6}, 6/m, 622, 6mm,$ $\bar{6}2m, 6/m2/m2/m$	P
Cubic	$a \sim b \sim c$ $\alpha \sim \beta \sim \gamma = 90°$	Four triad axes	$23, 2/m\bar{3}, 432, \bar{4}3m,$ $4/m\bar{3}2/m$	PI F

P primitive (points only at the corners), except rhombohedral.

R primitive rhombohedral.

In addition to the seven crystal systems and the 14 Bravais lattices there are also 32 crystal classes. These crystal classes, and there may be one or more in each crystal system, are related to symmetry within the unit cell. Point groups can be classified by the Herman Mauguin notation (see table 2.1).

Crystal classes are determined by various symmetry transformations and relate the position of an atom in the unit cell to a periodic repeat pattern. There may be more than one molecule within the unit cell and the positions of the atoms in one molecule relative to those in another molecule can be described by an arrangement of the symmetry elements. These symmetry elements can be classified as: a centre of symmetry ($\bar{1}$), a mirror plane (m), glide planes (a, b, c, d, n), rotation axes (2, 3, 4, 6), screw axes (2_1; 3_1, 3_2; 4_1, 4_2, 4_3; 6_1, 6_2, 6_3, 6_4, 6_5), and inversion axes ($\bar{3}, \bar{4}, \bar{6}$). These symmetry elements can be combined in groups, known as space groups. The International Tables for X-Ray Crystallography (Henry & Lonsdale, 1969), classify 230 space groups. This information on symmetry within the unit cell is important if the scientist wants to study intermolecular contacts when more than one molecule is contained in the unit cell. A knowledge of the symmetry enables him to locate exactly the positions of adjacent molecules from a single set of atomic coordinates.

2.3.2 Crystal and orthogonal axes

Natural crystal axes are defined by the lengths of the edges of the unit cell a, b, c and the interaxial angles α, β, γ. The angles may not be orthogonal to each other and so the axes system needs to be transformed

Figure 2.5. Filling space with regular arrangements of 3-, 4-, 5-, and 6-sided figures. Note pentagons cannot be arranged regularly.

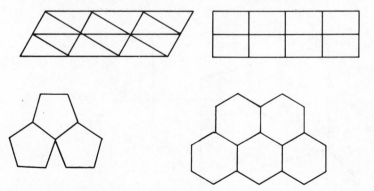

into a Cartesian system before it is possible to extract bond lengths, bond angles and torsion angles. We can consider firstly the transformation of a monoclinic lattice to an orthogonal system. Let **a** and **b** be two non-orthogonal vectors and the vector **c** is perpendicular to the plane described by **a** and **b**; the interaxial angles are $\alpha = \gamma = 90° \neq \beta$. Place vector **b** on the y-axis and vector **a** on the x-axis (figure 2.6); the fractional x- and z-coordinates need to be transformed into a Cartesian system. The transformation is

$$\begin{pmatrix} x' \\ y' \\ z' \end{pmatrix} = \begin{pmatrix} a \sin \beta & 0 & 0 \\ 0 & b & 0 \\ a \cos \beta & 0 & c \end{pmatrix} \begin{pmatrix} x \\ y \\ z \end{pmatrix} \tag{2.20}$$

for the monoclinic crystal system described.

In the triclinic system no natural axis is orthogonal to any of the others and the transformation matrix is more complicated. The transform from a natural system of axes to a Cartesian system is (Rollett, 1965):

$$\begin{pmatrix} x' \\ y' \\ z' \end{pmatrix} = \begin{pmatrix} a \sin \beta \sin \gamma^* & 0 & 0 \\ -a \sin \beta \cos \gamma^* & b \sin \alpha & 0 \\ a \cos \beta & b \cos \alpha & c \end{pmatrix} \begin{pmatrix} x \\ y \\ z \end{pmatrix} \tag{2.21}$$

where

$$\cos \gamma^* = (\cos \alpha \cos \beta - \cos \gamma)/(\sin \alpha \sin \beta)$$

and

$$\sin \gamma^* = (1 - \cos^2 \gamma^*)^{1/2}$$

Figure 2.6. The conversion of a monoclinic lattice system (full lines) to an orthogonal coordinate system (dashed lines); the interaxial angles α and γ are 90°, vector **b** is coincident with the y-axis and **c** with the z-axis. Point Q has to be rotated to Q'.

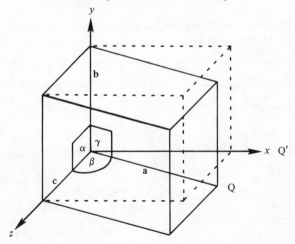

2.3.3   Structure determination

The objective in this subsection is to provide the reader, who has no knowledge of crystallography, with a simple glossary of crystallographic terms. Important textbooks on structure determination have been written by Blundell & Johnson (1976), Woolfson (1978), and Wheatley & Jaffrey (1981).

X-ray crystallography as a science is built round the study of the scattering of x-rays by atoms. Incident radiation in the form of a parallel beam is absorbed by the atoms and the energy is re-emitted in all directions. The distribution in re-emitted energy, or the scatter pattern, is related to the atomic structure of the scatterer. Thus, from a knowledge of the intensity pattern of scatter it is possible to extract the relative positions of the atoms. The x-rays are scattered by the electrons of the atoms and since the electrons are not points but are distributions then the scattering can be related to the electron distribution of the constituent atoms. The scattering of x-rays allows the crystallographer to form a spatial picture of the electron distribution. Atoms with more electrons are thus easier to locate than those with a smaller atomic number.

X-rays have wavelengths of the same order as the spacing of atoms in a crystal (1 Å) which makes them ideal for studying crystal molecular structure by diffraction. Diffraction waves interfere constructively when their amplitudes are in phase and destructively when their amplitudes cancel. If we consider two parallel rays P and Q travelling through a medium with reflecting layers (figure 2.7) then the difference in path length for the rays from adjacent layers is

$$AB + BC = 2d \sin \theta \tag{2.22}$$

Figure 2.7. The path difference for two rays P and Q which are reflected from adjacent layers; the layers are separated by a distance d.

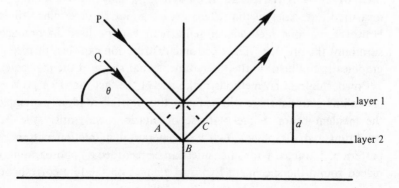

where d is the perpendicular distance between the layers and θ is the angle between the plane and the incident ray. If the ray is a wave then constructive interference will occur when the difference in path length is a multiple n of the wavelength λ, ie $AB + BC = n\lambda$. Thus we have the Bragg relationship

$$n\lambda = 2d \sin \theta$$

and so d can be calculated from the relationship by measuring θ; x-rays are reflected off the atoms in the plane in this layer. If we consider a unit cell then the plane would have an intercept a/h, b/k, and c/l with the three edges of the unit cell; h, k, and l are indices of reflection and are three integers. The distance d_{hkl} between the planes is defined by

$$\lambda = 2d_{hkl} \sin \theta$$

If there are N atoms in the unit cell and the fractional coordinates of atom j are x_j, y_j, z_j then the structure factor F_{hkl} can be defined as

$$F_{hkl} = \sum_j^N f_j \exp \{2\pi i(hx_j + ky_j + lz_j)\} \tag{2.23}$$

where f_j is a measure of the scatter of atom j and is approximately proportional to the number of electrons in the atom. The problem for the crystallographer is to analyse the intensity of the diffraction pattern and relate it to the structure factors so that he can extract the atomic arrangement in the unit cell.

In a molecular structure, the scattering is proportional to the number of electrons contained in a small volume of the unit cell; the structure factor is then obtained by integrating through all volume elements in the cell. It is possible by Fourier synthesis to relate the measured intensities to the electron density distribution. This, however, is where the central problem in crystallography lies: the measured intensity is proportional to $|F_{hkl}|^2$, the sign of F_{hkl} is unknown. Moreover the complex number $|F_{hkl}| \exp (i\phi)$ is used where ϕ is the phase. Although $|F_{hkl}|$ may be determined from measured intensities of the reflexions, ϕ cannot be determined directly. However, various methods of refinement can be used in practice to surmount the problem. The electronic distribution can then be displayed graphically to illustrate the three-dimensional shape of the molecule.

Powder diffraction methods make use of the fact that in a powder of crystalline material there will be many tiny crystals randomly oriented to the incident beam. Some of these orientations will satisfy the Bragg condition and the planes (hkl) will generate a diffraction pattern. The pattern of diffracted x-ray intensities can be recorded on photographic film placed round the specimen. Exposure images on the film correspond to cones of diffracted rays. The image cone then has to be related to a particular plane (hkl); this process is called indexing the reflexion. For crystals with large symmetry, image lines may overlap thus making it difficult to index the reflexions without ambiguity. Therefore in structure determination, data obtained from a single crystal is preferred.

In single crystal work the x-ray beam is monochromatic and finely collimated before being passed through a mounted crystal in the goniometer head. The crystal can be rotated round its three body axes and the scattered beam recorded either by a camera or, more usually, by a diffractometer to count the x-ray photons. Modern diffractometers allow very precise measurements of diffracted x-ray intensities and enable the collection of accurate data. Many counters are linked to a computer so that data collection and rotation of the crystal can be controlled precisely and automatically. Output from the diffractometer can be supervised electronically so that errors in the collection of data may be searched for and corrected.

The Patterson function is an aid to computing the electron density in the unit cell; the function makes use of shifting the origin of one unit cell and superimposing the electron density on another cell and so staggering the distributions of electron density. Regions of overlap may be found that can then be identified. This method is useful if one of the atoms in the cell is heavy, since the pattern of intensities can then be related to the heavy atom. The disadvantage of the method is that the complexity of the function increases in proportion to $\frac{1}{2}N(N-1)$ for N atoms in the unit cell. Another aid in determining crystal structures is the method of isomorphous replacement. It is sometimes possible to replace one or more atoms in a group by another atom, for example a methyl group could be substituted by a chlorine atom, and at the same time maintain a crystal cell and diffraction pattern which is very similar in its essential characteristics. Patterson synthesis can then be used to determine the structure. Isomorphous replacement is particularly useful in investigating the structures of large molecules; mercury derivatives may be prepared that are isomorphous. Stereochemical problems in structure determination can arise in the study and separation of isomers; the heavy atoms method can be used to simplify the problem. The heavy atom is relatively simple to locate provided that there is a balance of light atoms to prevent the intensities of the heavy atom overshadowing the diffraction pattern.

Crystal structure determination is in part a trial and error procedure and once a trial structure has been found it has to be compared with the observed structure factors. Thus the trial structure yields some approximate atomic positions in the unit cell; from these positions and from the symmetry parameters structure factors can be calculated. An index of reliability, or residual, is given by

$$R = \left(\sum \left| |F_o| - |F_c| \right| \right) \bigg/ \sum |F_o| \tag{2.24}$$

where F_o and F_c are observed and calculated structure factors. For a random distribution $R \sim 0.586$ for a non-centrosymmetric structure and 0.828 for a centrosymmetric arrangement. If the value of R determined is less than 0.3 or 0.4 for each respective structure, then the trial coordinates are adequate. A cyclical process of refinement is now embarked upon. An electron density map is re-computed using magnitudes from the observed structure factors and phases from the calculated structure factors. This new map will generate modified atomic positions and R can be re-calculated. The process is repeated until R converges to a low value which depends on the molecular complexity in the unit cell. An R value of < 0.05 indicates a high precision determination whereas $R > 0.15$ suggests that although the heavy atom positions are well defined the analysis needs further refinement. Hydrogen atoms are poorly defined in electron density maps; in a Fourier refinement they are omitted from the process until the final stages. Sometimes hydrogen atom coordinates are too poorly defined by x-ray crystallography to be quoted in structure papers.

Atoms arranged at the minimum energy position tend to form the molecular structure around which the unit cell is formed. However, since the crystal exists at temperatures above absolute zero, there will be atom movements by bond vibrations away from the equilibrium position. If the vibrational motion is significant this will alter the diffraction pattern, since each photon of energy emitted from an atom will occur at a slightly different position. The diffraction pattern will therefore be a time-averaged picture of multiple discrete reflexions. The displacement from the mean position can be incorporated into the structure factor calculation for different temperatures through the formula

$$(F_{hkl})_T = F_{hkl} \exp \left(-B_j \sin^2 \theta / \lambda^2 \right) \tag{2.25}$$

where B_j is the temperature factor of the atom and is expressed as

$$B_j = 8\pi^2 \bar{U}_{\perp j}^2 \tag{2.26}$$

and $\bar{U}_{\perp j}^2$ is the mean square displacement of the atom from its mean position. Thus the observed intensities are reduced by thermal vibrations. It is possible to calculate the atom vibrations and plot the probability distribution as thermal ellipsoids.

2.3.4   Neutron diffraction

Neutron diffraction shares some similarities with x-ray diffraction but it does have important differences. Neutrons are uncharged particles generated by nuclear reactions from an atomic pile. If the temperature of the reactor is T then the wavelength of the neutron, λ, is given by

$$\lambda^2 = h^2/3mkT \tag{2.27}$$

where h is Planck's constant, m is the mass and k is Boltzmann's constant. Therefore, at 373 K, $\lambda = 1.33$ Å. This wavelength is of the same order as that of x-rays used in crystal studies and so neutrons can be used in interference studies. A monochromatic beam is obtained by reflexion from an instrument crystal before being used for scattering studies.

X-rays passing through a crystal lattice are scattered by their interaction with the electrons in the atoms. The electron density cross-section of an atom is approximately the size of an x-ray wavelength. The scattering amplitude is related to the atomic number of the scattering atom. Furthermore, in x-ray diffraction the spatial distribution of the electrons produces an angular distribution of the scattering.

Neutron diffraction shows two marked differences from x-ray diffraction. Firstly, there are short-range interactions between the thermal neutron and the atomic nucleus; scattering is isotropic since the nucleus, which is much smaller than the atom, acts as a point scatterer. Scattering amplitude is not directly proportional to the atomic number of the scatterer as in x-ray diffraction. Thus neutron diffraction is useful for locating hydrogen positions in a crystal and gives reliable values for C—H bond lengths of 1.08 Å because the hydrogen nucleus is the scatterer, not the associated electrons. Secondly, neutrons are scattered by the interaction between the magnetic moment of the neutron and the spin and orbital magnetic moments of the atom. For drug-research the great importance of neutron diffraction has been in studying hydrogen bonds. It is possible to make detailed investigations of positional relationships such as acceptor–donor bond lengths and internal angles within the hydrogen bond. In many cases hydrogen bonds between a drug molecule and a receptor determine the spatial specificity since hydrogen bonding has an important directional component. Neutron diffraction studies are essential for examining the geometrical relationships between hydrogen-bonded molecular structures (Taylor, Kennard & Versichel, 1984).

2.4   The use of crystal data

2[(Hydroxyimino)methyl]-1-methylpyridinium chloride, with the trivial name pralidoxime, is an effective antidote to the anticholinesterase nerve gases. The drug was developed by rational principles of drug design. The backbone of the design of pralidoxime was based on a prolonged study of the biochemistry of cholinesterase inhibition by reversible inhibitors and by organophosphate inhibitors. The hydrolysis of acetylcholine is a two-step reaction; after binding of the substrate, choline is hydrolyzed slowly to regenerate the enzyme. With the nerve gas inhibitors, acetylcholinesterase is phosphorylated at a serine hydroxyl group near the active site and the subsequent hydrolysis is extremely slow and the enzyme is effectively irreversibly inhibited. Phosphate ester hydrolysis can be speeded up by using hydroxylamine as a nucleophile, but as hydroxylamine is too toxic for use *in vivo*, careful structural studies of hydroxylamine derivatives were undertaken to maintain a good nucleophile for phosphate ester hydrolysis. A pyridine moiety with a quaternary group was selected and placed at a calculated distance from the hydroxyimino so that optimal binding and orientation would be expected. Pralidoxime is one of the most successful derivatives developed and can be used clinically against poisoning by organophosphate anticholinesterases. The crystal structure of pralidoxime has recently been elucidated (Van Havere *et al.*, 1982); we can illustrate the manipulation of crystal data from this paper.

Two resonance forms of pralidoxime are possible (figure 2.8); the molecule crystallizes into a triclinic system with a point group $P\bar{1}$ with two molecules in the unit cell. The group $P\bar{1}$ shows that the unit cell contains a primitive lattice with an inversion centre. The atomic fractional coordinates for the second molecule in this symmetry group are therefore: $1-x$, $1-y$, $1-z$. Unit cell parameters are given in table 2.2 together with

Figure 2.8. Pralidoxime: (*a*) numbering scheme; (*b*) resonance forms.

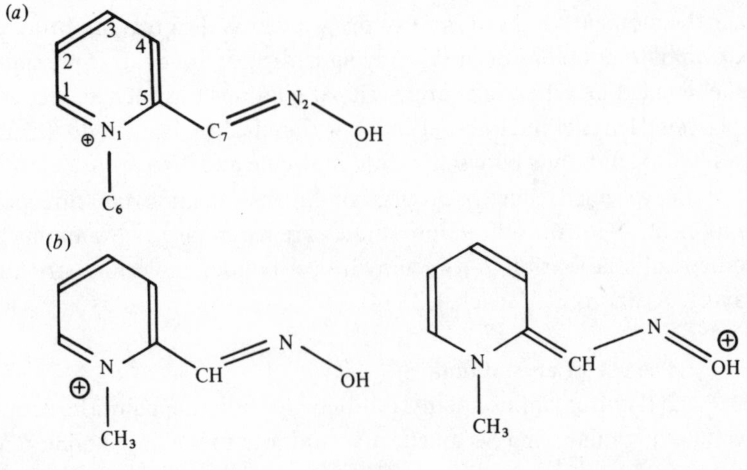

the crystal density, d_c, the x-ray wavelength, $\mu(\text{MoK}_\alpha)$, and the refinement parameter, R_w. The fractional crystal coordinates of all atoms of one molecule in the unit cell are shown in table 2.3; these are the coordinates in the natural coordinate system and must be transformed to Cartesian coordinates before the geometry can be extracted. An inspection of the

Table 2.2. *Unit cell parameters for pralidoxime chloride*

Formula	$C_7H_9N_2O^+ \cdot Cl^-$
Crystal system	triclinic
Crystal group	P$\bar{1}$
a	7.110 Å
b	7.165 Å
c	8.861 Å
α	76.52°
β	85.62°
γ	65.81°
z	2
d_c	1.432 Mg m^{-3}
$\mu(\text{MoK}_\alpha)$	4.205 mm^{-1}
R_w	0.039 for 2435 reflections

Table 2.3. *Fractional coordinates for atoms of pralidoxime*

	x	y	z
C1	0.04351	0.27738	0.23507
O	0.5463	0.7602	0.6689
N(1)	0.1961	0.7489	0.2547
N(2)	0.5520	0.7481	0.5162
C(1)	0.1778	0.7436	0.1052
C(2)	0.3282	0.7462	0.0018
C(3)	0.5052	0.7491	0.0523
C(4)	0.5223	0.7549	0.2034
C(5)	0.3656	0.7573	0.3069
C(6)	0.0230	0.7468	0.3569
C(7)	0.3794	0.7663	0.4686
H(1, O)	0.678	0.742	0.692
H(1, C1)	0.058	0.731	0.075
H(1, C2)	0.300	0.746	−0.099
H(1, C3)	0.612	0.743	−0.013
H(1, C4)	0.635	0.760	0.238
H(1, C6)	−0.062	0.881	0.352
H(2, C6)	0.075	0.678	0.460
H(3, C6)	−0.029	0.652	0.333
H(1, C7)	0.259	0.789	0.537

Data taken from van Havere *et al.* (1982).

fractional coordinates in table 2.3 reveals that the hydrogen atom positions are only defined to three significant figures; this ten-fold drop in precision is due to the difficulty of locating hydrogen atoms accurately by x-ray diffraction. Apart from the chloride ion and the hydrogen atoms of the methyl group all the other atoms can be seen to lie in a plane perpendicular to the y-axis in the unit cell.

The transformation matrix T, obtained by substituting $a, b, c, \alpha, \beta, \gamma$ into equation (2.21) is

$$T = \begin{bmatrix} 6.4817 & 0.0 & 0.0 \\ 2.8658 & 6.9676 & 0.0 \\ 0.5430 & 1.6702 & 8.8610 \end{bmatrix} \tag{2.28}$$

The corresponding Cartesian coordinates x', y', z' are obtained from

$$\begin{pmatrix} x' \\ y' \\ z' \end{pmatrix} = T \begin{pmatrix} x \\ y \\ z \end{pmatrix} \tag{2.29}$$

and are given in table 2.4.

Interatomic distances can be computed from the Cartesian coordinates by equation (2.3); bond angles can be obtained from equation (2.4) and

Table 2.4. *Orthogonal Cartesian coordinates for pralidoxime; positions are given in Å units*

	x	y	z
Cl	0.2821	2.0568	2.5691
O	3.5423	6.8624	7.4935
N(1)	1.2715	5.7800	3.6142
N(2)	3.5793	6.7944	6.1233
C(1)	1.1529	5.6907	2.2707
C(2)	2.1281	6.1398	1.4405
C(3)	3.2758	6.6672	1.9889
C(4)	3.3867	6.7567	3.3468
C(5)	2.3706	6.3243	4.1828
C(6)	0.1491	5.2693	4.4223
C(7)	2.4601	6.4266	5.6382
H(1, O)	4.3963	7.1130	7.7393
H(1, C1)	0.3761	5.2595	1.9170
H(1, C2)	1.9452	6.0576	0.5316
H(1, C3)	3.9683	6.9308	1.4581
H(1, C4)	4.1174	7.1152	3.7231
H(1, C6)	−0.4020	5.9608	4.5569
H(2, C6)	0.4863	4.9390	5.2492
H(3, C6)	−0.1880	4.4598	4.0239
H(1, C7)	1.6794	6.2397	6.2168

torsion angles from equation (2.5) with the correct sign from equation (2.6). These internal coordinates are displayed for the molecule in table 2.5. There is some redundant information in the cation: atoms N(1), C(5), C(7), N(2) dihedral angle, bond angle and bond length; atoms O, N(2), C(7), C(5) dihedral angle and bond angle; atoms H(1, O), O, N(2), C(7) dihedral angle. These atom sequence internal coordinates are not necessary to describe the pralidoxime moiety; 3N-6 coordinates only are needed. The dihedral angles show that the pyridinium ring is coplanar with the oxime moiety. Crystallographic comparisons with other congeneric oximes suggest that both resonance forms of figure 2.8(*b*) contribute to the structure. An interatomic distance of 2.99 Å between OH and Cl suggests hydrogen bonding to the chloride anion.

The example of pralidoxime chosen to illustrate the handling of crystal data was a high-precision determination of a crystal structure. Not all structure determinations locate hydrogen atoms and so their corresponding fractional coordinates are omitted. This is a disadvantage for the medicinal chemist, wanting to study the whole molecule. However, he can approximate his structure by adding standard bond lengths, bond angles and dihedral angles for the hydrogen atoms to the Cartesian

Table 2.5. *Internal coordinates for pralidoxime*

Atoms				Dihedral angle ABCD°	Bond angle ABC°	Bond length AB (Å)
A	B	C	D			
N(1)	C(1)	C(2)	C(3)	1.615	121.537	1.352
H(1, C1)	C(1)	C(2)	C(3)	−175.316	120.441	0.956
C(1)	C(2)	C(3)	C(4)	−1.722	118.817	1.357
H(1, C2)	C(2)	C(3)	C(4)	178.193	125.910	0.931
C(2)	C(3)	C(4)	C(5)	0.177	119.267	1.377
H(1, C3)	C(3)	C(4)	C(5)	−177.868	119.904	0.911
C(3)	C(4)	C(5)	N(1)	1.473	121.353	1.365
H(1, C4)	C(4)	C(5)	N(1)	−179.785	117.919	0.897
C(4)	C(5)	C(7)	N(2)	−7.690	122.294	1.385
N(1)	C(5)	C(7)	N(2)	171.655	119.784	1.352
C(5)	C(7)	N(2)	O	179.626	116.942	1.462
H(1, C7)	C(7)	N(2)	O	1.086	121.662	0.990
C(7)	N(2)	O	H(1, O)	178.599	111.765	1.274
H(1, O)	O	N(2)	C(7)	178.598	104.734	0.923
O	N(2)	C(7)	C(5)	179.622	116.942	1.372
H(1, C6)	C(6)	N(1)	C(1)	−91.169	106.487	0.894
H(2, C6)	C(6)	N(1)	C(1)	149.173	109.063	0.952
H(3, C6)	C(6)	N(1)	C(1)	39.544	109.327	0.963
C(6)	N(1)	C(1)	C(2)	179.621	117.068	1.474

coordinates. The hydrogen atom is added to the atom sequence A, B, C to give A–B–C–H. For a non-colinear sequence the procedure is as follows:

(a) Translate the origin of the sequence A–B–C to atom C by the translation matrix equation (2.13).

(b) Rotate about the z-axis to bring atom B into the x–z plane and with a positive x-coordinate equation.

(c) Rotate about the y-axis to place atom B on the positive x-axis.

(d) Rotate round the x-axis to place atom A on the x–y plane with positive y-coordinate.

The new coordinates of the atoms A–B–C are: $A_{(x, +y, 0)}$, $B_{(+x, 0, 0)}$, $C_{(0, 0, 0)}$, and H can now be added to the frame by using the three internal coordinates of the dihedral angles $\phi(A, B, C, H)$, the bond angle $\theta(B, C, H)$, and the bond length $R(C–H)$.

The hydrogen coordinates are:

$$H_x = R \sin \theta$$
$$H_y = R \sin \theta \cos \phi$$
$$H_z = -R \sin \theta \sin \phi$$

These coordinates are in the local frame of atoms A–B–C–H and now need to be transformed back to the original molecular Cartesian coordinate system by the reverse transformation of steps (a)–(d) using equation (2.14). Table 2.6 shows a variety of standard bond lengths with hydrogen atoms

Table 2.6. *Standard bond lengths for bonds to hydrogen atoms frequently encountered in problems of medicinal chemistry*

Atom link	Bond length to hydrogen atom (Å)
—CH$_3$	1.096
C—CH$_2$—C	1.073
\—C—H (trisubstituted) /	1.070
—NH$_3$	1.032
—NH$_2$	1.024
C—NH—C	1.038
C—OH	0.970
C—SH	1.329
=C—H	1.055
C=CH$_2$	1.083
C=CH—C	1.085
C—H (aromatic)	1.084

and these can be incorporated easily into a routine to add missing hydrogen atoms to Cartesian coordinates of a crystal structure. Standard routines of this type are very important in model building from crystal data of polypeptides where hydrogen atoms are not resolved. Each amino residue can have its hydrogen atoms internal coordinates stored in a look-up table to be added automatically as the residues are processed.

An identical geometrical method can also be applied if we wish to add substituents to a parent molecule with known crystal structure but without a structural determination for the substituent modifications. Model building programs are available to compute the Cartesian coordinates in a stepwise generation of a larger molecule according to a set of well-defined rules. MBLD (Gordon & Pople, 1969) is a model building program which can generate atomic Cartesian coordinates on the input of a molecular formula. The program has a small database of standard bond lengths taken from Interatomic Distances (Sutton, 1958) between different atom types and different bond types (single, double, triple); routines assign local geometries to connected atoms from a connectivity attribute (eg tetrahedral, trigonal, etc.). Bond lengths, bond angles and dihedral angles are generated before conversion to Cartesian coordinates; ring closure can be handled automatically. The standard model assumes bond rotational geometry to be *trans* but special options allow an explicit geometry to be generated. There are special routines that handle ambiguous valence structures that might arise with a radical or an ion. Programs of this type are very versatile and invaluable for medicinal chemists wanting to examine molecular fitting of putative drug molecules to known structural binding sites.

A reasonably accurate geometry of the ligand molecule can be generated and distance calculations made between its binding site before the molecule is synthesized and experimentally tested. This approach saves a lot of time in eliminating putative molecules with geometries unacceptable to the binding surface. The method is more accurate than building solid space-fitting models and using the chemist's eye to evaluate the fit.

2.5 Databases of molecular structures

With the development of computers built from very large-scale integrated circuits, the time to resolve a crystal structure has been reduced proportionately with the speed of the machine. A natural consequence of this expansion in computing power is the increase in number and size of new crystal structures being tackled. About 40 000 small molecule crystal structures are known to have been solved between 1935 and 1983. In the latter year 4000 new structures were reported; by comparison it is interesting to note that the total world output of structures up to 1970 had been about 4000. Thus we are in an explosive growth phase of data generation; parallel array processes will markedly reduce the time taken to resolve new structures. Therefore there has to be an efficient method of keeping track of an exponentially growing mass of data. A bibliographic retrieval system is only one part of the problem; if structural comparisons are to be made then the database must contain coded geometrical information that is complete in terms of description and at the same time accessible by existing and future software. Two databases are becoming standard: the Cambridge Structural Database (CSD) for organic molecules and the Brookhaven Protein Data Bank for macromolecules. Updated files of these databases are held in major crystallographic centres throughout the world. Both databases are vital for the medicinal chemist investigating structural relationships between ligands and macromolecules. A brief outline of how the databases are constructed will be given.

2.5.1   Cambridge Structural Database (CSD)

The database began at the Cambridge Crystallographic Data Centre (CCDC) in 1965 and is confined to x-ray and neutron diffraction studies of organo-carbon compounds; these are organic compounds, organo-metallic and metal complexes. Each entry is compiled in three categories: chemical connectivity which describes the linkages between atoms; numerical information such as coordinates and crystal cell data; and a bibliographic entry which handles the compound name, authors on structure papers, and date of publication. The CCDC also produces annually a volume entitled *Molecular Structures and Dimensions* which contains index information of chemical names, molecular formulae, authors and journals, for all crystal structure determinations of the previous year: this volume is essentially bibliographic information.

The files of CSD are: BIB, a bibliographic information file; CONN, a chemical connectivity file; and DATA, a numeric data file. Each master file is linked by an eight-character reference code REFCODE. Software programs carry out search, retrieval, analysis and display in response to queries. The BIBSER file handles items in BIB, and CONNSER handles the CONN file. Primitive grammar is included in the search procedure by use of the AND, OR and NOT operations with a field or a character string. There are 12 search fields used by BIBSER. For example, a query might be to collect bibliographic information on penicillin crystal structures containing chlorine but not published in *Acta Crystallographica* B.

Q * COMPND 'PENICILL' AND * ELEMENT 'Cl' NOT * CODEN '107'

where COMPND, ELEMENT and CODEN are search fields. With the three operators all search combinations are possible and an exhaustive literature search of all published crystal papers is possible, Allen *et al.* (1979).

Of more use to the medicinal chemist are the connectivity searches by the CONNSER program. This scans atoms by atom, and bonds by bond, for a coded structure, or structural fragment, contained in the CONN file. Each atom in the molecule forms a node with a number (n) and element (symbol); the atom is connected to nca (non-hydrogen atoms) and/or nh (hydrogen atoms) with a net atomic charge (nch). A bond matrix has three entries: atoms i and j and a bond type (bt); seven bond types are specified. With a correct question coding, the search through the CONN file can locate all structures that comply with the specified question. Thus it is possible to locate particular structural groupings and, via the REFCODE, output the bibliographic information, or use the REFCODE to link with the DATA file.

The DATA file is a massive database of numerical information. It contains the inter- and intra-molecular geometry, ie crystal coordinates and crystal cell parameters. All the information in the files has been automatically checked and discoverable mistakes corrected. A geometry retrieval routine GEOM 78 can be run on the DATA file to extract internal coordinates, centroids, vectors, planes and statistics on defined data subsets. This is an extremely useful facility for making geometrical comparisons between molecules. It is now used extensively in statistical analysis of molecular dimensions. PLUTO 78 is a set of drawing programs that can be used to draw the molecules as stick, ball-and-stick or space-filling models, in mono or stereo, and crystal packing diagrams. Anyone considering becoming a user of CSD should read the CCDC user manual; affiliated datacentres throughout the world are given in the appendix to the paper of Allen *et al.* (1979).

2.5.2   *Computer analysis of data files of molecular geometry*

The definitive papers on using the CSD are those of Murray-Rust & Motherwell (1978a, b) and Murray-Rust & Bland (1978). With 40 000 crystal structures in the CSD file DATA, and with the file expanding annually at an exponential rate, it is now possible to examine variations in molecular geometry from numerous entries containing the moiety selected for study. The variation reflects three factors: the crystal environment, the experimental errors in determination and the variation which is produced by neighbouring atoms. These separate causes of variation between molecular moieties can be analysed. Having decided on the moiety to be examined a CONNSER search is carried out. This lists all molecules with the particular atom connectivities. However, there are weaknesses in CONNSER; it cannot pick out stereochemical information since it is essentially a topological file. Stereochemical screening can be carried out using the supplementary GEOM 78 package to compute torsion angles from atomic Cartesian coordinates. This step is necessary if isomers are to be studied. These combined searches mean that it is possible to scan the database for particular connectivities, and at the same time to apply geometrical constraints to select isomers where necessary. A data system subfile is generated from which a statistical analysis is required. Since particular bond lengths, bond angles and torsion angles are the result of a complex quantum-mechanical force field, the relationships between variables are going to be non-trivial. It is possible that we do not, at the onset, even have a model framework in which we can analyse the data. Therefore, from the beginning it is essential to use multivariate analysis on the molecular crystal data. The statistical package for social sciences (SPSS) is used to carry out the statistical analysis.

Before relationships between members in a data system subfile can be scrutinized, errors in the crystallographic determination have to be separated out from the total variance to leave structural variation (due to crystallographic and chemical effects) as relationships to be resolved. In most cases this process cannot be done separately from the multivariate analysis. However, it is useful to consider this as a separate conceptual first part of the problem. Errors in crystallographic determination can be classified into random errors and systematic errors. Random errors may be the result of catastrophic errors in crystal structure determination, such as placing the molecule at the wrong position in the cell, mistyping of cell dimensions and atomic coordinates, or refinement on inadequate data. Fortunately, these gross errors are largely singled out by CCDC checking routines and flagged before entering the DATA file as an error-free set. Other random errors are encountered in calculating the estimated standard deviation (esds) from experimental data. Not all esds are held in the DATA file. Systematic errors fall into two groups: those which do not affect the variance and those generating pseudovariance. Nuclear position and the centre of electron density do not always coincide; this is most marked with H atoms and can result in C—H bond lengths varying by about 0.1 Å. Thermal motion of the atoms gives rise to a pseudorandom error but it is usually small, <0.05 Å, and the standard deviation from thermal motion effects is about 0.003 Å. Disorder can contribute to pseudorandom errors and must be examined. A rough estimate of these errors is essential before looking for structural variance. Esds are only given in DATA for C—C bonds; we have to assume that other esds in other lengths and angles are in proportion. If the esds are similar in magnitude to data variance then there is unlikely to be any detectable structural variance. However, if the esds are much smaller than the structural variation then we can begin to analyse the geometry statistically.

Once we have a reliable indication of structural variation not related to determination errors, the problem is then: how do we interpret the variation? Factor analysis can be used to reveal whether a small set of variables or factors can account for the underlying variation. The factors have to be identified. We could, of course, have a model framework to operate on and have decided in advance which factors to examine. The advantage of carrying out factor analysis is that it avoids preconceptions which might bias our analysis of the data. Factor analysis makes our model formulation rigorous.

As an example of the method we shall take the analysis of the molecular geometry of the β-1'-aminofuranoside fragment studied by Murray-Rust & Motherwell (1978b) and look at torsional variation in the ring. The nomenclature of the ring is given in figure 2.9. CONNSER searches of CSD yielded 99 hits and after stereochemical screening 78 ribose fragments were analysed. The means, standard deviation and variance of the torsion angles are given in table 2.7. The deviation from the mean is expressed as a z-score.

$$z_{ij} = (p_{ij} - \overline{p_j})/\sigma_j \tag{2.30}$$

These scores form the data matrix $\mathbf{Z}$. The correlation matrix $\mathbf{R}$ is calculated, where

$$\mathbf{R}_{mm} = (1/n)\mathbf{Z}_{mn}^{\mathrm{T}}\mathbf{Z}_{nm} \tag{2.31}$$

Figure 2.9. Nomenclature for the β-1'-aminofuranoside ring.

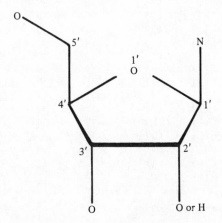

The eigenvalues λ and eigenvectors $\mathbf{E}$ of $\mathbf{R}$ are used to generate the m factors $\mathbf{F}$

$$\mathbf{F}_{mm} = \lambda^{1/2} \mathbf{E}_{mm} \qquad (2.32)$$

also

$$\mathbf{F}\mathbf{F}^{\mathrm{T}} = \mathbf{R}_{mm}$$

and the matrix of factor scores $\mathbf{S}$ is computed from the relationship

$$\mathbf{S}_{np} = \mathbf{Z}_{nm}\mathbf{F}_{mp}$$

where $\mathbf{F}_{mp}$ contains only significant factors. Note that in table 2.7 T08 and T09 belong to different populations; the eigenvalues of the correlation matrix are given in table 2.8 together with the corresponding factor. Three factors account for 99.7% of the variance. These three factors can then be tabulated for each variable (table 2.9). Factor 3 is related to torsion angles T14 and T15 whereas factors 1 and 2 are related in a complex way to ring puckering. The factor scores can be plotted as a scattergram in figure 2.10; the problem now is how to interpret the factors and relate them to molecular deformations. Murray-Rust & Motherwell were then able to show that the factors 1 and 2 were related, firstly to $(\tau)_{\mathrm{max}}$ – an amplitude for

Table 2.7. *Torsion angle statistics for β-1'-aminofuranosides*

Torsion angle number	Atoms				Mean (°)	Standard deviation (°)	Variance
T01	C(1')	C(2')	C(3')	C(4')	−3	34	1155
T02	C(2')	C(3')	C(4')	O(1')	−2	30	906
T03	C(3')	C(4')	O(1')	C(1')	6	19	356
T04	C(4')	O(1')	C(1')	C(2')	−8	18	333
T05	O(1')	C(1')	C(2')	C(3')	7	30	870
T06	N	C(1')	O(1')	C(4')	−130	19	386
T07	N	C(1')	C(2')	C(3')	126	29	856
T08	O(2')	C(2')	C(1')	N	54	210	44 379
T09	O(2')	C(2')	C(3')	C(4')	−111	183	33 550
T10	O(3')	C(3')	C(2')	C(1')	115	37	1373
T11	O(3')	C(3')	C(4')	O(1')	−120	34	1137
T12	C(5')	C(4')	C(3')	C(2')	−122	30	906
T13	C(5')	C(4')	O(1')	C(1')	130	19	371
T14	O(5')	C(5')	C(4')	C(3')	72	71	5072
T15	O(5')	C(5')	C(4')	O(1')	−46	71	5045

Data from Murray-Rust & Motherwell (1978b).

an out-of-plane deformation – and secondly to P – a phase angle used by Altona & Sundaralingham (1972) in a description of pseudorotation of five-membered rings. Research of this type will be useful in conformational studies associated with molecular fitting between drug molecules and binding site moieties. Conformational energy barriers could be assessed by

Table 2.8. *Torsion angle data for β-1′-amino-arabinosides. Factor analysis of the data to obtain the eigenvalues of the correlation matrix*

Factor	Eigenvalue
1	7.98
2	3.03
3	1.95
4	0.02
5	0.009
6	0.004
7	0.002
8	0.002
9–13	<0.001

Data from Murray-Rust & Motherwell (1978b).

Table 2.9. *Factor values tabulated against torsion angles*

Torsion angle	Factor 1	Factor 2	Factor 3
T01	−1.00	0.06	−0.01
T02	0.97	0.22	0.04
T03	−0.66	−0.73	0.13
T04	−0.53	0.84	−0.15
T05	0.93	−0.35	0.06
T06	−0.48	0.86	−0.13
T07	0.93	−0.36	0.06
T10	−1.00	0.04	0.01
T11	0.97	0.23	−0.6
T12	0.98	0.20	−0.04
T13	−0.67	−0.72	−0.13
T14	0.03	0.26	0.96
T15	0.03	0.26	0.97

Data from Murray-Rust & Motherwell (1978b).

computer searches of the type outlined here. Moreover, in complex conformational rearrangements, it might be possible to work out the pathway by factor analysis.

2.5.3   The Brookhaven Protein Data Bank

The Brookhaven Protein Data Bank started in 1971 with the objective of providing machine-readable archives of macromolecular structural information from crystallographic coordinates (Bernstein *et al.*, 1977). The data is collected by direct deposit from those scientists who solve the crystal structures, while the data for small molecules is generally extracted from the literature. The reason for this is that, with macromolecular structures, coordinate lists are too long for publication. Deposited data is put into standard format consisting of a header which would be a title, the compound name, source of protein, author and journal (see figure 2.11). A series of remarks about the crystal solution method and resolution is followed by the sequence of the residues. A few footnotes comment on the sequence. This is followed by a structural analysis which assigns residues to helix, sheet or turn structures. Substructural types are also coded. Crystal cell data with the position of the origin and the scale precede the atomic coordinate data. Each atom contains a line of information: the atom number in the molecule, the atom type, the residue type, residue number, x-, y- and z-coordinates, the occupancy and temperature factor. The ordering of atoms within the amino acid residue is $N(1)$, $C\alpha(2)$, $C'(3)$, $O(4)$, $C\beta(5)$ .... Useful information ends with connectivity records which enable the identification of hydrogen bonds and salt-bridges. The Data Bank has now been extended to incorporate polynucleotide data.

Figure 2.10. Plot of the scores for factors 1 and 2 with autoscaling. (Re-drawn from Murray-Rust & Motherwell, 1978b.)

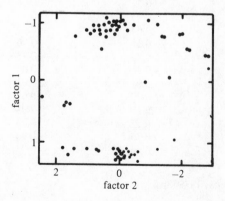

A limited number of programs is available to analyse data records. These include programs to enable model building to be undertaken from atomic coordinates. Geometrical programs check chirality, extract torsion angles and calculate interatomic distances. Graphics programs are available for the whole, or selected portions, of the molecule. In October 1985 there were about 350 macromolecular structures held in the Data Bank.

The atomic coordinate data of macromolecules is used extensively in the investigation of protein structures. Numerous studies are available for sequence homologies and their structural correlations between proteins; if marked relationships do exist between tertiary structure and sequence then these may be useful in developing methods for resolving protein structures.

Figure 2.11. The Brookhaven database entry for DHFR.

Header	Oxido-reductase
compnd	Dihydrofolate reductase (EC.1.5.1.3) complex with NADPH
compnd	Methotrexate
source	(Lactobacillus casei),
source	dichloromethotrexate-resistant strain
Author	D.J. Filman, D.A. Matthews, J.T. Bolin, J. Kraut.
:	
JRNL	REF J.Biol.Chem 257, 13650, 1982.
:	
:	
Remark	7 The side-chain of LYS 51 is missing four atoms.
:	
Seqres	.1 162 THR ALA PHE LEU TRP ALA GLN ASN ARG ASN GLY LEU ILE
:	
FTNOTE	2 See remark 7
:	
Helix	1 HB LEU 23 THR 34 1
:	
TURN	1 TA ASN 8 GLY 1/
:	
CRYST1	71.860 71.860 93.380 90.000 90.000 120.000 P61 6
:	
ATOM	1 N THR 1 −.313 18.726 33.523 1.00 21.00
:	
TER	1299 ALA 162
:	
CONECT	27 26 1353
:	
END	

Indeed, one of the long-term aims of structure studies is to predict three-dimensional structure from amino acid sequences (see section 4.6). However, the most important aspect of the Data Bank for the medicinal chemist is that he has available coordinates of some important enzyme active sites.

2.6   Protein structures

2.6.1   Chemical composition

Proteins are polypeptides formed from the interaction of more than 20 amino acids. Each amino acid has the structure NH_2—CHR—COOH where R is a side group. The $C\alpha$ carbon is central to a tetrahedral arrangement with N of NH_2, C of COOH and H and R. Atoms of the side group R are labelled progressively with letters of the Greek alphabet. Naturally occurring amino acid residues in proteins are in the L-configuration. This L-configuration can be observed by looking in the direction from H to $C\alpha$; the COOH, R, and NH_2 are then found in that order in clockwise positions (the acronym CORN aids the memory) (figure 2.12). Polypeptides are formed by the condensation of the amino group from one amino acid with the COOH of an adjacent residue to form the peptide bond, $Ci\alpha$—CO—NH—$C(i+1)\alpha$ linking the residues. At neutral or physiological pH the α carboxyl or α amino groups are charged and can form a zwitterion for the isolated amino acid. Some amino acids possess other acidic or basic groups in the side-chain and these also can be ionized at physiological pH. However, in many amino acid side-chains there are non-polar groups and these tend to be hydrophobic. Hydrophobic residues are usually oriented to lie internally within the polypeptide three-dimensional structure and the hydrophilic residues are predominantly found at the surface. This disposition of residues helps to maintain the stability of the three-dimensional structure (sections 4.5, 6.4).

Figure 2.12. The stereochemical configuration of L-amino acids.

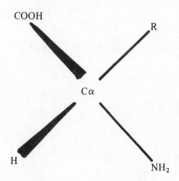

2.6.2 Protein crystals for x-ray diffraction

The solubility of proteins in water is dependent on the ionic strength, the pH, the presence of counter ions, the presence of other organic solvents, eg ethanol which decreases the dielectric constant of water, and temperature. Compared with the determination of small molecular structures of <80 atoms, which use very small crystals, protein structure determination requires bigger crystals since the unit cell is much larger and the intensity of the reflexions is approximately proportional to the number of unit cells in the crystal and inversely proportional to the size of the unit cell. Growing suitable protein crystals, therefore, is a delicate problem. The initial solution of protein has to be concentrated enough to reach supersaturation before a favourable thermodynamic condition enables crystal growth to be triggered by nucleation from the container wall or by artificial seeding. In the growing protein crystal, much solvent is included and it may be as high as 50% by weight. Solvent may be structured and relatively immobile near the macromolecular surface whereas solvent can be free to move in channels between subunits. Most of the solvent, structured and free, is retained in the crystal used for diffraction studies. The presence of solvent in the channels within a protein crystal makes isomorphic replacement possible by diffusing a heavy atom into the crystal. The heavy atom may then bind to the protein surface either in the position where replacement has occurred, or, more commonly, at some other additional site. Measurements of the x-ray intensities from the natural protein and heavy atom derivative can then be used to determine the phase. The preparation of heavy atom derivatives is non-trivial and is often achieved by a trial and error procedure. An alternative method can be used where the crystal is formed from cocrystallization of protein with a cofactor or inhibitor. In this case a heavy metal derivative of the small molecule may be prepared and thereby introduced into the crystal. However, the presence of the small molecule may cause a conformational change in the protein and care is needed when making structural extrapolations to the free protein.

2.6.3 Electron density maps

The x-ray determination of macromolecular structures is necessarily based on the same basic procedure as that used for small molecules, but the interpretation of the x-ray data is much more complex. In part, this complexity is related to the greater size of the macromolecule but, perhaps more important, the resolution is limited by the amount of disorder in the molecules due to their increased flexibility by vibration. The resolution of the electron density maps is quoted in terms of their interplanar spacing.

In low-resolution maps the spacing is >4.5 Å and only general features of gross molecular structure can be identified. Helices can often be identified as rods of electron density. β-Pleated sheets cannot be identified; the resolution is too poor to allow the identification of side-chains. Historically the low-resolution maps were used to build models of the general shape of the protein, using balsa wood or polystyrene sheets. Medium-resolution electron density maps with an interplanar spacing of 2.5–3.5 Å enable one to identify the polypeptide chain in ordered parts of the protein, particularly by following the carbonyl peaks in electron density; α helices and β-pleated sheets can be identified. Various amino residues can be discerned at closer spacings and high-resolution maps can be obtained to 1.9 Å. At high resolution β carbon atoms can be precisely located along with aromatic side-chains.

Electron density maps are contours in a single plane and have to be stacked on top of each other to represent the density in the whole molecule. In the early days of protein structure determination, the identification of a residue with electron density was an extraordinarily complicated process and involved building wire frame models to fit the density maps. Maps and models had to be drawn or built to a precise scale. Orientation of the backbone could be achieved by superimposing images of the model by a series of mirrors onto the stacked density maps. The amino acid sequence can be determined directly by crystallography but the procedure is hazardous and ambiguous assignments can easily be made. It is better to determine the sequence chemically by tryptic digestion of the protein into overlapping fragments where this is possible. If the sequence can be established in this way then solving the x-ray structure is a much quicker procedure. The peptide backbone is constructed by maintaining a planar transconfiguration for the peptide bond. Rotations are made round $C\alpha$–C and $C\alpha$–N bonds. A good starting point for structure determination is not necessarily at a chain terminus since the vibrational freedom is larger in that region and structural identification is poor. In many cases a marker residue, such as cysteine, can be used if it has been labelled with a heavy atom.

The early manual methods have now been superseded by computer graphics and automated fitting procedures (Barry & North, 1972). Model proteins can be built up, at will, from a set of residues with bond lengths and angles and dihedral angles that can be varied. An electron density for the model is computed and overlayed onto the electron density determined crystallographically. A density difference map is obtained and a goodness of fit between them can be calculated. The best least squares fit can then be found by varying the model geometry and minimizing the electron density difference until the calculated and determined electron densities are optimally matched.

2.6.4 Structure refinement

Atomic coordinates of heavy atoms derived from electron density maps will have standard errors of at least 0.3 Å. For detailed structural studies these errors are unacceptable and the protein model must be refined further if details about intramolecular contacts are required. Incorrect structural assignments can arise when the electron densities on the maps are weak for a side-chain due to a large degree of freedom, or in regions where electron densities overlap between side-chains across polypeptide chains. Bond lengths and bond angles are therefore imprecise. Furthermore, variation of dihedral angles might lead to interatomic contacts that would be unacceptable due to high repulsion energies. Energy refinement procedures can be used to minimize these problems. An empirical potential with terms for bond lengths, bond angles, torsion angles and non-bonded interactions is employed. This function can be minimized by changing the atomic coordinates in the direction of the steepest descent. Energy refinement is most useful in eliminating unacceptable non-bonded contacts. The calculation of inter- and intramolecular energies is described in the next chapter.

3

Intra- and intermolecular forces

This discussion will not consider covalent bond formation. The only intramolecular forces to be considered are those which control the molecular geometry. It is important that the pharmacologist is aware that structural geometry can be modified in molecular regions distant and beyond those where different moieties are interchanged. Thus if we study the geometry of the pharmacophore in a congeneric series by crystallography, the molecular geometry of the skeleton can show differences in bond lengths and bond angles as well as the more obvious torsion angle changes. These geometrical variations are increasingly being studied with the aid of databases of molecular structure and should provide a rich hunting ground for shape mapping and correlating with group modification. Small changes in intramolecular shape are very important in some biochemical reactions. For example, in oxygen binding to haemoglobin a very small shift is needed in the iron atom to lead to oxygen coordination. Rather more gross changes in macromolecular shape are common features of substrate binding to enzymes. However, in drug–receptor interaction, and particularly in the recognition step, intermolecular forces are the controlling forces to be investigated. What are these forces and how can they be calculated? How does the field of forces generated by one molecule modify the orientation and approach of a second molecule? Once a productive collision has occurred between a drug and its receptor can we assess the stabilization energy and partition it into component parts? How can we represent and display these forces to provide us with a detailed understanding of the interaction so that we can apply their properties in drug design? In this chapter we shall attempt to answer some of these questions.

If we take the simplest case of two neutral atoms, A and B infinitely separated so that the electron systems show negligible interaction, then the total energy E_{tot} is given by the sums of the energies of each atom E_A and E_B.

$$E_{tot} = E_A + E_B \qquad (3.1)$$

Now if these atoms are brought together with a separation distance between them of r, then an interaction occurs and the equation for the total energy has to be adjusted,

$$E_{tot} = E_A + E_B + U(r) \qquad (3.2)$$

where $U(r)$ is the intermolecular pair-potential energy function

$$U(r) = E_{tot}(r) - E_{tot}^{\infty} \qquad (3.3)$$

$$= E_{tot}(r) - E_A - E_B \qquad (3.4)$$

$U(r)$ is the work performed in bringing the two atoms from infinite separation to r. The force between the atoms $F(r)$ is given by

$$F(r) = -dU/dr \qquad (3.5)$$

and the sign $(-)$ denotes attraction and $(+)$ repulsion.

The potential energy function $U(r)$ has to be partitioned into components with physical meaning and extended to the case for molecules without simple spherical symmetry. The relationship between $U(r)$ and r is schematically shown in figure 3.1. The potential energy is negative at long range, which signifies that the forces are attractive, and becomes positive with repulsive forces at short range. The minimum energy $-e$ is found at a

Figure 3.1. The potential energy of two atoms $U(r)$ plotted as a function of separation r. The minimum energy e is found at a separation r_{min}; σ is the closest separation for which the energy is zero.

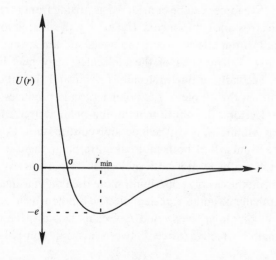

separation r_{min} and σ is the closest separation at which the energy is zero. If the interacting partners A and B are two molecules without spherical symmetry then the intermolecular pair-potential energy function is dependent firstly on the relative positions of their centres of mass R_{AB}, which is a function of R, θ, ϕ in a spherical coordinate system, and secondly on the relative orientations governed by the Euler angles of the two molecules $\omega A, \omega B$ (α, β, γ). Since the energy of interaction for two non-linear polyatomic molecules is orientation-dependent it is often useful to consider the average potential between the partners, where the averaging is carried out over all relative positions. This averaging can be computed using a Boltzmann distribution and hence it has an entropic part and a free energy component. Furthermore, if the interaction shows this geometrical dependence, it is quite likely that other contributions to the interaction arise from internal vibrational and rotational components in the bonds of each molecule.

In intermolecular interactions, the only forces operating are electromagnetic forces and these are electrostatic in origin between the protons in the nuclei and the electron charge density for the electrons in the molecules. Any calculation of the forces between the molecules presupposes that we can calculate the electron charge density. Fortunately, recent advances in quantum chemistry yield methods for determining the charge, provided that the positions of the nuclei are fixed. The simple objective of any theory of intermolecular interactions is to be able to describe the forces in terms of the electron distribution and parameters assigned to the different atomic nuclei. Not surprisingly, this difficult task has not been completely achieved and we have to use various approximations in building a picture of the whole problem.

The long-range attractive forces can be considered as arising from three components. Firstly, electrostatic components, U_{el}, exist in the interaction between polar molecules through their dipole moments and are strongly dependent on the relative orientations of the molecules; they may be attractive or repulsive. Secondly, if one molecule is polar and the other molecule is non-polar, then the dipole of the polar molecule produces a field which distorts the electronic distribution in the non-polar molecule to induce a dipole moment within it. This is the induction contribution, U_{ind}, to the long-range force. Similarly if both molecules are polar there is a mutual inductive effect. Thirdly, even if both molecules do not possess a permanent dipole moment, the electron distribution can oscillate causing a temporary electronic imbalance and a transient electric dipole which can induce an instantaneous dipole in the second molecule. This dispersion force, U_{disp}, is the main attractive force between neutral non-polar molecules. The long-range intermolecular energy can be summed in the equation

$$U = U_{el} + U_{ind} + U_{disp} \tag{3.6}$$

Short-range repulsive interactions arise from the interaction between the electron orbitals of one molecule and those of the other molecule. This overlap contributes to the repulsion. Moreover, when there is electronic overlap, then the electron distribution in the internuclear regions may be distorted to reduce the screening between the nuclei, with a further consequent electrostatic repulsion. Short-range forces are best calculated by quantum-mechanical procedures and are too complex to be represented by a simple analytical function. The energy of interaction in the region of the minimum of the potential function is then treated as a simple sum of the short-range and long-range interactions.

In drug–receptor interaction analytical potential functions are commonly used to estimate the interaction energies; they have the advantage that they are computationally rapid and can be used for large molecular systems. These methods are particularly useful for assessing an order for binding affinities within a congeneric series. A full quantum-mechanical treatment of drug–receptor interactions can only be performed when the total number of atoms in the system is small. However, quantum-mechanical methods have the advantage that they are more accurate and can reveal subtleties of the interaction that are not catered for by analytical pair-potential methods; the disadvantage is that they are costly computations.

3.1 Methods to calculate different forces

3.1.1 The electron distribution

Intermolecular forces are electrostatic in origin and are a combination of effects resulting from fixed nuclear charges and orbiting electrons. There is no way that we can measure the molecular charge at any desired position in a molecule because the electron distribution would be perturbed by the measuring instrument. We can only calculate the distribution from quantum theory. Even then, the computations only provide a probability distribution of charge density throughout three-dimensional space. For simplified approximate computations of intermolecular forces, the charge density needs to be converted into point charges to make the equations less cumbersome. This can be achieved by contracting the charge down onto the atomic nuclei to give atomic charges and this procedure is adequate for the study of long-range effects in intermolecular interactions. Atomic charges are less useful when close-range interactions are considered because orbital overlap may be a significant factor. Nevertheless, atomic charges are widely used in the theoretical studies of drug–receptor interaction and have provided many insights into the recognition mechanism as well as being useful for determining certain parameters used in quantitative structure–activity relationships (QSAR).

The simplest way of thinking about electron distributions is to start by taking the case of a single electron in an orbital $\psi(r)$. Then the probability of finding the electron in a small volume element dv at the position r is

$$|\psi(r)|^2 \, dv \tag{3.7}$$

so that the amount of charge found there is $-e|\psi(r)|^2 \, dv$. Now the electron will have associated with it a spin quantum state of $+1/2$ or $-1/2$. Thus we can determine the probability of finding the electron with a particular spin in the volume dv as $|\psi(r, \sigma)|^2 \, dT$, where $dT = dv_1 \, d\sigma_1$ and σ is the spin variable. For N electrons, the probability of finding electron 1 in volume dT_1 is

$$N \int |\psi(r_1\sigma_1 \ldots r_N\sigma_N)|^2 \, dT_1 \ldots dT_N = \rho(r_1\sigma_1) \tag{3.8}$$

where $\rho(r_1\sigma_1)$ is the charge density function. The total charge density $P(r)$ is obtained by integration

$$P(r) = \int \rho(r, \sigma) \, d\sigma \tag{3.9}$$

and can be calculated from the wave function for the N-electron system. Numerous quantum chemistry programs are available to compute accurately the wave function and they have been summarized by Cook (1978).

Values for the charge density surrounding a molecule can be mapped by contouring in a plane through the molecule. These contour maps show that the charge is spread round the molecule but it can also be partitioned into particular regions. Usually the greater concentrations of charge are found along the bonds of covalent molecules. With ionic bonds there is a greater concentration of charge round the anion. Charge density may also be partitioned into the molecular orbitals. These may be localized molecular orbitals where the charge resides predominantly on one atom or delocalized where the charge spreads to atoms further away. Localized orbitals often produce a conventional picture for the electron distribution in the molecules where charge density is concentrated in regions of the lone pairs and along the bond axes. A description of the charge density in terms of molecular orbitals can provide a good picture of how the molecule is built but it does not tell us how much charge is associated with each atomic nucleus in the molecule. Effective atomic charges have to be determined by a procedure called population analysis, first developed by Mulliken (1955). The wave function is expressed in terms of atomic orbitals associated with each nucleus.

For a diatomic molecule AB the molecular orbital density function $P^i(r)$ is

$$P^i(r) = |\psi_i(r)|^2 = P_A^i(r) + P_B^i(r) + P_{AB}^i(r) \tag{3.10}$$

where ψ_i is the molecular orbital found from a combination of two atomic orbitals. The quantities P_A^i and P_B^i represent the atomic densities associated respectively with atoms A and B in the molecular orbital i; P_{AB}^i is the overlap density between the atoms which can be thought of as forming the bond (Steiner, 1976). Similarly for a polyatomic molecule

$$P^i(r) = \sum_\alpha P_\alpha^i(r) + \sum_{\alpha > \beta} \sum P_{\alpha\beta}^i(r) \tag{3.11}$$

where α and β are summations over atoms and pairs of atoms in the molecule. The total molecular charge density function is obtained by adding together all the molecular orbital contributions i. Integration of the orbital densities in the molecule gives the total net atomic population $n(\alpha)$ and the total overlap population $n(\alpha\beta)$.

$$n(\alpha) = \int P_\alpha(r)\, dv \tag{3.12}$$

$$n(\alpha\beta) = \int P_{\alpha\beta}(r)\, dV \tag{3.13}$$

The electronic charge associated with each atom can then be apportioned by splitting the overlap population between each atom and adding that portion to the atomic population of the atom in the atom pair. In the Mulliken procedure the overlap population contributes a half to each atom. Thus for N_i orbitals

$$N_i(\alpha) = n_i(\alpha) + \frac{1}{2} \sum_{\beta \neq \alpha} n_i(\alpha\beta) \tag{3.14}$$

The total gross population on atom α is

$$N(\alpha) = \sum_i N_i(\alpha) \tag{3.15}$$

If the nuclear charge is $Z_\alpha e$ then the net charge on the atom is

$$q(\alpha) = [Z_\alpha - N(\alpha)]e \tag{3.16}$$

This is the value of atomic charge often used as the basis for computation of long-range forces. However, since the method depends on how the molecular orbitals are computed, one would expect that computed charge distribution would vary according to the computational method employed. If differences in charge distributions are to be investigated then they should only be compared between different states using the same procedure and with the same basis set. Theoretical methods employing charges computed by Mulliken population analysis should be treated with caution; qualitative trends in the data will be useful but quantitative differences may prove to be a trap for the unwary.

3.1.2   Electrostatic energy

For large intermolecular separations between two molecules, that is, where there is insignificant overlap of electron orbitals, the net atomic charges can be considered as an array of fixed point charges. The electrostatic interaction then has charge–charge, charge–dipole, dipole–dipole and other multipole components. These various terms can be understood by considering the potential generated by a diatomic molecule at a point P with charges Q_1 and Q_2 (figure 3.2). The distances r_1 and r_2 lie between the charges and P. Then the electrostatic potential is given by

$$V = (1/4\pi\varepsilon_0)(Q_1/r_1 + Q_2/r_2) \qquad (3.17)$$

where ε_0 is the permittivity ($4\pi\varepsilon_0 = 1.112 \times 10^{-10} \, C \, V^{-1} \, m^{-1}$, or 1 esu). The potential can then be partitioned into its various contributions. For simplicity let the origin of the coordinate system, O, lie at the centre of mass with the atoms on the z-axis; the point P makes an angle θ to the axis. We may now write the potential as

$$V = (1/4\pi\varepsilon_0)(Q/r + \mu \cos \theta/r^2 + \Theta(3\cos^2 \theta - 1)/2r^3 + \cdots) \qquad (3.18)$$

Figure 3.2. The calculation of the electrostatic potential at a point P generated from two fixed charges Q_1, Q_2. The centre of mass is situated at O.

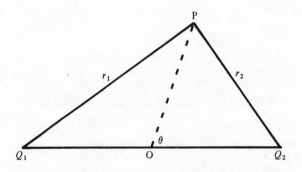

Q is the total charge $Q = Q_1 + Q_2$, μ is the dipole moment, $\mu = Q_2 Z_2 - Q_1 Z_1$, and Θ is the quadrupole moment $\Theta = Q_1 Z_1^2 + Q_2 Z_2^2$. Thus if the molecule is neutral the potential is generated by the dipole and quadrupole terms. Multipoles arise from the imbalance in the arrangement of partial charges in the molecule. There are $2n + 1$ independent multipole moments of degree n. The electric potential of the multipole decreases with distance r by the factor $1/r^{n+1}$, and the electric field, since it is the derivative of the potential, decreases by $1/r^{n+2}$. Monopoles have order $n = 0$, dipoles $n = 1$, quadrupoles $n = 2$, octupoles $n = 3$. Thus monopoles exert the largest effect in long-range forces. Monopoles can be thought of as an effective point charge; dipoles are two charges of opposite sign separated by a line, and quadrupoles are charges distributed at the corners of a quadrilateral. Molecules which possess symmetry may lose some of their multipole moments; for example, CO_2 has no monopole or dipole moment although it does possess a quadrupole moment. The dipole moment is a vector and has components

$$\mu_x = \sum_i Q_i x_i; \quad \mu_y = \sum_i Q_i y_i; \quad \mu_z = \sum_i Q_i z_i$$

and a total magnitude

$$\mu = \sum_i Q_i r_i \tag{3.19}$$

Two opposite charges separated by 1 Å have a magnitude of about 3.3×10^{-30} C m (commonly abbreviated to 1 Debye). The convention for representing dipoles is by an arrow with the head at the negative end. For molecules with a net charge the magnitude of the dipole is dependent on the origin of the coordinate system; for a neutral molecule the magnitude and direction are independent of the coordinate system. Crudely calculated molecular dipole moments use the summation procedure for a set of point charges. However, this approach is inadequate for two reasons; firstly, there is often an asymmetry dipole due to bonding orbitals being of different size; secondly, atomic hybridization dipoles can arise where the centre of charge on the atom does not coincide with the nuclear position. These two effects can make a significant contribution to the total dipole moment for a molecule, and are frequently employed in quantum-mechanical calculations of molecular dipoles.

Suppose we now rotate position P to be coincident with the z-axis and place a charge Q_P at that point, then the interaction energy between the diatomic charge distribution and the point is ΔE

$$\Delta E = (1/4\pi\varepsilon_0) \left[Q_P/r \sum_{i=1}^{2} Q_i + Q_P/r^2 \sum_{i=1}^{2} Q_i z_i + Q_P/2r^3 \sum_{i=1}^{2} Q_i (3z_i^2 - r_i^2) \right] \tag{3.20}$$

or in general terms for any position P at large distances

$$\Delta E = (1/4\pi\varepsilon_0)\left(Q\sum_i Q_i/|r_i - r|\right) \qquad (3.21)$$

This equation can be developed further to calculate the electrostatic interaction energy between two molecules A and B having i and j atoms respectively; it is a drastic simplification

$$\Delta E = (1/4\pi\varepsilon_0)\sum_i\sum_j Q_i^A Q_j^B (r_{ij})^{-1} \qquad (3.22)$$

where r_{ij} is the interatomic distance between atoms i and j.

The interaction energy is therefore orientation dependent in the dipole–dipole interaction term and, depending on the orientation, may be attractive or repulsive between two neutral molecules. This orientation dependence can be very important in drug–receptor recognition because the drug molecule in the biophase is usually free to rotate (see section 7.5). The alignment of the dipole is governed by the local electric field of the receptor. It is possible to calculate the significance of dipole–dipole orientation effects by computing the probability of a particular orientation from the Boltzmann factor, $\exp\{-\Delta E(\omega_1\omega_2)/kT\}$ and normalizing through the orientation space; $\omega_1\omega_2$ are the orientations expressed around the Euler axes. The average interaction energy $\langle U_{el}\rangle$ is

$$\langle U_{el}\rangle = \frac{\displaystyle\iint \Delta E(\omega_1\omega_2)\exp\{-\Delta E(\omega_1\omega_2)/kT\}\,d\omega_1\,d\omega_2}{\displaystyle\iint \exp\{-\Delta E(\omega_1\omega_2)/kT\}\,d\omega_1\,d\omega_2} \qquad (3.23)$$

Thus for freely rotating neutral molecules there is always a net electrostatic interaction and the dipole–dipole term, if it is large enough, can exert a significant steering effect on the drug molecule at the receptor site (section 7.5.4).

3.1.3 Induction energy

The distribution of electrons in an isolated molecule has been shown earlier to be controlled by the spatial arrangement of the nuclear charges. If the molecule is no longer isolated but placed in an electric field, the electron distribution is perturbed and we say that the molecule has been polarized by the electric field. This polarization induces a dipole moment μ_{ind} in the molecule proportional to the field strength

$$\mu_{ind} = \alpha\mathscr{E} \qquad (3.24)$$

where α is a constant called the polarizability. The energy of induction can be obtained by integration

$$E_{ind} = -\int_0^{\mathscr{E}} \mu_{ind} \cdot d\mathscr{E} = -\alpha \mathscr{E}^2/2 \tag{3.25}$$

The electric field is a vector and can be obtained by partial differentiation of the electrostatic potential

$$\mathscr{E} = -\partial V/\partial \mathbf{r} \tag{3.26}$$

In the discussion of equation (3.18) it was clear that the electrostatic potential could be generated by a neutral dipolar molecule through its dipole moment. Therefore, it is possible for the dipole of one molecule to induce a dipole in a second molecule located nearby. So if α' is the polarizability of the second molecule, the induction energy is $-\frac{1}{2}\alpha'\mathscr{E}^2$ and is attractive for all orientations. If r is the separation between the centres of mass of the two molecules then the average induction energy, weighted for all orientations is

$$\langle E_{ind} \rangle = -\mu^2 \alpha'/r^6 (4\pi\varepsilon_0)^2 \tag{3.27}$$

Furthermore, if both molecules are polar they can induce dipole moments in each other; the average induction energy is then found by adding the two terms

$$\langle E_{ind} \rangle_{\mu\mu'} = (\mu^2\alpha' + \mu'^2\alpha)/r^6 (4\pi\varepsilon_0)^2 \tag{3.28}$$

Equations can be developed in an analogous manner for higher-order moments that contribute to the electrostatic potential.

The induction energy may be significant near a receptor binding site where a strong electric field can be generated by charged residues and strongly dipolar regions. However, a systematic evaluation of induction energies at drug binding sites has not been carried out. In part the reason for this lack of information lies in the difficulty of obtaining adequate values for the polarizability; in many cases polarizability is non-isotropic. Polarizabilities can be measured by dielectric methods or can be computed for quantum-mechanical calculations using the finite field procedure applied to equation (3.24) (Amos, 1979). The latter method appears to be very accurate and would seem to offer an ideal approach to determining the polarizability parameter whilst, at the same time, providing a handle to examine the anisotropy of the parameter.

3.1.4 *Dispersion energy*

With non-polar molecules there are no direct electrostatic or induction forces contributing to the long-range interaction energy. The only long-range force producing attraction is the dispersion force. Even in a non-polar molecule fluctuations in the electron distribution can occur, giving rise to an instantaneous dipole; consequently a temporary electric field is generated. This field can affect a neighbouring second molecule and induce a temporary molecular dipole there (figure 3.3). The second molecule now generates a field which feeds back on the first molecule to give a net attraction. As with the induction forces, the effect is dependent on the polarizabilities of the two molecules. Dispersion effects are not confined to non-polar molecules, they also exist between polar molecules. In order to calculate the dispersion energy between two molecules, the dipole resonance frequency needs to be determined. This is a non-trivial matter and beyond the scope of this book; a review is given by Buckingham (1978). A formula for the dispersion energy can be given as

$$E_{disp} = C_6/r^6 \qquad\qquad (3.29)$$

where the coefficient C_6 has to be computed quantum-mechanically. In studies of drug–receptor interaction, dispersion energy contributions are almost always neglected because they are difficult to assess. Instead, dispersion is represented by a choice of values for C_6 as an incorporated term in an intermolecular pair-potential function.

3.1.5   Short-range energies

Short-range forces arise when the electron distribution in one molecule overlaps significantly with that in another molecule and a repulsion energy is generated. At close intermolecular separations the perturbation of the electron distribution leads to an inadequate screening between the nearest atoms on each molecule. This partial exposure of the nuclei results in a strong nuclear repulsion. In tandem the close proximity of the electron orbitals also causes repulsion in the non-bonding case. Therefore, the wave function for the bimolecular system is drastically modified so that a simple analytical function to describe the interaction is not possible. Short-range forces can only be examined by computation of

Figure 3.3. A model for dispersion between two identical non-polar molecules 1 and 2. Let q^+ be positioned at the centre of mass of each molecule and the molecules have a separation r; the negative charge q^- oscillates along the z-axis to create a dipole with a magnitude qz, for molecule 1. This temporary dipole can induce a dipole in molecule 2.

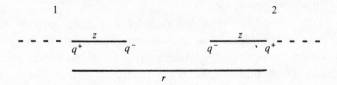

the wave function. However, the energy has been modelled by a function of the form

$$U_{rep} = E(r_{min}/r)^{12} \tag{3.30}$$

where the parameters E, r_{min} and r are defined in figure 3.1. This equation is the repulsion term in a Lennard–Jones (12−6) potential. There are numerous other functions available to model the repulsion energy.

3.1.6 Interaction energy calculations

An analytical potential function should be able to be expressed in terms of long-range and short-range interactions, the desire being to have a simple method for calculating the intermolecular energy. One of the most widely used approaches is the Lennard–Jones method of the $n - 6$ potential

$$U(r) = \varepsilon \left\{ \frac{6}{n-6} \left(\frac{r_{min}}{r} \right)^n - \frac{n}{n-6} \left(\frac{r_{min}}{r} \right)^6 \right\} \tag{3.31}$$

where the minimum in the potential function is ε at a separation r_{min}; the exponent n occurring in the repulsion term is often 12 but can be varied. Values for ε are obtained from experimental data or by calculation. There is no coulomb term in this equation. The Stockmayer potential adds a dipole–dipole interaction term together with an orientation function to take into account the electrostatic interactions.

$$U(r, \theta_1, \theta_2, \phi) = \varepsilon \left\{ \left(\frac{r_{min}}{r} \right)^{12} - 2 \left(\frac{r_{min}}{r} \right)^6 \right\} - \frac{\mu^2}{4\pi\varepsilon_0 r^3} \zeta(\theta_1, \theta_2, \phi) \tag{3.32}$$

where $\zeta(\theta_1, \theta_2, \phi)$ provide the orientations of the dipoles. A large number of variant analytical potential functions have been used; Maitland *et al.* (1981) catalogue 27 types.

3.1.7 Atom–atom pair-potentials

The main difficulty with analytical potential functions described previously is two-fold; firstly getting the most appropriate equation to describe the problem, and secondly obtaining satisfactory parameter constants to insert into the selected equation. Historically the choice of function has been to some extent arbitrary. An exciting and very promising new development has been made by Clementi and colleagues (Bolis & Clementi, 1977; Clementi, Cavallone & Scordamaglia, 1977; Scordamaglia, Cavallone & Clementi, 1977; Carozzo, Corongiu, Petrongolo & Clementi, 1978; Ragazzi, Ferro & Clementi, 1979). In an interaction between N molecules in a system there are a number of terms into which the interaction can be decomposed; these are two-body terms, three-body terms, and ... N-body terms. Two-body terms are simply additive whereas the higher terms are not. However, in a molecule the atoms are usually arranged into distinct groups which confer specific chemical properties. For example, the oxygen atom in COOH, CHO, CHOH, R_2CO etc. behaves differently, but in a group specific way. Can we divide all atoms in a molecule into classes and then parameterize each atom class in a simple analytical pair-potential function? Can we then obtain an accuracy in the atom–atom pair-potential similar to that obtained by *ab initio* quantum-chemical calculations?

The following scheme is used to examine these equations. A simple potential function is used having a Lennard–Jones term together with a coulomb interaction term

$$E = \sum_i \sum_{j \neq i} (- A_{ij}^{ab}/r_{ij}^6 + B_{ij}^{ab}/r_{ij}^{12} + C_{ij}^{ab} q_i q_j/r_{ij}) \tag{3.33}$$

where the fitting constraints are A_{ij}^{ab}, B_{ij}^{ab}, C_{ij}^{ab} between atom i of one molecule and atom j of the other molecule; r_{ij} is the interatomic distance, $q_i q_j$ are the atomic net charges. The superscripts a and b distinguish the atom type *and* its class on the two molecules. The r^{12} term represents atom–atom repulsion and the r^6 term dispersal attraction. Each atom is classified according to the following scheme and its class code used as an identifier:

Class ID, Z, NL, MOVS, ΔMOVS, NC, ΔNC, A1, A2, A3, A4

where class ID is the atom class code, Z is the atomic number, NL is the number of atoms bonded to the atom being classed, MOVS is the molecular orbital valence state energy which has been computed and is an average value taken from a histogram of computed values for this class, ΔMOVS is the variance of the MOVS value, NC is the average net charge for the atom together with its variance ΔNC (again taken from a histogram of computed values), and A1 ... A4 are the atoms bonded to the classified atom. The objective of the classification scheme is to reduce the number of possible parameters in a pairwise interaction to a relatively small number since many interactions fall into equivalent groups. The idea of atom class is a breakthrough in potential function work since it incorporates neighbouring atoms, that is, the electronic and atomic environment of the atom, into the function. Thus the constants A, B and C are not only dependent on atoms i and j but on their local environments as well. Thirdly we have to compute the values of the parameters A, B and C. These are obtained from quantum-mechanical calculations of a large number of complexes at different positions and relative orientations. A sample size of around 2000 computed points was used to fit the potentials of hydrogen atoms in water to an interaction with phenylalanine. A library of atom–atom pair-potential constants can be built up and used to compute the interaction energy between two molecules. The transferability of the parameters A, B and C can then be checked with specific cases computed by full *ab initio* computations. Checks show a remarkable comparability between the two methods of calculation. Agreement is within about $4 \, \text{kJ} \, \text{mol}^{-1}$ in attractive regions and up to $21 \, \text{kJ} \, \text{mol}^{-1}$ in regions showing repulsion.

The objective of this atom–atom pair-potential method has been to derive a simple potential function which can be used in a Monte Carlo computation as a rapid and reliable method for evaluating the interaction energy. A large-scale study of water binding to amino acids and nucleotide bases has been carried out to map water binding regions round interesting biomolecules (section 6.3.2).

3.2   Molecular dynamics

The crystal coordinates for a particular molecular structure relate to the mean positions of the atomic nuclei. The actual positions of the nuclei are in continual motion as the bonds vibrate. Packing forces restrict the movement of atoms in crystals to small bond vibrations and small shifts in torsion angles. In the absence of crystal packing forces, the constraints on nuclear movement are diminished and greater flexibility is possible due to the energy of thermal motion. Molecular dynamics attempts to simulate this intramolecular motion. In many respects molecular dynamics is the offspring of molecular mechanics. Molecular dynamics simulations allow the observer to monitor intramolecular movement on special graphics visual display units on a real-time scale. This simulation process is a very powerful tool in drug research, particularly in the study of drug binding sites and docking interactions between a drug molecule and its receptor. Intramolecular movements in binding sites can be followed and compared with the static coordinates obtained by crystallography. The drug designer can then see how much leeway he has in constructing a putative ligand by combining conformational, thermodynamic and dynamic considerations into his rationale for design.

The first step in any molecular dynamics simulation is to derive a potential function for a molecular mechanics calculation which will accurately reflect the static crystal structure of the molecule. Thus, in a protein with many atoms, the best function will give the smallest root mean square (rms) coordinate deviation from the x-ray conformation. After obtaining an approximate potential function, a dynamic component is added by including the equations of motion for the thermal energies. An application of molecular dynamics to bovine pancreatic trypsin inhibitor has been elegantly presented by Levitt (1983a, b), and that study will be used as an outline illustration of the method in this section.

The potential functions used by Levitt are shown in table 3.1 and include terms for: bond stretching, angle bending, torsion angle twisting, non-bonded interactions and hydrogen-bond terms (modelled either by an electrostatic formula or by a directional method). Approximate force constant parameters were taken from previous work and used as transferable functions which could be employed in any protein dynamics simulation. First derivatives of the potential function, used in energy minimization, can be calculated by analytical differentiation, eg for bond stretching

$$U(b) = K_b(b_i - b_0)^2, \quad \partial U(b)/\partial b_i = 2K_b(b_i - b_0) \tag{3.34}$$

and combined in the chain rule to give

$$\frac{dU}{dx} = \sum_i \frac{\partial U(b)}{\partial b_i} \frac{\partial b_i}{\partial x} + \sum_i \frac{\partial u(\theta)}{\partial \theta_i} \frac{\partial \theta_i}{\partial x} + \sum_i \frac{\partial u(\phi)}{\partial \phi_i} \frac{\partial \phi_i}{\partial x} + \sum_i \frac{\partial u(d)}{\partial d_i} \frac{\partial d_i}{\partial x} \tag{3.35}$$

Energy minimization was carried out by a conjugate gradient technique for up to 3000 evaluations so that the energy change over the last 800 steps was less than $0.84 \, \text{kJ mol}^{-1}$.

The equations of motion are taken from classical dynamics and are solved for each atom; temperature adjustments can readily be made to the system. Consider a force, f, acting on an atom of mass, m, in the x-direction, then the acceleration, a, in the x-direction is

$$a = d^2x/dt^2 = f/m \tag{3.36}$$

Table 3.1. *Potential energy functions employed by Levitt (1983a, b) in molecular dynamics simulation of bovine trypsin inhibitor*

Bond stretching	$k_b(b - b_0)^2$
Bond angle bending	$k_\theta(\theta - \theta_0)^2$
Van der Waals' interaction	$A/r^{12} - B/r^6$
Hydrogen-bond energy	$(A/r^{12} - B/r^6) \exp(-\theta_{O \ldots N-H}/\sigma^2)$
	$\quad + (A'/r^{12} - B'/r^6)[1 - \exp(-\theta^2_{O \ldots N-H}/\sigma^2)]$
	where A and B are for the $O \ldots H$ bond, and A' and B' are derived from van der Waals' parameters, $\sigma = 20°$, $\theta_{O \ldots N-H}$ is the linearity of the hydrogen bond
Torsion angle twisting	$K[1 + \cos(n\phi + \delta)]$ where ϕ is the torsion angle
Backbone (ϕ, ψ) potential	$\sum_i^4 E_{ig}(\phi - \phi'_O, \omega'_\psi)g(\psi - \psi'_O, \omega'_\psi)$

In practice the force is a complicated function of time t for all atoms in motion. However, if integration is performed over a very short period, it can be assumed to be constant or a slowly varying function of t, and a has a constant value a_0. Then

$$d^2x/dt^2 = a_0 \qquad (3.37)$$

and integrating

$$dx/dt = V(t) = V_0 + a_0 t \qquad (3.38)$$

$$x(t) = x_0 V_0 t + a_0 t^2/2 \qquad (3.39)$$

where V_0 is the velocity at $t = 0$ in the x-direction, and x_0 is the initial x-coordinate. If we now take a sequence of steps, $n, n+1$ separated by a time interval, Δt, then

$$V_{n+1} = V_n + a_n \Delta t \qquad (3.40)$$

$$x_{n+1} = x_n + V_n \Delta t + a_n(\Delta t)^2/2 \qquad (3.41)$$

Similarly taking the time interval $n-1$ to n, then the velocity midway is

$$V_{n-1/2} = (x_n - x_{n-1})/\Delta t \qquad (3.42)$$

$$V_n = V_{n-1/2} + a_n(\Delta t/2)$$
$$= (x_n - x_{n-1})/\Delta t + a_n(\Delta t/2) \qquad (3.43)$$

V_n is now expressed in terms of x and can be substituted in equation (3.41)

$$x_{n+1} = x_n + \{(x_n - x_{n-1})/\Delta t + a_n(\Delta t/2)\} \Delta t + a_n(\Delta t)^2/2$$
$$= 2x_n - x_{n-1} + a_n(\Delta t)^2 \qquad (3.44)$$

Therefore, we can calculate the x-coordinate for the $n+1$ step in terms of the coordinates for n and $n-1$ steps and the acceleration at the n step. Furthermore, the velocity at step n is

$$V_n = (x_{n+1} - x_{n-1})/2 \Delta t \qquad (3.45)$$

The time step, Δt, affects the performance of the algorithm and a value of 0.5×10^{-15} s appears to be optimal. Equations (3.44) and (3.45), or Beeman's (1976) variant on them, are used in the molecular dynamics simulation.

In running a molecular dynamics calculation care must be exercised in heating up the system from the start. Temperature changes, to be modelled mathematically, need to be introduced slowly to provide gradual relaxation of the molecule from initial strain. If this precaution is not taken 'hot spots' can appear in the molecule and the simulation can be grossly distorted thus affecting subsequent behaviour. Heating is achieved by selecting random pairs of atoms and assigning them equal and opposite changes in momentum. For an atom pair ij given a temperature change ΔT

$$\Delta V_i = (R_{nd} k_B \Delta T/m_i)^{1/2} \qquad (3.46)$$

where k_B is Boltzmann's constant, m_i is the mass of atom i and R_{nd} is a random number between 0 and 1, also

$$\Delta V_j = -\Delta V_i m_i / m_j \tag{3.47}$$

After n_e steps adding energy to N_a atom pairs the energy of the molecule is increased by ΔE

$$\Delta E = n_e N_a k_B \, \Delta T / 2 \tag{3.48}$$

and the temperature of the system ΔT_{sys} is raised by

$$\Delta T_{sys} = n_e \, \Delta T / 6 \tag{3.49}$$

If the desired temperature of the system is 300 K, and $n_e = 3000$, then $\Delta T = 0.6$ K and will take place over a time span of 6 ps.

Once an equilibrium temperature has been reached then the dynamics simulation is ready to follow. In Levitt's study, heating up to equilibration took 6 ps and protein movement was followed over 132 ps (60 000 time intervals). All coordinate information was stored on magnetic tape before detailed analysis. The results can be assessed qualitatively by making a movie film using each conformation as a still in the kinetic sequence. This film gives an overall impression of how the geometry of the molecule changes and highlights certain gross features such as a major conformational flip. A quantitative assessment of structural change is more complicated because of the huge amount of information generated by the simulation. Various methods for this detailed analysis will be dealt with more fully in the next chapter (section 4.2); but for now we shall itemize a few of the changes that can be followed. One of the simplest parameters that can be computed to give a global indication of change in molecular shape is the radius of gyration

$$R_g = \left\{ \left(\frac{1}{N}\right) \sum_{i=1}^{N} (r_i - \bar{r}) \right\}^{1/2} \tag{3.50}$$

where r_i is the position of atom i and $\bar{r}$ is the molecular centroid. This parameter gives an idea whether the molecule is condensing and becoming more globular, or the converse. In tandem with a change in radius of gyration is a fluctuation in accessible surface area. Area changes can be studied for the whole molecule or in particular regions and provide very useful information about whether these regions have restricted access in the dynamic state. Hydrogen-bond formation and breaking can easily be monitored by computing interatomic distances of the participating atoms. Movement of the backbone of the protein is revealed by difference distance matrix methods as well as by following torsion angle (ϕ and ψ) changes between adjacent pairs of residues. Even a superficial analysis by these methods reveals a remarkable phenomenon in intramolecular movements in protein, namely, cooperative shifts in conformation that generate distinct conformational states for the whole structure. The molecular geometry seems to oscillate about particular conformational means and the pathway between states can be broadly followed through the conformational energy hypersurface. More powerful techniques are now being developed to study in greater detail the atomic shifts that lead to a conformational transition. An automated analysis of these cooperative dynamic changes is essential for a full understanding of the mechanism of geometrical rearrangement. A promising method is the correlative technique employed in multi-dimensional scaling. This method enables the observer to follow changes in n-dimensional data (that is, mathematical n dimensions, rather than the limit of three spatial dimensions) by computing clustered data points and correlating changes to the clustered points with other clusters. A general summary of multivariate analysis is given by Mardia, Kent & Bibby (1979).

A few words of caution are needed to temper the success of molecular dynamics simulations in macromolecules. The method is only as good as its potential functions. Heating to equilibration must be slow and evenly spread, otherwise the conformation can be drastically altered – akin to denaturing the real protein. Current calculations omit solvent and counter ions which can result in the burying of charged groups within the structure, a phenomenon which is not likely to occur in solution. Nevertheless, these problems are not insurmountable. The application of the method to drug–receptor interaction promises to be an exciting development and may reveal new classes of drug compound. Obvious classes to look for are those which bind to cooperative sites regulating conformation changes; these may then enhance, or restrict, natural transmitter or pharmacophore access to its normal binding site. The practical results of Levitt's work are developed in section 4.7.

3.3   Intramolecular forces

The molecular geometry is stabilized by the balance of directional forces exerted by all atoms on each other. These forces are entirely electrostatic in origin and are generated by nuclear and electronic charges. An exact solution of this problem cannot be achieved analytically because a polyatomic molecule yields a many-body problem. An approximation has to be introduced by keeping the nuclei fixed and allowing the electrons to move to the minimum energy level. This is the Born–Oppenheimer approximation in quantum mechanics; the nuclear forces on the electrons can then readily be calculated for that particular nuclear configuration by assuming them to be point charges in the Hamiltonian of the Schrödinger equation. The Schrödinger equation has the form

$$H\psi = E\psi \tag{3.51}$$

where ψ is the wave function, H is the Hamiltonian and E is the ground state energy. The Hamiltonian is given by Steiner (1976):

$$H = \frac{h^2}{8\pi^2 m_e} \sum_{i=1}^{N} \nabla_i^2 - \sum_{i=1}^{N} \sum_{\alpha=1}^{\nu} \frac{Z_\alpha e^2}{4\pi\varepsilon_0 r_{i\alpha}} + \sum_{i>j=1}^{N} \frac{e^2}{4\pi\varepsilon_0 r_{ij}} + \sum_{\alpha>\beta=1}^{\nu} \frac{Z_\alpha Z_\beta e^2}{4\pi\varepsilon_0 R_{\alpha\beta}} \tag{3.52}$$

where we have a molecule composed of ν nuclei each with a nuclear charge $Z_\alpha e$, there are N electrons of mass m_e and the electron charge $-e$; h is Planck's constant, ε_0 is the permittivity, $r_{i\alpha}$ is the distance between electron i and nucleus α, r_{ij} is the distance between electron i and electron j, $R_{\alpha\beta}$ is the internuclear distance between nuclei α and β and ∇_i is the Laplacian for each electron i. Thus the ground state energy of the molecule can be calculated for the nuclei held in fixed positions. The solution of this equation, however, is non-trivial and there is a wide choice of methods available. The bottleneck in the computation occurs in trying to solve the integrals employed in the representation of the molecular orbitals. Theoretical methods for the solution of the molecular wave function are surveyed by Steiner (1976) and the application of quantum chemistry to drug research has been adequately documented by Richards (1983).

The procedure outlined would give the ground state energy for a particular nuclear geometry. However, the geometry input may not be that for the global minimum energy conformation. If we want to examine the global minimum energy, then numerous geometrical inputs are needed for each cycle of computations. The energy of the molecule is a function of its internal coordinates, bond lengths, bond angles and dihedral angles; all would need to be varied by a minimization process to obtain the minimum energy geometry. Quantum-chemical calculations can, at the moment, only be performed on small molecules or on molecular fragments because of the practical limitations imposed by current computers. A full geometry optimization by quantum-chemical methods is not yet feasible, therefore other approximate methods are necessary to study intramolecular variations in geometry. The force field method provides an attractive alternative for studying geometry and molecular energies.

The force field method attempts to take into account bond stretching, bond angle variation, torsion angle rotations and non-bonded interactions by using simple potential functions optimized for different atom combinations. The potential energy is then given by the equation

$$V = \sum V_{\text{stretch}} + \sum V_{\text{bend}} + \sum V_{\text{torsion}} + \sum V_{\text{nbi}} \tag{3.53}$$

which, if we consider in terms of atom–atom vibrations, can be written in the form

$$V = V_0 + \sum_{i=1}^{3n} \left(\frac{\partial V}{\partial x_i}\right)_0 \Delta x_i + \frac{1}{2} \sum_{i,j=1}^{3n} \left(\frac{\partial^2 V}{\partial x_i \, \partial x_j}\right)_0 \Delta x_i \, \Delta x_j$$
$$+ \frac{1}{6} \sum_{i,j,k=1}^{3n} \left(\frac{\partial^3 V}{\partial x_i \, \partial x_j \, \partial x_k}\right)_0 \Delta x_i \, \Delta x_j \, \Delta x_k \tag{3.54}$$

where the molecule contains n atoms with three coordinates x_i, x_j, x_k, V_0 is the global minimum in the potential energy with coordinates x_0, Δx_i is the displacement from this position. If the energies are measured relative to V_0 and V_0 set to zero then only the third term is important for small displacements and it can be simplified to

$$V = \frac{1}{2} \sum_{i,j=1}^{3n} f_{ij} \Delta x_i \, \Delta x_j \tag{3.55}$$

where f_{ij} are the force constants. Thus a valence force field can be represented by

$$V = \frac{1}{2} \sum_i f_{r,i}(r_i - r_{0i})^2 + \frac{1}{2} \sum_k f_{\theta,k}(\theta_k - \theta_{0k})^2 + \frac{1}{2} \sum_l f_{\omega l}(\omega_l - \omega_{0l})^2 \tag{3.56}$$

where r is the bond length, θ the bond angle and ω the torsion angle. However, this simple force field will not take into account electrostatic, dipole or non-bonded interactions and terms including these need to be added to generate a more representative force field. In many respects the type of force field that needs to be used is dependent on the molecular problem being investigated. As well as trying to employ the best equation to suit the problem there is the further question of selecting appropriate parameter constants. The force constants are empirically determined by spectroscopy, structural or energy data from a large variety of compounds. Therefore the parameters need to be optimized by a least squares procedure. A variety of programmed molecular mechanics computations are available for different classes of problem.

Geometry optimization is included in most molecular mechanics computations since the cost of each calculation for a single geometry is very small; many calculations can be carried out with ease. The optimization methods utilize standard minimization techniques by iterative calculations. The principles of geometry optimization are easy to follow, but there are pitfalls to avoid which are directly related to the optimization method. For example, consider the elementary case of a diatomic molecule A—B in which the bond length is to be varied (see figure 3.4). The energy of the molecule is a function of the separation x and has an energy $V = f(x)$. If the initial position at the beginning of the iteration is x_0 and x_0 is not at the maximum energy, then the minimum can be approached by moving in the direction of the gradient $\mathrm{d}V/\mathrm{d}x$ until the first derivative is zero. So the new position x, after the first iteration, is $x_0 + \Delta x$ where Δx is the gradient of $f(x)$ at x_0. The gradient at x_1 is computed and x_2 found. This is repeated successively until $\mathrm{d}V/\mathrm{d}x = 0$ although in practice the minimum is found to some desired level of accuracy. This is an example of a first derivative technique and the demonstration diatomic molecule could be expanded to a polyatomic molecule by varying each bond length successively. To make the algorithm efficient the atom moving step must incorporate a scale factor, otherwise the procedure will be unnecessarily slow in converging to the minimum in the function $f(x)$. There is one obvious pitfall of the optimization; the minimum found is the one that lies on the steepest descent pathway from the starting position. If we were studying rotation round bonds, this local minimum may not necessarily be the global minimum and other lower minima could well be missed. If the other minima need to be calculated, then a more rational set of starting positions for the problem is required.

The steepest descent technique outlined here can be improved further by utilizing second derivatives in the well-known Newton–Raphson method. Starting at position x_0 the first and second derivatives, $f'(x_0)$ and $f''(x_0)$, are computed. The new position x_1 is then given by

$$x_1 = x_0 - f'(x_0)/f''(x_0) \tag{3.57}$$

Figure 3.4. Energy minimization of a function $V = f(x)$ by the steepest descent, where x is a bond length which can vary, x_0 is an arbitrarily chosen starting position. After the first iteration a new position x_1 is found; iteration is repeated until the local minimum $x_{\min}$ is generated with $\mathrm{d}V/\mathrm{d}x \doteq 0$.

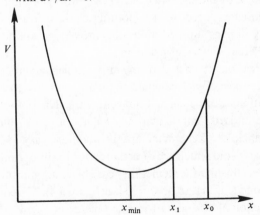

The advantage of the Newton–Raphson method is that it converges much faster than the first derivative method since second derivatives can be calculated analytically with simple potential functions.

In the foregoing example of optimization we have used stretching of one bond only. However, for a polyatomic molecule there will be $3n - 6$ internal coordinates to consider in a full optimization of the molecular geometry. The problem is, therefore, rather more complex than the elementary case. The force constants are handled as a matrix of second derivatives and can be manipulated in the optimization procedure in a variety of ways. A survey of mathematical methods for geometry optimization is given by Burkert & Allinger (1982).

The method of molecular mechanics is extensively used in drug research for the purpose of geometry optimization of both drug and receptor binding site structures. An empirical conformational energy program for peptides (ECEPP) has been written by Browman *et al.* (1975) to examine only torsion angle changes in a polypeptide. The potential function contains a coulomb term, non-bonded interactions, hydrogen-bond energies, torsional energies and loop closing energy terms. The program MM2 designed by Allinger (1977) is for general molecular mechanics calculations. An example of the use of these computation methods in the study of drug–receptor interaction will be given in section 7.6. Perhaps an even more important development has been the incorporation of a dynamic component to the calculations to enable the drug designer to simulate molecular movement in a drug molecule or in a region of the binding site. By the methods of molecular dynamics it is possible to observe molecular movement in scaled real time at a visual display unit and follow the docking manoeuvre from an initial approach trajectory to the formation of a drug–receptor complex.

3.4 Hydrogen bonds in drug–receptor interaction

Hydrogen bonds have the general structure X—H ... Y where X is the donor group, conventionally thought to be covalently bonded to the hydrogen atom, and Y is the acceptor group which, in the isolated state, does not have that hydrogen atom attached to it. A hydrogen bond occurs between the neighbouring atoms X and Y when the distance X ...Y is less than the sum of the van der Waals radii of X and Y. This general structure is termed a two-centre hydrogen bond; there are also three-centre and four-centre bonds. Hydrogen bonds are ubiquitous in biomolecular structures. They are found in nucleic acid structures and are responsible for base-pair formation across the strands of DNA and RNA. In proteins they are major determinants of the three-dimensional architecture of the proteins and hold together α helices and β-pleated sheets. Why do they hold such an important position in structural molecular biology? What is their role in drug–receptor interaction? The answer to the first question is that hydrogen bonds are energetically significant and commonly of the order $20\,kJ\,mol^{-1}$ and, perhaps more important, they are geometry dependent. The effect of an array of acceptors and donors on one molecule is to form a three-dimensional recognition site capable of interacting specifically with another molecule bearing a complementary molecular structure. It is this facility to form complementary geometrical structures which accounts for their importance in structural biology and in drug–receptor interaction. Hydrogen bonds are an essential, but not the only, ingredient in recognition mechanisms.

3.4.1   Geometries of hydrogen bonds

The existence of the Cambridge Structural Database has provided an enormous resource of hydrogen-bond data immediately accessible to computer analysis. Very detailed examinations of different types of hydrogen bond have started to emerge. One particular type of hydrogen bond commonly found in biomolecular structures is N—H . . . O=C where O=C may be a carbonyl or part of a carboxyl group. A complete survey of this type of bond has been carried out by Taylor, Kennard & Versichel (1983, 1984a, b).

Figure 3.5. A geometrical arrangement for the participation of a lone-pair of a carbonyl oxygen atom in hydrogen bonding with an amine. The atoms A, B, C, O lie in the x–y plane.

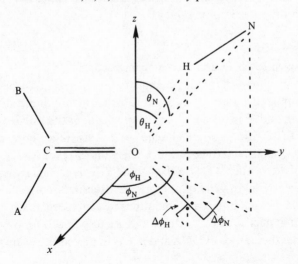

The coordinate system for the N—H . . . O=C bond is shown in figure 3.5. An idealized lone-pair, showing sp^2 hybridization on the oxygen atom, is placed in the $x–y$ plane at $30°$ to the x-axis. The positions of the hydrogen atom and nitrogen atom with respect to the lone-pair can then be described; the distribution is shown in figure 3.6. Hydrogen bonding appears to be in the plane ABC and close to the direction of the lone-pair. There are differences between intermolecular and intramolecular hydrogen bonds; intramolecular hydrogen bonds have a larger θ_H and smaller ϕ_H. The positions of the nitrogen atom show a similar distribution to that of the hydrogen atom (figure 3.6). However, the difference in position for the hydrogen atom relative to the lone-pair $\Delta\phi_H$ and $\Delta\phi_N$ for the nitrogen atom show that the hydrogen atom lies closer to the lone-pair axis than the nitrogen atom. The nomenclature for two-, three- and four-centre hydrogen bonds is shown in figure 3.7. About 20% of all hydrogen bonds are three-centre; four-centre bonds are rare ($< 1\%$). It seems possible that these multiple hydrogen bonds are related to steric forces in crystal packing. Their possible relevance in structural molecular biology is unknown.

Bond lengths and angles can be studied; the nomenclature is shown in figure 3.8. The mean value for $r(H—O)$ is 1.921 ± 0.143 Å. There are significant differences between inter- and intramolecular bond lengths and the nature of the donor or acceptor group has a significant effect on bond length. The angle $\alpha(N—HO)$ corrected for geometrical factors is $180 \pm 16°$; $\alpha(N—H . . . O)$ is inversely correlated with $r(H . . . O)$. The distance $r(N—H)$ is 1.030 ± 0.016 Å. The $r(C=O)$ length is of course dependent on the environment of the acceptor group, ie whether it is a carbonyl, carboxylate, ketone or amide group. A knowledge of hydrogen-bond geometry is important for drug design because the bond, although

Figure 3.6. An approximate distribution for the hydrogen-bond angles θ_H and ϕ_H.

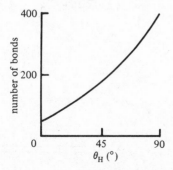
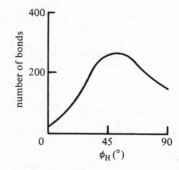

geometrically directed, does show a statistical deviation from the mean. For example, if we were attempting to design a molecule with an N—H moiety putatively hydrogen-bonded to a C=O group at the receptor, then the nitrogen atom nuclear position could lie in a polar cap of the sphere of radius $r(N=O)$ with a diameter of 0.56 Å without appreciably altering the hydrogen bond. It is this leeway in hydrogen-bond geometry which makes them so useful in biomolecular structures.

Similar studies of the geometry of hydrogen bonding have been carried out on particular classes of molecular structures such as water (Pedersen, 1974), polysaccharides and amino acids (Jeffrey, 1982; Jeffrey & Maluszynska, 1982). A list of common hydrogen-bond lengths is given in table 3.2; these bond lengths are approximate. In general, hydrogen-bond donors are: for the —OH group, alcohols, phenols, carboxylic acids, oximes and hydroperoxides; for the —NH group, primary and secondary

Figure 3.7. Multicentre hydrogen bonding: (*a*) two-centre; (*b*) three-centre; (*c*) four-centre hydrogen bonds.

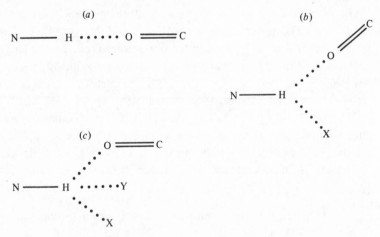

Figure 3.8. Hydrogen-bond length and angle relationships. (Re-drawn from Taylor, Kennard & Versichel, 1983.)

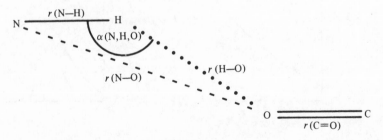

amines, pyrroles, amides and carbonates; for —SH groups, thiols and thiophenols; for C—H chloroform and acetylenes. Hydrogen-bond acceptors have unshared electron pairs with the exception of nitro groups. π-Electron systems can also form acceptor groups; for example, ethynes and aryl rings.

Accurate hydrogen-bond energies for different types of interaction are not widely available but where calculations have been performed the energy of the bond is very dependent on the donor and acceptor groups. In water dimers the hydrogen bonds have energies in the range $16-25\,\text{kJ mol}^{-1}$, whereas ion complexes form very strong hydrogen bonds, eg $\text{F}^-\text{H}_2\text{O}$ has an energy of $96-109\,\text{kJ mol}^{-1}$ (Schuster, 1978). A recent attempt to obtain a potential function for hydrogen-bond geometries has been provided by Vedani & Dunitz (1985). Their function employs the geometrical description drawn in figure 3.5; force field parameters include polarizability, the number of outer shell electrons and the van der Waals radius. One consequence of hydrogen-bond energies being significant is the tendency to maximize the number of hydrogen bonds in the formation of a crystal complex. This phenomenon is particularly marked in carbohydrate crystals where water molecules may also be included in the hydrogen-bonded network. Furthermore, cooperativity in hydrogen-bond formation seems possible; in water a chain of hydrogen bonds has a greater stabilization energy then the sum of the component parts (Del Bene & Pople, 1970). Substantial reviews of hydrogen bonding are given by Vinogradov & Linell (1971) and Schuster, Zundel & Sandorfy (1976).

Table 3.2. *Hydrogen-bond length and geometries*

Bond	Angle (°)	Interatomic distance (Å)	
N—H...O=C	α(NHO)160	r(N—H)	1.03
Two centre		r(H...O)	1.86
		r(N—O)	2.90
Three centre		r(H...O)	2.10
H_2O...N		r(O...N)	3.14
		r(H...N)	2.24
C=O...H—O—P		r(O...H)	1.21
		r(O...O)	2.42
N—H...O=P		r(H...O)	1.99
O—H...O	α(O—H...O)167	r(H...O)	1.82
P—O—H...OH_2		r(H...O)	1.53
		r(O...O)	2.58

3.5   Molecular electrostatic potentials

The molecular electrostatic potential is the potential generated by the charge distribution of the molecule. It can be expressed in the analytical form by equation (3.18) as the sum of the multipole moments of the molecule. The electrostatic potential is a scalar quantity whereas its derivative, the electrostatic force, is a vector. Computations of the electrostatic potential round drug molecules and binding sites are now commonplace. For polar molecules the electrostatic potential is anisotropic and in many cases indicative of certain types of chemical reactivity; for example, we can visually locate nucleophilic or electrophilic sites of putative hydrogen-bonding regions by inspecting contour maps of the potential. Electrostatic potentials are now being used as a parameter in QSAR studies of drug design.

3.5.1   Calculation of the electrostatic potential

The electrostatic potential $V(r)$ at the point r can be calculated from the formula

$$V(r) = \sum_{\alpha} \frac{Z_{\alpha}}{|R_{\alpha} - r|} - \int \frac{\rho(r')\,dr'}{|r' - r|} \qquad (3.58)$$

where $\rho(r')$ is the electron density function at position r', Z_{α} is the charge on the nucleus α found at R_{α}. There are two terms to the equation: the first is the nuclear charge term and the second is due to the electron distribution. If at position r there was located a separate single charge Q, then the electrostatic interaction energy would be $QV(r)$. If this charge is a proton then $Q = +1$ and the interaction energy is $V(r)$ with the energy expressed in kJ mol^{-1}. It is a widespread convention to quote the electrostatic potential in units of energy as though it were an interaction energy with a proton. The electrostatic force $F(r)$ is

$$F(r) = -\nabla[\pm QV(r)] \qquad (3.59)$$

and has components in x, y and z of a Cartesian coordinate system. The electrostatic potential computed from equation (3.58) is a static potential; the charge distribution is that for the isolated state and not perturbed in any way. However, if the molecule is interacting with another molecule then $\rho(r')$ needs to be re-calculated to take into account possible polarization and charge transfer effects which may occur. This can be done in quantum-mechanical computations by computing $\rho(r)$ from the wave function

$$\rho(r) = \sum_{i} N_{i} \sum_{j} \sum_{k} \sum_{l} \sum_{m} C_{il} C_{im} \psi_{l}(r) \psi_{m}(r) \qquad (3.60)$$

where l and m are atomic orbitals on atoms j and k, N_i is the number of electrons in molecular orbital ψ_i, C_{im} is the coefficient of the atomic orbital ψ_m and r is the position of the nucleus; $\rho(r)$ can then be inserted into equation (3.58) and the interaction performed. Generally the integrals are solved in a three or six Gaussian expansion. Other approximations for the wave function can be used profitably if qualitative information is desired, with the advantage of a considerable reduction in cost of computing. With the semi-empirical methods the overall shape of the potential distribution in unchanged although the location of the minima and values close to the molecular surface will be different. The semi-empirical approximation methods have been reviewed by Politzer & Daiker (1981). A further gross approximation can be used by considering the electronic charge to be located at the nuclear positions as a point charge. The accuracy of this approach is dependent on the set of atomic partial charges used. However, the point charge model is very useful for examining the electrostatic potential round polymers where partial charges are available for the monomer units. The electrostatic potential round carboxypeptidase (Hayes & Kollman, 1976a, b) and nucleic acids (Dean & Wakelin, 1979, 1980a) have been examined by these point charge methods (section 7.3).

3.5.2 Display of molecular electrostatic potentials

The anisotropy of molecular electrostatic potentials makes it difficult to display them on paper since there are four variables, three spatial and one magnitude of potential. No single display method on a paper surface is adequate; currently available graphical display methods are surveyed by Dean (1984). Isoenergy maps in a single plane such as that shown in figure 3.9 for the antibiotic actinomycin D are the most common method of display. Its advantage is that the map is quick to compute but it may need to be drawn in numerous planes to provide detail of the potential surrounding groups out of the graphical plane.

Representation of electrostatic potential throughout three-dimensional space can be achieved in a limited way by two methods. The 'chicken wire' type of diagram connects three-dimensional grid points with the same potential value. This method is widely used in computer graphics terminals with fast refresh facilities so that the potential can be turned on inspection. The method is less useful when drawn on paper since overdrawing of lines can be very confusing; for example, figure 3.10 displays the potential for the diabetogenic drug alloxan at different contour values. The second method overcomes the limitations of a single graphical plane by drawing the potentials round the molecule in stereo-pairs. This gives a very good demonstration of the contour positions relative to each atom in the molecule (figure 3.11). Different potential values can be represented by colours and thus it is possible to build up a clear picture of the distribution of molecular electrostatic potential. The potential surrounding a drug molecule can be visually compared with that of its receptor and complementarity searched for (Dean & Wakelin, 1980a, b). For many interactions we are probably more interested in studying the potential on a particular surface, such as the steric surface surrounding a drug binding site. In this case the potential can be mapped on to a grid drawn on the surface and coded either by a grey-scale (figure 3.12) or by a colour code (Dean, 1983).

Figure 3.9. The molecular electrostatic potential surrounding the antibiotic actinomycin D. The isopotential map is constructed in the plane of the phenoxazone ring, contour values are given in kJ mol^{-1}. (Taken from Dean & Wakelin, 1980b.)

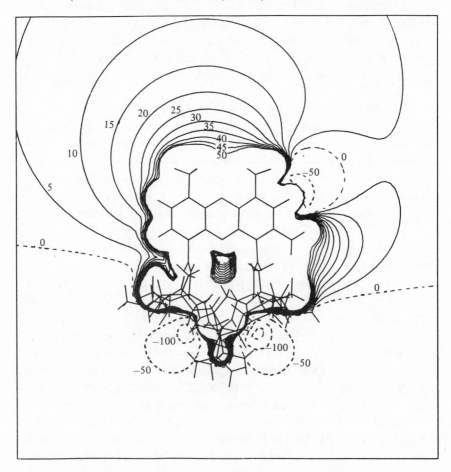

3.5.3 The use of molecular electrostatic potentials

Electrostatic potential calculations are widely used in theoretical chemistry; they point to the location of electrophilic and nucleophilic sites, and where the calculations are of a high order of accuracy they can indicate the preference for electrophilic or nucleophilic attack on particular atoms in the molecule. For example, in guanine, a base in naturally occurring DNA, there are two electrophilic sites associated with N3 and N7. The local minimum of N7 is deeper than at N3, indicating that N7 is preferentially protonated or alkylated. Experiment confirms the observation (Bonaccorsi *et al.*, 1975). In many respects the use of molecular electrostatic potentials has been confined largely to spotting correlations in isopotential maps between differing molecular structures. However, this

Figure 3.10. A 'chicken wire' diagram of the molecular electrostatic potential surrounding the diabetogenic drug alloxan. The mesh is drawn through three coordinate planes. Contour values: (a) $200\,\mathrm{kJ\,mol^{-1}}$, (b) $-50\,\mathrm{kJ\,mol^{-1}}$, (c) $0\,\mathrm{kJ\,mol^{-1}}$. (Taken from Dean, 1984.)

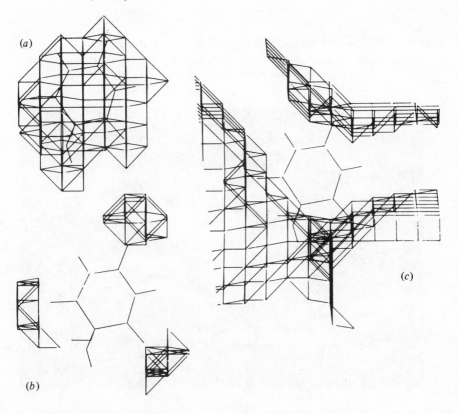

(a)

(b)

(c)

Figure 3.11. Stereoscopic drawing of the molecular electrostatic potential round A-DNA with the alternating sequence poly(dG-dC)·poly(dG-dC), contour value $-2717\,kJ\,mol^{-1}$. (Taken from Dean & Wakelin, 1980a.)

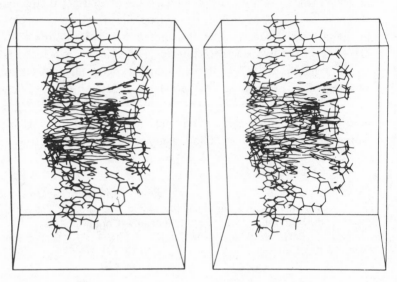

Figure 3.12. The electrostatic potential mapped onto a three-dimensional surface and expressed as a grey-scale. The surface drawn is for one intersubunit binding site on lactate dehydrogenase; scale marks are in Å. (Taken from Dean, 1984.)

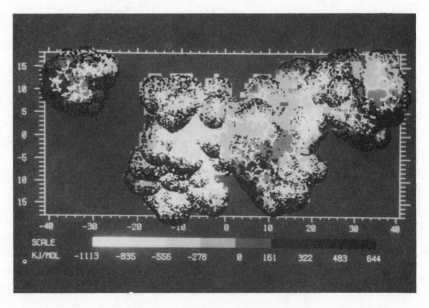

approach is fraught with difficulties concerning which plane to select for comparison. No simple three-dimensional method of comparison has yet been used successfully in QSAR studies.

A further development in electrostatic potential studies has been to take derivatives of the field and examine the electrostatic force at particular positions (equation (3.59)) (Dean, 1981a, b; Lavery, Pullman & Pullman, 1982). The electrostatic field derived from the potential is a vector and the direction of the resultant shows the path taken by a proton at that point. These electrostatic forces are orienting forces and could exert a torsional rotation energy on a molecule located nearby (section 7.5). It is quite likely that the orientation of an approaching drug molecule is controlled, in part, by the receptor electric field. With polar drug molecules the orientation of the molecule at the minimum electrostatic energy is correlated with the direction of the molecular dipole and the direction of the electric field generated by the receptor projected to the centre of mass of the drug molecule. Electric field vectors can be conveniently displayed by an arrow with hidden line suppression (figure 3.13). Arrows may be scaled for magnitude and coded to represent regions of negative or positive electrostatic potential. Critical points can occur where the direction component of the field is nil. These critical points would be a region of orientational instability and could have a profound effect on drug binding by acting as pivot points in particular electrostatic regions which control drug orientation.

The correlation of best orientation of a drug molecule with the direction of its dipole and the local direction field is not necessarily exact even in electrostatic terms since the potential is derived from a multipole expansion,

$$V(r) = \sum_{n=0}^{\infty} (Q_n / r^{n+1}) \tag{3.61}$$

where Q_n is the strength of the 2^n-pole at position r. Other multipoles may play a major role in the orientation mechanism. The realization of this led Weinstein *et al.* (1981) to describe the drug molecule by an orientation vector that includes these higher multipole terms. The orientation vector is the vector sum of all the multipole moments.

The electrostatic potential considered so far has been that computed *in vacuo* and has a qualitative use in the study of drug–receptor interaction. In practice, however, we need to compute the potential in solution if we are to use the potential in a quantitative theory to study electrostatic binding between a variety of drug molecules and their receptor sites. The great problem in transferring computational methods for the *in vacuo* state to that for solution is, how can the dielectric effect be handled? Anisotropy in the dielectric for water occurs near to a protein surface and gives rise to a very complicated pattern of electric fields and multipole moments. This problem would be greatly simplified if the molecular charge distribution was regular in a Cartesian framework; the Poisson equation

$$\nabla\{\varepsilon(r)\,\nabla\phi(r)\} = -\rho(r) \tag{3.62}$$

could then be solved analytically; $\phi(r)$ is the electrostatic potential at the vector position (r), $\varepsilon(r)$ is the dielectric constant and $\rho(r)$ is the charge density. A novel numerical solution to this problem has been proposed by Rogers & Sternberg (1984) and Edmonds, Rogers & Sternberg (1984). The irregularly spaced charge distribution of a molecule is re-distributed to the eight vertices of a surrounding regular cubic grid system. The single point charge q in figure 3.14 is re-distributed to the corners of the cube to give fractional charge $w_i q$ according to a function of the form

$$w_i = (x_i + a)(y_i + b)(z_i + c)/(8x_i y_i z_i) \tag{3.63}$$

Figure 3.13. An electric field map in a plane passing through an intersubunit binding site of lactate dehydrogenase. Each arrow indicates the direction of the force on a proton at each position. (Taken from Dean, 1983.)

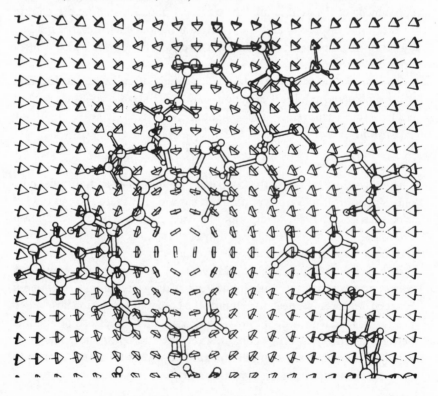

where x_i, y_i, z_i are the coordinates of the *i*th vertex of the cube and a, b, c is the position of charge q. The molecular electrostatic potential is then computed by a numerical finite difference iterative procedure.

This novel strategy takes into account monopole, dipole and quadrupole moment contributions to the potential as well as the dielectric variation. The computational method was tested by comparing it with exact calculations based on the original irregular charge distribution for a small α-helix model protein. The accuracy of the approximate method depends, as expected, on the grid size used for redistributing the charge. With a 1 Å grid, the electrostatic potential outside the van der Waals radii is accurate to 95% of that computed by the exact calculations.

3.6   Hydrophobic interactions

Hydrophobic interactions are referred to sometimes as hydrophobic bonds. The latter expression can be misleading. In a chemical bond there is a sharing of electrons; in hydrophobic interactions no electrons are shared. Thus they are not chemical bonds in the strict sense and here the term hydrophobic interactions will be used. The importance of hydrophobic interactions in structural biology has been stressed now for many years (Kauzmann, 1959); however, it has to be admitted that hard evidence in support of their role is not as strong as one might imagine. In many cases the term has been abused by being invoked to explain anomalies in structural hypotheses. A further reason for this nagging worry

Figure 3.14. The distribution of charge from the point q to the corners of a cube as a method for solving the Poisson equation.

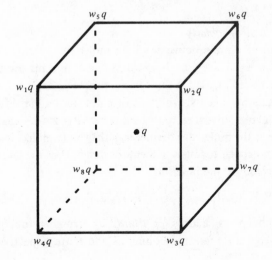

is two-fold: the definition of a hydrophobic interaction is not unique (there are two ways of viewing the problem), and the theory of the interactions is not as well advanced as for other intermolecular interactions. Hydrophobic interactions between two molecules only have meaning in the presence of a third ingredient – aqueous solvent; we are therefore dependent on a detailed knowledge of water structure and solute–solvent behaviour before we can appreciate the importance of hydrophobic effects as a factor in biomolecular interactions.

As far as drug–receptor interaction is concerned, hydrophobic properties of the drug molecule can be separated into two components with pharmacological function. Firstly, hydrophobic properties of the drug are intimately related to the pharmacokinetic parameters in transport of drug molecules through lipid barriers to the site of action of the drug. This factor forms an essential component in Hansch analysis in QSAR studies and the design of more 'potent' compounds. We shall ignore this pharmacokinetic component. Secondly, hydrophobic interactions are believed to be a significant component in the binding energy linking the drug molecule with its receptor. In this section we shall examine this second factor, intermolecular binding. Curiously, these two factors reflect the two ways of looking at hydrophobicity; one is seen as the free energy of transfer of solute between aqueous solvent and a lipid-like environment; the second approach looks at the tendency of two molecules to form dimers and aggregate in aqueous solution. A further complicating factor in drug–receptor interaction is that only part of the binding energy may be hydrophobic in origin, since even at the microenvironment of the receptor binding site only a subregion may exhibit hydrophobic molecular groupings.

3.6.1   Evaluation of hydrophobicity

The formation of dimers between solute molecules in aqueous solvent has been studied extensively by Ben-Naim (1980) and his formal description of the process will be used in this section to give an explanation of hydrophobic interactions. Let two solute molecules, ss, be placed at positions 1 and 2 in aqueous solvent so that the distance R_{12} is very much greater than σ, where σ is the molecular diameter of the solute molecule. If we now bring the two molecules together to form the dimer, then the Gibbs free energy change is given by the function

$$\Delta G(\sigma) = G(T, P, H_2O, R_{12} = \sigma) - G(T, P, H_2O, R_{12} > \sigma) \qquad (3.64)$$

where $\Delta G(\sigma)$ is the change in energy produced by bringing the two molecules from wide separation to form a dimer, T and P are respectively temperature and pressure and are constant, and H_2O is the solvent. This free energy can be viewed as the difference between the free energies of solution of the monomer and the dimer plus the intermolecular potential between the two solute molecules,

$$\Delta G(\sigma) = \Delta \tilde{\mu}_D - 2\Delta \tilde{\mu}_s + U_{ss}(R_{12} = \sigma) \tag{3.65}$$

where $\Delta \tilde{\mu}_s$ is the free energy of solution of the monomer (s) and $\Delta \tilde{\mu}_D$ is an analogous quantity for the dimer (D), and U_{ss} is the intermolecular potential between the two solute molecules *in vacuo*. U_{ss} is constant and at low solvent densities the free energy change $\Delta G(\sigma)$ approaches U_{ss}. If the free energy change is now studied with respect to separation (R) between the solute molecules, then the force between the two solute molecules $F_{ss}(R)$ is

$$F_{ss}(R) = -\partial \Delta G(R)/\partial R \tag{3.66}$$

The change in the free energy of the hydrophobic interaction at separation (R) is given by

$$\Delta G(R) = U_{ss}(R) + \delta G^{HI}(R) \tag{3.67}$$

where $\delta G^{HI}(R)$ is the energy change of the hydrophobic interaction and $U_{ss}(R)$ is independent of the solvent.

The problem facing us is how to determine $\delta G^{HI}(R)$, because there are at least two physicochemical components. Firstly there is a solvation effect whereby water interacts with the solute and the solvated solute changes the properties of the solvent close by; in short, the solute deforms the three-dimensional net of hydrogen bonds in water. Secondly there is the free energy component of solvent plus solvated solute. Furthermore, since solvation round many molecules is anisotropic there will be an orientation component to the calculations (section 6.4). Not surprisingly there is no universally satisfactory method yet for determining $\delta G^{HI}(R)$. However, each of the parts of the problem can be investigated separately and combined in a Monte Carlo simulation with approximations for computing the pair-potentials. A model calculation for the hydrophobic interaction between two methane molecules in water has been carried out by Dashevsky & Sarkisov (1974). They took a cubic block of space in which they placed two methane molecules at different separations; water molecules were placed in the block to give the density of water. The water molecules were then moved in a Monte Carlo simulation and the energies of the system calculated. For an explanation of the Monte Carlo method, see section 6.3.3. The potential function for water–water interaction included hydrogen-bond potentials. The solvation energy of methane in water could be calculated using statistical mechanics and the hydrophobic interaction between methane molecules in water estimated as the dimer formed. At 4 Å separation the $\delta G^{HI}(R)$ had a value of $\sim -6.3\,\mathrm{kJ\,mol^{-1}}$ whereas the $U_{ss}(R)$ was about $-1.3\,\mathrm{kJ\,mol^{-1}}$. The hydrophobic interaction is therefore attractive and is the dominant term in $\Delta G(R)$ for methane in water; the interaction extends over about 7 Å separation.

3.6.2   Molecular shape and hydrophobic interactions

A characteristic feature of the study of x-ray crystallographic structures of proteins is a directional preference for polar and non-polar groups. In general, polar groups are orientated towards the aqueous solvent and non-polar regions are found within the three-dimensional structure away from the solvent regions. Hydrophobic interactions are important determinants in the folding of polypeptide chains (section 6.4.2).

Consider a molecule composed of four atoms arranged in a *trans* or a *cis* configuration as in figure 3.15(a) and (b). Let the interaction energy of the four atoms be $U(a)$ and $U(b)$ for the two geometries, then the probability P of the occurrence of the two forms *in vacuo* is given by the ratio

$$P(a)/P(b) = \exp\{-U(a)/kT\}/\exp\{-U(b)/kT\} \tag{3.68}$$

If we now place these two molecules as solutes in water then the hydrophobic interaction can be expressed as

$$\delta G^{HI}(a) = \sum_{i<j} \delta G^{HI}(R_{ij}^a) + \phi_a \tag{3.69}$$

$$\delta G^{HI}(b) = \sum_{i<j} \delta G^{HI}(R_{ij}^b) + \phi_b \tag{3.70}$$

Figure 3.15. The van der Waals surface of a molecule in (a) *trans* and (b) *cis* configurations together with the respective solvent accessible surfaces.

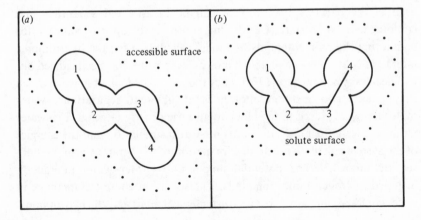

where R_{ij} is the interatomic distance between atom i and j, and ϕ_a and ϕ_b are non-additive functions in the hydrophobic interaction. If the intramolecular potentials $U(a)$ and $U(b)$ are equal, the ratio of the probability of the conformations is

$$P(a)/P(b) = \exp\left(-\{\delta G^{HI}(a) - \delta G^{HI}(b)\}/kT\right) \tag{3.71}$$

and by substitution

$$P(a)/P(b) = \exp\left(-\{\delta G^{HI}(R_{14}^a) - \delta G^{HI}(R_{14}^b)\}/kT\right) \exp\left\{-(\phi_a - \phi_b)/kT\right\}$$
$$\tag{3.72}$$

since we have the equalities $R_{ij}^a = R_{ij}^b$ for i and j not equal for 1 and 4. The difference in the hydrophobic interaction between the *trans* and *cis* conformers is dependent on the distance between atoms 1 and 4. It can readily be shown that this difference is related to the solvent accessible surface of the conformations (Ben-Naim, 1980).

For simplicity, the molecule shown in figure 3.15 can be considered to be formed from four hard spheres, each of radius σ_1. If a solvent molecule, also considered as a hard sphere of radius σ_2, is brought to the surface of the solute molecule and rolled along the surface, the centre of the solvent will map a further surface denoted by the dotted lines in figure 3.15 (section 4.3, 4.8). If the position vector of the solvent molecule is denoted by **R** then the ratio of the probabilities of the two conformations is

$$\frac{P(a)}{P(b)} = \frac{\displaystyle\int \exp\left(-\{U(a) + U(2|a)\}/kT\right) d\mathbf{R}}{\displaystyle\int \exp\left(-\{U(b) + U(2|b)\}/kT\right) d\mathbf{R}} \tag{3.73}$$

where $U(2|a)$ is the interaction energy between the solvent and solute at position **R** for confirmation (a), and a similar factor for conformation (b). The integration is performed throughout the volume of the block V. If the integrand has the property

$$\exp\left\{-U(2|a)/kT\right\} = \begin{cases} 0 & \text{if closest separation } < (\sigma_1 + \sigma_2)/2 \\ \text{or} \\ 1 & \text{if closest separation } > (\sigma_1 + \sigma_2)/2 \end{cases}$$

then the integral is 0 when the solvent centre lies within the accessible surface marked by the dotted line in figure 3.15, and 1 when the solvent lies outside the accessible surface. Thus we have two excluded volumes $V_{ex}(a)$ and $V_{ex}(b)$. The ratio of the probabilities of the two conformations is

$$P(a)/P(b) = \{V - V_{ex}(a)\}/\{V - V_{ex}(b)\} \tag{3.74}$$

thus, if $V_{ex}(a) > V_{ex}(b)$, then

$$P(b) > P(a)$$

and conformation (b) would be more favourable in terms of the hydrophobic interaction with solvent than conformation (a), that is, if $U(a) = U(b)$ and if the solute molecule is non-polar.

The idea of the accessible surface area surrounding a molecule having an effect on hydrophobic interactions has become increasingly important in drug research. Computing methods for determining the accessible surface area and excluded molecular volumes will be described in the next chapter. A clear relationship is emerging between water–lipid solubility and the surface area of the cavity occupied by a hydrophobic molecule in solvent (Amidon et al., 1975; Hermann, 1972). At the moment the relationships are empirical although refined enough to predict, with a promising level of accuracy, the solubility in water of tetrahydropyran and cholesterol. Furthermore a relationship is emerging between the cavity surface area of different molecular moieties in a drug molecule and the overall hydrophobic constant used in QSAR parameters (for a review of this development see Franke, 1984). These empirical relationships need to be placed on a firmer theoretical base by combining water–solute pair-potential calculations with equation (3.73) and computed solubilities with equation (3.65). This should make it possible to assign group contributions to hydrophobic interactions between two molecules in solution. The scheme outlined earlier by Ben-Naim (1980) for studying hydrophobic interactions opens up an exciting new area for quantifying hydrophobic effects in drug–receptor interaction.

4

Characterization of molecular shape

4.1   Electron density distribution and shape

The intramolecular factors that determine molecular shape are extremely complicated for the non-specialist to understand. However, a simple description can give an insight into how the atoms are held together. In general we conceptualize the arrangement of atoms as being linked by bonds. Bonds are really only topological entities that define the line of maximum attractive force between adjacent nuclei when the nuclei are arranged in some stable configuration. The direction of these bonds is determined by the electron density distribution associated with each atom and the nuclear charges. For a polyatomic molecule this presents a many-body problem and it cannot be solved analytically by the Schrödinger equation since both the nuclei and the electrons are free to move. Fixed geometrical positions between the nuclei are generated by the energy gradient between them and the electron distribution when a balance between the two parameters occurs. This fixing of positions, however, is not absolute and rotations round the internuclear axes may produce equivalent conformational energy stages. A re-distribution of the electron density permits these rotational changes. Concomitant with a change in electron distribution is a change in shape of the electron density in regions associated with each atom. These local regions of electron density can be defined by the topological properties of the local gradient field of the scalar electron distribution. Figure 4.1 illustrates diagrammatically the topological properties of the field generated by the electron density distribution round a diatomic molecule. Contours of charge density encircle the two nuclei and spread to the internuclear regions. The gradient paths are indicated by lines perpendicular to the density contours with the direction shown by the arrows. The collection of gradient paths which terminate at a nucleus defines the volume of space associated with that nucleus. In terms of topology, the nucleus is a parabolic critical point. The boundary between two atoms is, therefore, the separatrix which runs directly towards the hyperbolic critical point where the field vanishes. A full quantum topological description of molecular charge distribution is given by Bader, Anderson & Duke (1979) and Bader, Nguyen-Dang & Tal (1979).

With nuclei in fixed positions, the charge density can easily be calculated by *ab initio* quantum-mechanical methods. If the charge density is plotted out, say in the molecular plane, then the separatrix for the gradient field can be drawn where the gradient is a minimum and moving towards a critical point. In general, the separatrixes lie perpendicular to a bond. An example of this procedure is given for the pharmacologically interesting molecule tetramethylammonium (TMA) in figure 4.2. TMA is used as a model for the cationic head of acetylcholine. Integration of the electron density in regions defined by this topology, for successive planar sections through the molecule, yields the charge associated with each atom (Barrett, 1983). Charge distributions in two moieties can be compared by superimposing the atoms to be inspected and constructing a density difference map. Comparison between TMA and neopentane, an uncharged analogue, by density difference maps indicate that the cationic charge of the quaternary ammonium ion is entirely delocalized to the outer surface of the molecule and on all 12 hydrogen atoms.

Figure 4.1. Electron density contours for a diatomic molecule (dotted lines). Electric gradient paths and their directions are indicated by the full lines and the arrow heads. The dashed line is the direction of the bond between the two atoms.

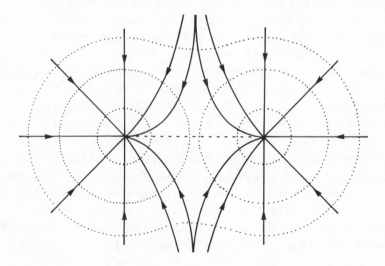

4.2 Distance matrices (DMs) and data manipulation

The shape and absolute configuration of a molecule composed of n atoms can be expressed by using $3n-6$ internal coordinates of bond lengths, bond angles and torsion angles; alternatively it can be represented by $3n$ Cartesian coordinates and $n-1$ bonds. Both of these methods require a three axes scale for graphical representation. Another convenient method for representing shape is by the DM. This is a square matrix with $n \times n$ elements and thus a two-dimensional representation. Each element in the DM is found by computing the interatomic distance d_{ij} between atom i and j from coordinate information

$$DM_{(ij)} = d\{i_{(x_i, y_i, z_i)} - j_{(x_j, y_j, z_j)}\} \tag{4.1}$$

The matrix is symmetric with diagonal elements of zero, therefore only one half of the matrix is needed to give the relevant distance information and the lower triangular matrix is conventionally used to describe the molecule with $\sum_{i=1}^{n} i$ elements. The matrix is not affected either by rotation or translation of the coordinates. However, it must be remembered that handedness information is lost; the DM of two mirror image isomers will be identical.

The pattern of values, but not the values themselves, in the distance matrix is dependent on the atomic numbering scheme. If the numbering scheme follows some logical topological order, such as the sequential numbering of residues along a polypeptide backbone, then a difference distance matrix between two conformations can be portrayed graphically by a contour plot, for example for the drug tetrodotoxin in figure 4.3. The

Figure 4.2. Sketch of the electron density contours for TMA. The plane passes through two carbon atoms with the nitrogen atom at the centre. Separatrixes partition the molecule into atomic regions.

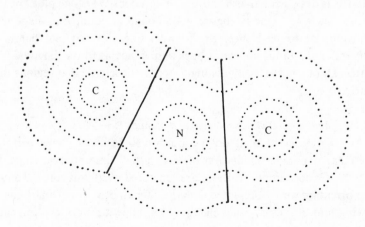

difference distance matrix (DDM) is obtained by subtraction of the corresponding elements of two distance matrices. The DDM elements $DDM_{(ij)}$ are

$$DDM_{(ij)} = |DM1_{(ij)} - DM2_{(ij)}|$$ (4.2)

Figure 4.3 shows the DDM for the two forms, A and B, of tetrodotoxin as a contour plot and as a surface drawing. This method of displaying differences in structural geometry has been used extensively in the comparison of protein structures for different conformations and is employed routinely in the analysis of data from molecular dynamics calculations. The statistical difference, s, between the two distance matrices is given by Padlan & Davies (1975)

$$s = (1/N)\left\{\sum_i \sum_j DDM_{(ij)}^2\right\}^{1/2}$$ (4.3)

Other types of DM have been developed for different requirements. For example, one might want to examine certain structural features such as helices and sheets between similar proteins. These subregions can be expressed as submatrices and a partitioned difference matrix calculated; if comparisons need to be made, a partitioned DDM can be evaluated in a similar way to the DDM together with its corresponding matrix statistic. This method has proved very useful in studies of protein structure where correlations are sought in sequence homologies between proteins with similar geometrical conformations (Padlan & Davies, 1975; Nishikawa & Ooi, 1974). The method can also be used to study the differences in geometry between two known structural binding sites for drug molecules obtained from proteins of different species.

The DM for Cα–Cα distances in polypeptides often shows interesting patterns for Cα–Cα distances close to each other. If the matrix, scaled for distance, is drawn either as a contour or as a density shaded plot, then two features show up readily (figure 4.4); bonds lie along the diagonal or perpendicular to it. Helices are found along the diagonal bands and antiparallel β-strands appear as lines perpendicular to the diagonal of the matrix. Domains are easily distinguished as closely related regions in the matrix.

4.2.1   Animated DMs in molecular dynamics

Consider a small polypeptide of 50 residues; if we perform a molecular dynamics simulation of, say, 50 000 time interval steps, then 2.5×10^6 Cα coordinate positions are generated. With this colossal amount of information we are presented with the difficult problem of analysing it. A moving film of the molecule changing its shape does not circumvent the

Figure 4.3. DDM for two forms of tetrodotoxin displayed as a
contour plot and as a surface drawing.

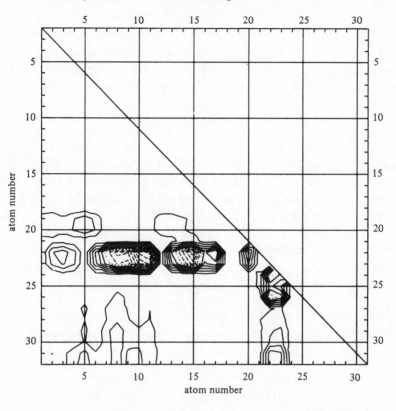

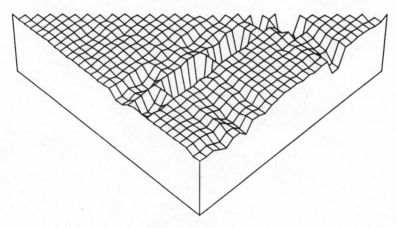

difficulty of perception; relative information about atom positions is forgotten as the film is played through. One method for following relative position changes precisely is the animated distance matrix method developed by Todd & Gillett (1983) and Morffew (1983). After the molecule has progressively been 'heated' in the simulation to the desired equilibrium temperature, the DM is calculated as the initial master matrix, $DM1_{(ij)}$; each successive step, n, has a calculated distance matrix $DMn_{(ij)}$. A difference distance matrix $DDMn$ is then calculated by the expression

$$DDMn_{(ij)} = a\{|DM1_{(ij)} - DMn_{(ij)}|\}^c \tag{4.4}$$

where a is a scaling variable and c is an inverting variable ($c = -1$). The scaling variable is used to generate visible differences on a graphics screen and is constant throughout the run. The DDMs are stored on a Winchester graphics system and converted to a surface plot of 'heights' on a grid system like that in figure 4.3. Each DDM is measured relative to the master matrix. When the DDMs are drawn as successive frames on a graphics device then movement relative to the initial position is portrayed as rising or falling of peaks in elements of the DDM. With a rational numbering system, such as a sequence of successive Cα atoms, then complete shifts in

Figure 4.4. A DM for haemoglobin α chain (Cα–Cα distances). Close to the diagonal the interatomic distance is < 10 Å (double density shading). The contour for distances between 10–13 Å travels out perpendicular to the diagonal (single density shading). Contoured regions perpendicular to the diagonal are characteristic of antiparallel β-strands.

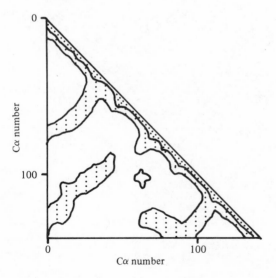

the chain can be followed. These appear as moving waves in the animated DDM, see figure 4.5. Cooperative movements can be followed as the waves move through the DDM and critical nodes can be identified. Sometimes waves move parallel to each other through the DDM and suggest a rhythmic flexing of α helices. The behaviour of selected regions in the molecular structure can be studied more closely by animating an analogous partitioned difference distance matrix.

The application of animated DDMs to drug–receptor interaction is an exciting possibility. Dynamic structural changes at the binding site can, in principle, be observed and need to be correlated with the docking manoeuvre; site accessibility could be handled through the DM.

4.2.2 Molecular superposition

The DDM can easily be used to produce a statistical assessment of the similarity between two geometric structures by computing the rms value for the DDM elements using equation (4.3). Since the matrices are rotationally invariant, they cannot readily be used to superimpose one geometric structure over another to give the best fit. Optimal superpositioning is frequently required in drug design, particularly using

Figure 4.5. One line from an animated DDM for ten atoms at three separate time intervals. When the whole matrix is plotted as a surface plot and the picture is re-drawn for each time interval, the surface peaks move. Coordinated movement can sometimes be observed as the peaks and troughs move in unison.

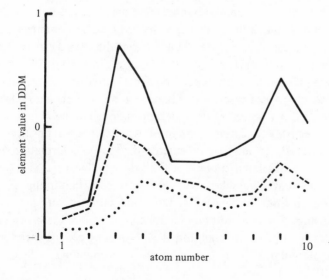

molecular graphics, to observe similarities and differences in molecular shape. Superpositioning is often used to compare geometries between ligands and also conformational differences between homologous protein sequences. A more tedious application of superposition methods is the search for geometrical matches of a particular amino acid sequence throughout the Brookhaven Protein Database of all structurally resolved proteins. This latter type of project is very important for establishing rules of polypeptide folding which might be used in the generation of tertiary structure from simple sequence information. An ambitious project of this type is being carried out internationally and has been reviewed by Sternberg (1983). Large-scale searches of this nature have led to the development of highly efficient superposition matching algorithms that can be routinely applied to 10^6 superpositions.

The algorithm developed by McLachlan (1982) simply calculates a rotation matrix and a translation to minimize the rms values of the fit between two structures. For example, consider two molecular fragments A and B, each containing N atoms, then the atoms will have position vectors $\mathbf{a}_n, \mathbf{b}_n$ $(n = 1, \ldots, N)$ and a statistical weight w_n for each atom. An orthogonal rotation matrix, $\mathbf{R}$, has to be found together with a translation vector, $\mathbf{t}$, that converts the coordinates a_{in} $(i = 1, 2, 3)$ to

$$r_{in} = \sum_j R_{ij} a_{jn} + t_i \tag{4.5}$$

and at the same time minimizes the residual

$$E = \frac{1}{2} \sum_n w_n (r_n - b_n)^2 \tag{4.6}$$

The translation vector $\mathbf{t}$ is simply computed from the centroids between A and B. The rotation matrix $\mathbf{R}$ is computed by an iterative minimization procedure for a succession of finite rotations about the axes. Usually the iteration is complete to a difference of 10^{-10} rad in 5–8 steps.

If matching and eventual superimposition are required in a systematic search of a large number of cases, say of the order 10^3–10^7, then a DDM method can be used as a scan prior to superimposition. This combination has been used effectively to search for similarities in geometrical positions, of all possible atoms that could be involved in hydrogen-bond formation to a receptor, for two dissimilar molecules (Danziger & Dean, 1985). All sequences of atom numbering schemes are generated combinatorially. The DDM is calculated from the two initial DMs and summed with further optimization using a branch-and-bound algorithm. The sum is a measure of the rms difference outlined in equation (4.3). Matching occurs when the rms value is a minimum. The numbering scheme corresponding to the rms minimum can then be used for superimposition at the final step by McLachlan's method. The saving in computing time can be as much as 100-fold by combining the two methods in the way outlined. Structural matching between the drugs saxitoxin and tetrodotoxin, using this method, are illustrated in figure 4.6. Positional matches are illustrated for four- and eight-atom correspondences. It can be seen that the quality of the match diminishes as more atoms are included in the search algorithm; this is not unexpected considering that the molecules are two largely dissimilar geometrical structures. The method is comprehensive; all possible match combinations are considered and it avoids visual subjective bias in spotting molecular matches by eye.

Figure 4.6. Stereo drawing for superimposed matched atoms of saxitoxin and tetrodotoxin (dashed lines). (*a*) Four-atom match; (*b*) eight-atom match. (Taken from Danziger & Dean, 1985.)

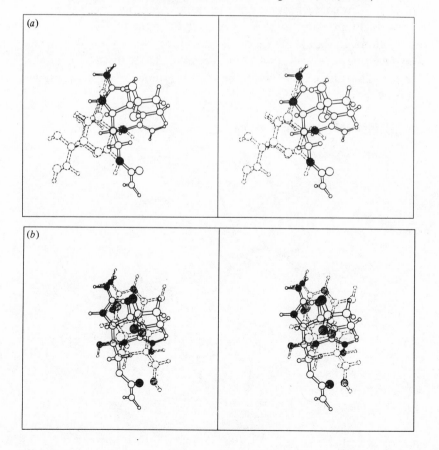

4.3   Molecular surfaces

Internuclear distances and geometries are used extensively in the calculation of intermolecular forces between molecules. But this type of formal treatment handles the atomic positions as only points in space whereas the electron distribution is more correctly approximated by spheres centred on the points. If we represent the spheres by atomic van der Waals radii then the spheres of linked atoms will cut each other through their circle of intersection. The surface area of an atom can then be calculated as that portion remaining after the cap is removed by the circles of intersection of adjacent atoms. However, the chemical properties of a molecule are largely determined by those atoms which lie at the surface interface with the solvent. This is particularly important for a large molecule like a protein where many atoms lie buried beneath the interface and are not available for interaction. Two descriptions of the external surface are possible (figure 4.7) and are related to the ability of a small probe, such as solvent, to touch the van der Waals surface of the solute molecule. The contact surface is the surface actually touched by the spherical solvent probe as it is rolled over the molecular surface. Some atoms, although they lie on the molecular surface, may not make contact with the probe because the path of the probe is deflected by adjacent atoms. The accessible surface, sometimes known as the steric surface, is the surface

Figure 4.7: Definition of molecular surfaces at an exposed interface.

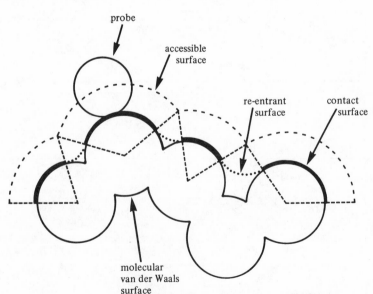

swept out by the centre of the probe molecule as the probe rolls over the contact surface and does not penetrate any other atoms of the molecule.

The accessible surface area of a surface can be calculated by the methods of Lee & Richards (1971) and Hermann (1972). A program MOLAREA is available for the calculation. Both methods use a numerical approximation. A different approach by Wodak & Janin (1980) is particularly applicable to accessible surface area calculations for proteins and an analytical approximation is used. This is based on the assumption that intersecting spheres are randomly distributed and that amino acid residues do not penetrate each other. The agreement of this method with that of Lee & Richards is good, and within 2 % for lysozyme. One caution to bear in mind is that both methods would include cavity surface areas in the calculation of accessible surface areas, even though the cavities may have no connecting channel to the exterior surface.

An exact analytical function for the computation of accessible and contact surface areas, using topology and differential geometry, has been provided by Connolly (1983a, b). The surfaces considered are divided into three regions (figure 4.8) as a spherical probe rolls over the van der Waals surface. Firstly, there is a portion on the van der Waals molecular surface where the probe may roll without touching an adjacent spherical region; this is the contact surface for an atom. The centre of the probe sweeps out the accessible surface related to the atom being considered. Secondly, when the probe rolls into a cleft formed by three atoms on the van der Waals surface, a triangle is drawn between the three points of contact with the

Figure 4.8. Partitioning the molecular surface for area calculations.

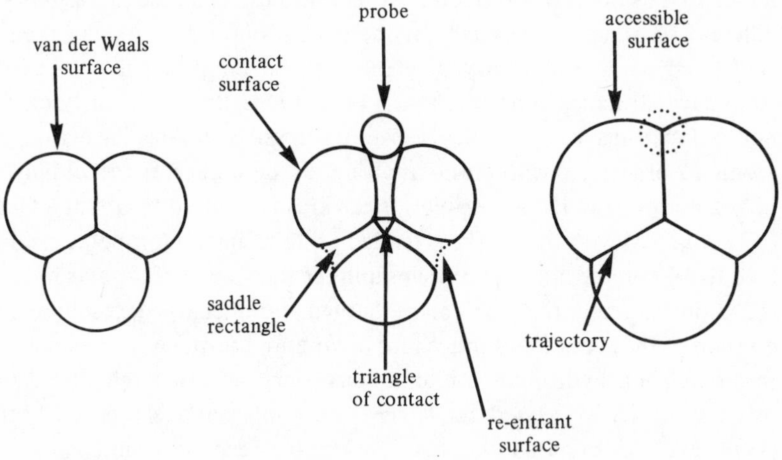

probe. Thirdly, when the probe is rolled along the crevice between two spherical surfaces it traces out a circle round each atom, a concave saddle region, which ends at the edge of the triangular region previously described. With a spherical probe this saddle region will have a circular cross-section with an arc radius equal to that of the probe. This saddle surface is called the re-entrant surface. Therefore the molecular contact surface is built from the contact surface and the re-entrant surface. The accessible surface is swept out by the probe centre. The intersection between adjacent accessible surfaces, the trajectory of intersection, runs in a spherical arc from triangular regions to nearest neighbouring triangular regions. All surfaces are now defined and areas can easily be calculated. Raster graphics can be used to display the surfaces; contact areas appear as isolated patches separated by re-entrant saddle surfaces. This method of partitioning surfaces into accessible surfaces and contact surfaces has proved to be of tremendous value in docking studies between a drug and its binding site. For example, in the binding of the thyroid hormone thyroxine to its binding site on pre-albumin, a pocket region is found adjacent to a phenylalanine residue. The dimensions of this pocket are such that it can be filled by a naphthyl ring. The binding affinity of a series of naphthyl analogues is related to the ability of the naphthyl ring to fit the pocket (Blaney et al., 1982). Contact area patches on the receptor surface should simplify the design of putative ligands by reducing the number of attachment points to precisely defined regions that can be viewed by molecular graphics.

The importance of accessible surface area calculations for drug–receptor interactions is that the surface can be divided according to the chemical properties of the local atoms. For example, the atoms in a moiety could be designated as polar or non-polar. Model studies of surface accessibility of different secondary structures, for the residues in a protein, can then be calculated to give a library of values for each residue in a particular secondary structure conformation (Lee & Richards, 1971). In lysozyme, non-polar atoms occupy 41 % of the total accessible surface; the majority of atoms are associated with the polar groups at the surface. In the folding of a polypeptide chain, the accessible surface area is reduced by about a factor of 3 compared with that of the extended protein chain. Non-polar residues tend to fold on the inside of proteins and the burying of non-polar residues is 2–3 times greater than for polar (charged and uncharged) residues. This division of the accessible surface into polar and non-polar groups can also throw light on hydrophobic interactions, since we saw in chapter 3 that hydrophobicity is related to solvent accessible surface areas. Chothia (1974) has investigated the relationship between accessible area and protein residue hydrophobicity; there is a linear relationship between the two parameters and the free energy change on transfer of the residue from organic solvent to water is $100 \, J \, mol^{-1}$ per $Å^2$ of accessible surface area. Thus it is possible to estimate the hydrophobic contribution to the free energy of dimer formation from the change in solvent accessible area. In haemoglobin the drop in total surface accessible area for α and β contact is $900 \, Å^2$ per monomer. This corresponds to a free energy change of $180 \, kJ \, mol^{-1}$ in dimer formation for the hydrophobic component in the interaction; other components in the interaction will, overall, reduce this value. In the contact region 68 % of the previously accessible surface is non-polar, 16 % polar and 16 % charged (Greer & Bush, 1978). Nevertheless, what is important is the handle on molecular interactions that surface area calculations provide.

4.3.1 Geometrical complementarity between molecular surfaces

Structural complementarity, analogous to a jelly fitting into its mould, seems to be widespread in macromolecular and drug–receptor interactions. Certainly the binding of subunits in protein assembly, and the interaction of DNA with DNA-binding proteins suggests that complementarity is a necessary structural phenomenon (Takeda *et al.*, 1983). A consequence of complementarity is that the possible accessible surface between two binding domains is reduced to a minimum over a large surface region. Thus one way of studying the molecular interaction is to examine the relative positions of two molecules that minimize their joint accessible surface. We thus have a geometrical problem related only to surface shape; energy calculations, which are very costly, can be omitted until final refinement is required. In other words, all non-surface atoms are not really relevant to the molecular interaction. Surface fitting procedures are a rapid short cut for getting close to the correct relative orientation between interacting molecular surfaces. These methods can also be used to search a macromolecular surface for a putative drug-binding site (Kuntz *et al.*, 1982), see section 5.3.

The fitting procedure for two molecular surfaces of molecules B and C in xyz-space is shown in figure 4.9. Each molecular surface is projected onto a yz-plane of grid points, the yz-planes being parallel for both molecules. Each grid point has an associated x-value at the molecular surface. The surface points are contained in the two matrices B_{ij} and C_{ij}. A matrix D_{ij} is then obtained by subtraction

$$D_{ij} = (B_{ij} - C_{ij} - m)/2 \tag{4.7}$$

where

$$m = \min_{ij}(B_{ij} - C_{ij}) \tag{4.8}$$

which brings the two surfaces together to touch at their closest point of contact. The matrix D_{ij} is a measure of the goodness of fit of the two surfaces and the maximum surface area in contact occurs when $\sum D_{ij}$ is a minimum. A value for the geometric fit can thus be computed rapidly. The relative positions of the two molecules can then be shifted by rotation and translation matrix operations to generate other possible fits between their surfaces. An algorithm has been developed by Santary & Kypr (1984) on these lines to search for the optimum interaction between two molecular surfaces. The results for searching and fitting in a model study of haeme–myoglobin interaction and thyroid hormone binding to pre-albumin by Kuntz *et al.* (1982) are very encouraging. They found that their algorithm generated a coordinate fit within 1 Å of that observed in the X-ray crystallographic structures. An obvious consequence of this type of work is that we can specify the molecular surface of a drug molecule and then search for a complementary binding site on a receptor surface even if we do not yet have a drug–receptor crystal complex.

4.4   Intramolecular flexibility and conformation

One of the triumphs of quantum chemistry, applied to drug research, has been the unequivocal demonstration that many drug

Figure 4.9. Section through two molecular surfaces of molecules B and C to measure their complementarity, distances b_{ij}, c_{ij} and d_{kl} are elements of the DMs B_{ij}, C_{ij} and D_{ij}. The dashed surface shows the translation of C to its nearest contact point on molecule B.

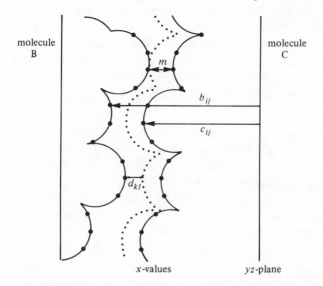

molecule
B

m

b_{ij}

c_{ij}

d_{kl}

molecule
C

x-values   *yz*-plane

molecules, and portions of their receptor, have highly flexible regions. The probability of the occurrence of particular conformations can be readily computed from a starting, rigid, crystal structure. This new tapped wealth of knowledge has provided a fresh impetus to structure–activity relationships within a congeneric series of drug molecules. Steric differences between structures are now being incorporated into QSAR techniques to improve rational methods for drug design (see section 8.3). With small molecules the conformational space can be mapped with relative ease. However, with macromolecular structures, conformational changes at the receptor binding site are less easy to follow and few have been studied in detail. Molecular dynamics calculations should provide a breakthrough method for examining spontaneous structural changes in complex macromolecular regions and hopefully we could build up a time-averaged set of dimensions to define the binding site. Since so much is known about the conformation of amino acid residue pairs in proteins, one can ask the question: is it possible to predict the secondary and tertiary structure of polypeptides from their amino acid sequence? A solution to the problem of structure prediction is crucial to the development of novel peptide drug molecules and this whole research area presents one of the most exciting challenges to drug designers.

4.4.1 Conformational energy computations

A bonded torsion angle is found from a sequence of three connected molecular bonds linking four atoms. The torsion angle convention for the sequence A—B—C—D for the bond B—C is illustrated in figure 4.10; the torsion angle of D is measured positive as a clockwise rotation relative to the fixed position of A when looking in the direction B → C. The torsion angle can be expressed in the range $0-360°$ or $0-180°$ in $+$ and $-$ directions. If atoms B and C have a tetrahedral distribution of bonds then rotation round B—C will give varying torsional energies with minima or maxima at $60°$ intervals. For example, with a 1,2-disubstituted ethane, the pattern of conformational energy changes will be similar to that represented in figure 4.11. When the substituted atoms are in the *cis* position, all atoms are eclipsed and conformationally unstable. A rotation by $60°$ to the *gauche* conformation will give a staggered arrangement and a local energy minimum. An eclipsed unstable conformation is produced by a $120°$ rotation. At $180°$ the substituted atoms are *trans* and staggered; this is the minimum in the conformational energy. Quantum-mechanical computations are ideal for determining torsional barriers. The calculations are of course limited to what we might expect in the *in vacuo* state. Accurate calculations can be made on small molecules, or fragments, by *ab initio* methods although some semi-empirical quantum-mechanical methods are ideally suited for rapid conformational analysis. The program PCILO (perturbative configuration interaction using localized orbitals) (Claverie *et al.*, 1972) has been used extensively for the study of drug molecule conformations, peptide (ϕ, ψ) torsion angles and nucleic acid backbone conformation. In general, the results from PCILO programs are substantiated by more accurate *ab initio* computations. An important insight into rotation barriers that has been revealed by quantum-

Figure 4.10. Torsion angle convention for two adjacent tetrahedrally arranged atoms B and C; B lies in front of C; ϕ is the angle between planes ABC and BCD. Bonds from C are shown by dashed lines.

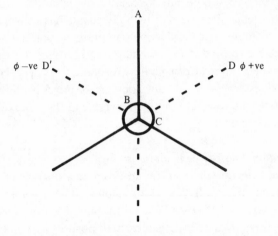

Figure 4.11. Rotation energy barriers for a 1,2-disubstituted ethane.

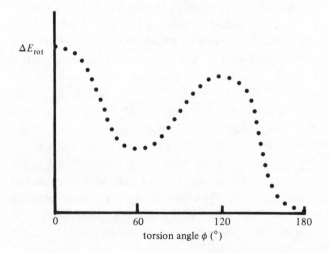

mechanical calculations is the change in orbital energies as the bond is rotated (Veillard, 1976). A more recent approach by Gresh, Claverie & Pullman (1984) has been to consider the molecule as composed of rotatable fragments that can be summed to give the correct intramolecular conformational energy. This method, SIBFA (sum of interactions between fragments computed *ab initio*), establishes a fragment library with *ab initio* computed parameters; the conformational energy is then summed for the contribution of the parameters. The results from this method are similar to those computed from full *ab initio* computations for choline and histamine.

The rotation around two adjacent torsional bonds can be followed by computing the rotational energies round one bond completely and repeating this rotation for incremental steps round the second bond. Conformational energies can be displayed as isoenergy contour maps. Conformational energy minima and rotation barriers can easily be identified by inspection and can be compared with particular conformations encountered in crystal structure determinations. For example, the conformational energy map for acetylcholine taken from Pullman & Courriere (1972) is shown in figure 4.12; two energy minima are found $\tau_1 = -180°, \tau_2 = -60°$ and $\tau_1 = 180°, \tau_2 = 60°$. The torsion angles in

Figure 4.12. Isoenergy maps for rotation of the acetylcholine backbone. Contours: (dashed) 20 kJ mol^{-1}, (dotted) 10 kJ mol^{-1} above the global minimum, the minima are indicated by a large dot. (Re-drawn from Pullman & Courriere, 1972.)

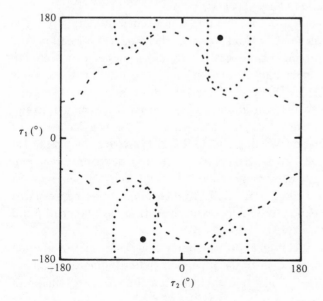

acetylcholine are defined as $\tau_1 = $ C6O1C5C4 and $\tau_2 = $ O1C5C4N1. The conformation in crystal structures of acetylcholine and related analogues prefers the region $\tau_1 = 60°$. However, the conformational energy map indicates that a large region of the total conformational space, about half, has an energy value less than $12.6\,\mathrm{kJ\,mol^{-1}}$ so that the possible conformations are not sharply restricted.

The probability of the occurrence of a particular conformation can be estimated by applying a Boltzmann distribution to the conformational energy map. For each regular grid point on the energy map a probability value is obtained from the equation

$$P'(\tau_1, \tau_2) = \exp\left\{-E(\tau_1, \tau_2)/kT\right\} \tag{4.9}$$

where k is the Boltzmann constant, $E(\tau_1, \tau_2)$ is the energy of the point at τ_1, τ_2 above the global minimum and T is the temperature. A normalizing factor Z then has to be obtained by summation

$$Z = \sum_{\tau_1 = -180°}^{+180°} \sum_{\tau_2 = -180°}^{+180°} P'(\tau_1, \tau_2) \tag{4.10}$$

The normalized probability P is then given as

$$P(\tau_1, \tau_2) = P'(\tau_1, \tau_2)/Z \tag{4.11}$$

These probability values can be contoured as a probability map to portray accurately the probability region in which a particular range of torsion angles will be found.

4.4.2   Conformation of small peptide units

A representation of a dipeptide is illustrated in figure 4.13. The peptide bond in general is planar and in the *trans* conformation, the exception being proline which occasionally has a *cis* conformation. *Ab initio* calculations show that the rotation barrier, ω, in the peptide bond is large $\sim 84\,\mathrm{kJ\,mol^{-1}}$. The C—N bond is short and has a large π contribution due to delocalization of charge from the carbonyl group. There are, therefore, only two torsion angles ϕ and ψ which determine the conformation between two adjacent residues. Moreover, the rotational properties round ϕ and ψ are related only to the residue spanned by ϕ and ψ and are not greatly affected by residues on either side. This observation suggests that, to a large extent, short-range interactions between the residue and the backbone atoms determine the local conformation of ϕ and ψ (Pullman, 1976).

The dipeptide model is an approximation which can be improved upon by extending the unit chain to the tripeptide and including two pairs of torsion angles (ϕ_1, ψ_1) and (ϕ_2, ψ_2). Conformational energy maps of

(ϕ_1, ψ_1) in general are similar to those taken from the dipeptide model with the exception that sometimes additional minima are found at set values of (ϕ_2, ψ_2). The construction of tripeptide models show that these further positions of stability are probably due to the formation of hydrogen bonds across the backbone to form a ring of ten atoms; these are known as C_{10} rings or U-turns (figure 4.14). U-Turns are commonly encountered in crystal structures of protein. Perahia & Pullman (1971) were able to predict 88 % of the turns crystallographically found in α-chymotrypsin by PCILO

Figure 4.13. The dipeptide unit is contained within the dotted lines and consists of the two torsion angles ϕ and ψ adjacent to the central amino acid residue.

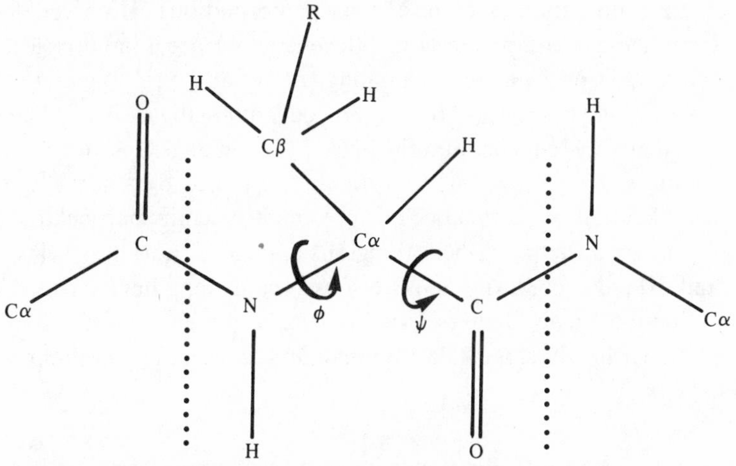

Figure 4.14. A U-turn in a polypeptide composed of a cyclical C_{10} ring hydrogen bonded between residues i and $i+3$.

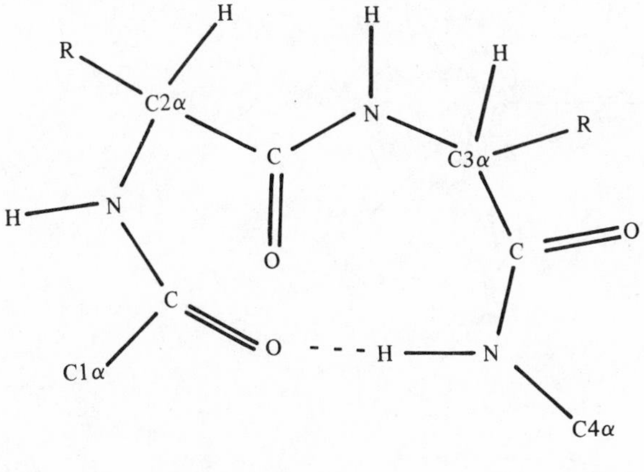

calculations. This type of conformational analysis can be extended to larger sequences of peptide. Other hydrogen-bonding patterns emerge; α-helical structures are generated between 1 . . . 5 residues. With the nonapeptides, a sheet structure with a central U-turn can be observed. Therefore, with higher peptide models, conformational energy calculations can reveal rudimentary features of protein secondary structure.

4.4.3   Graphical representation of multiple torsion angles

In a protein chain, the backbone conformation is described by two adjacent torsion angles across the peptide bond ϕ, ψ, see figure 4.15, to give a ribbon of twisted planes of atoms. Conformational energy plots of ϕ and ψ have already been illustrated for a pair of angles. Isoenergy plots have the advantage that the conformational energy surface can be scanned to reveal relative probabilities of particular conformations. However, for many torsion angles comprehension of the energy surface is impossible. One step round this problem for examining all residue ϕ, ψ angles is to use a Balasubramanian plot (Morffew & Todd, 1984) (figure 4.16). The residues are treated in sequence and the pair of torsion angles plotted with angle ϕ having a + symbol. The two points are joined by a short line to help identification. With practice it is possible to identify particular features of secondary structure: α-helical regions appear as short bars below the axis and extended conformations are observed as long bars across the axis.

Torsion angle changes in a molecular dynamics simulation can be plotted using the same Balasubramanian type of plot for each time interval,

Figure 4.15. Representation of the polypeptide backbone as a ribbon of twisted planes.

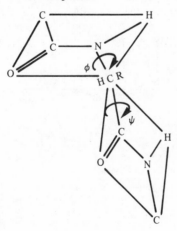

say 0.05 ps, and then changing the display at the end of the next time interval. This creates a graphic animation as the individual bars extend or contract. The method is able to simulate real-time motion of the polypeptide main-chain and cooperative movements can be looked for between large-scale conformational shifts. If it is desired, a static display for a torsion angle pair can be plotted against time (figure 4.17) to examine some critical dynamic torsional effect.

Figure 4.16. Representation of torsion angle variation along a polypeptide sequence. The angle marked + is ϕ, the angle for ψ is given at the other end of the bar. This form of representation allows patterns in ϕ and ψ to be picked out.

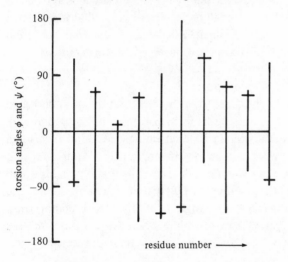

Figure 4.17. Torsion angle variation with time along the backbone. Angles are plotted for three different time intervals (full, dashed and dotted lines).

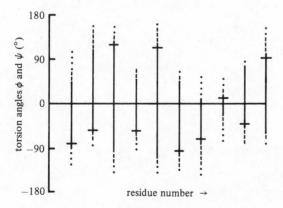

4.5   Macromolecular receptor structures

The only two examples of macromolecular structure that we shall consider are protein and nucleic acid geometry. There are of course other possible receptor structures such as polysaccharides, membrane lipids and glycoproteins, but too little is known about their structures to give a helpful survey. With proteins there are now ~ 140 x-ray structures known and so there is enough detail in molecular geometry to look for specific or classifiable patterns. Similarly there are now a number of polynucleotides, both of DNA and RNA, with elucidated crystallographic structures and different three-dimensional geometrical patterns, which have emerged to provide exciting models for certain types of drug–receptor interaction. The research area of macromolecular structure is still very much in its infancy compared with crystal data on small molecules where 40 000 well-defined geometries are known at present. Therefore, the classification of structural features presented here, whilst generally accepted, is still tentative.

4.5.1   Classification of protein structure

The polypeptide is formed by the condensation of amino acid residues. In general, the bond lengths in each residue of the peptide are invariant compared with those of the isolated amino acid. Bond angles also show little difference between the condensed and individual states. Torsion angles within the residue can vary considerably. The torsion angles which are most important are those along the polypeptide backbone. There are three torsion angles (figure 4.18). Angle ω is 180° in a planar *trans* configuration; the exception is with proline where the *cis* form ($\omega = 0°$) occurs in about 25% of cases. This torsion angle is therefore almost always

Figure 4.18. All the torsion angles of the amino acid residue that establish the tertiary structure of proteins.

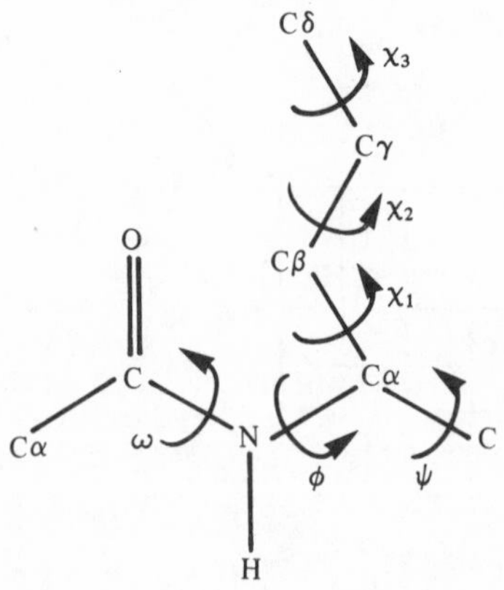

fixed, leaving the two torsion angles ϕ and ψ to determine the conformation of the backbone. Angle ϕ is defined by C_{n-1}—N—Cα—C and ψ is defined by N—Cα—C—N$_{n+1}$. Dihedral angles along the side-chain of the residue are $\chi_1, \chi_2, \ldots$ with χ_1 defined by N—Cα—Cβ—Cγ. For a given pair of residues there is often a region of conformational space that is forbidden due to atom collision between the residues; this greatly restricts the freedom for rotation of the polypeptide region.

The torsion angles (ϕ, ψ) are not the only determining factors for the shape of the polypeptide backbone. Probably more important is the ability of various groups to form hydrogen bonds. These hydrogen bonds stabilize the geometry and various arrangements are possible. α-Helical structures are formed by a loop of 13 atoms along the backbone and have a helix structure of 3.6 residues per turn. The torsion angles are approximately $\phi = -60°$, $\psi = -60°$ and represent a local conformational energy minimum even without the hydrogen bond. The hydrogen bond occurs between N—H ... O=C separated by three residues (figure 4.19). These α helices are right-handed (as in a right-handed screw thread). One frequent characteristic of α helices is that they tend to lie on the outside of protein structures and consequently polar residues are favoured. Charge dipoles point in the direction of the helix and can give considerable summations. Another helix can occur with a backbone of ten atoms in the sequence and

Figure 4.19. The hydrogen-bonding pattern of four consecutive amino acid residues that lead to the α-helix conformation. (Hydrogen atoms are omitted.)

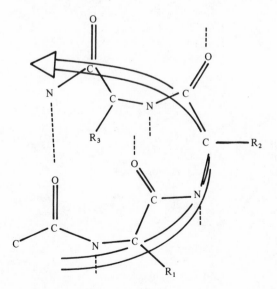

conformation angles are $\phi = -60°$, $\psi = -30°$; two residues are found between the hydrogen-bond groups.

Hydrogen-bonding patterns also give rise to another group of secondary structures, the β structures. These are extended chains with hydrogen bonds formed between chain segments. This gives rise to a stranded structure that can form sheets of strands. The direction of the adjacent strands may be parallel or antiparallel (figure 4.20). In general, sheets composed of many strands are usually of pure parallel or antiparallel type due to the geometrical arrangement of hydrogen bonds. Occasionally mixed sheets can be formed. The orientation of the residues is also important in β sheets. Antiparallel sheets have residues which tend to be either hydrophobic along the strand and point inwards, or hydrophilic and

Figure 4.20. Conformational arrangements of the β-sheet structure of proteins; (*a*) antiparallel alignment, (*b*) parallel alignment. (Hydrogen atoms are omitted.)

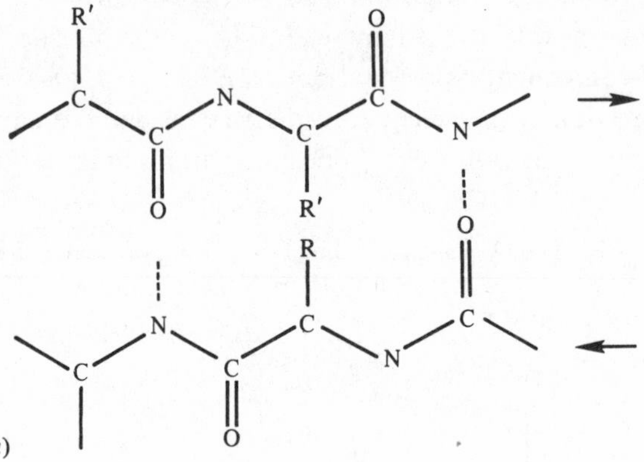

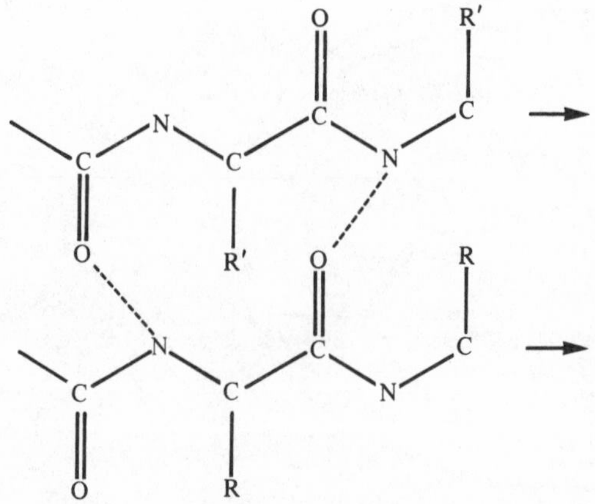

point out into the solvent. Similarly, hydrophobic groups on opposite strands are near to each other; alternating charged groups also lie opposite each other. A gross feature of β sheets is their lack of planarity. They have a characteristic twist and this can be pronounced generating a barrel structure.

If the sheets are formed from a linear polypeptide sequence, then parallel or antiparallel strands can be linked by a variety of topologies shown in figure 4.21. Interconnecting loops are of two types: hairpins and crossovers; x is used to designate a crossover and the absence of x is by definition a hairpin. A number defines how many strands the loop traverses; this number is preceded by a sign $\pm$ which gives the direction of the loop (above or below the sheet). The starting point for the topological description is the N-terminus. Most crossovers are right-handed loops and in many cases these loops have helical structures. However, turns of non-specific structure do occur.

The secondary structure of proteins can therefore be divided into two types – α helices and β sheets. There are four possible interactions between these two basic structures (Levitt & Chothia, 1976). These four classes are

(*a*) all α proteins with α-helical structure
(*b*) all β proteins with β-sheet structures

Figure 4.21. β-Sheet topologies for connecting loops in proteins. (Redrawn from Richardson, 1981.)

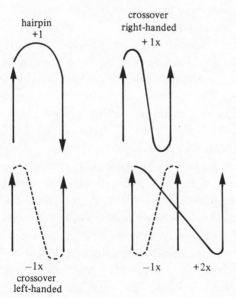

(c) $\alpha + \beta$ proteins with segregated regions of α helices and β sheets

(d) α/β proteins where α helices and β sheets approximately alternate. For two adjacent segments there are four secondary structures, $\alpha\alpha$, $\alpha\beta$, $\beta\alpha$, and $\beta\beta$, and for three adjacent segments eight secondary structures are possible, $\alpha\alpha\alpha$, $\alpha\alpha\beta$, $\alpha\beta\alpha$, $\beta\alpha\alpha$, $\alpha\beta\beta$, $\beta\alpha\beta$, $\beta\beta\alpha$, and $\beta\beta\beta$. These structures form the principal folding units and are subassemblies within protein domains. Each domain forms a complete globular protein and may have specific functions. A protein subunit may be formed from a number of domains. Good reviews of protein structure are given by Richardson (1981) and Chothia (1984).

4.5.2   Protein structure correlations with function

Our knowledge of protein structure and its molecular function is, in most cases, rudimentary. Research is hampered by a numerical lack of crystallographic structures and by the great complexity of chain folding. Comparison can be made with families of similar proteins but insertion or deletion of residues complicates the analysis. The nucleotide binding domain of a number of dehydrogenase enzymes does show a definite correlation of structure with function (Remington & Matthews, 1980). In this case there are no obvious homologous amino acid sequences between lactate dehydrogenase, glyceraldehyde-3-phosphate dehydrogenase and liver alcohol dehydrogenase at their respective nucleotide binding sites. Approximately 80 residues produce a three-dimensional structural similarity, in terms of backbone $C\alpha$ atom positions, at the site. Very few other structural binding sites have been analysed in this way.

We are used to considering structure–activity relationships in drug–receptor interactions where the structure of the drug molecule is the variable; with the nucleic acid binding sites we have the first converse study, namely the study of the structure of various receptor binding sites for a common drug molecule. The energetics of nucleotide binding to these variant sites have not been studied and would obviously be worthwhile.

4.5.3   Nucleic acid structure and conformation

Nucleic acids are formed by the polymerization of nucleotide residues. Each nucleotide has a purine or pyrimidine base attached to it. In DNA the bases are guanine (G) and cytosine (C) or adenine (A) and thymine (T). Cross links readily form by hydrogen-bonding between G and C, and A and T to form base pairs and produce a double helical structure. In RNA the pyrimidine base uracil (U) replaces thymine to form A–U base pairs. Each base in the nucleotide is attached to a deoxyribose phosphate in DNA or a ribose phosphate in RNA. The sugar phosphate forms the backbone of the polynucleotide. The structure of a nucleotide unit is shown in figure 4.22 and the 12 unit torsion angles are listed in table 4.1. A review of nucleic acid conformation and theoretical calculations of potential energies is given by Olson (1982).

The base pairs are usually held together in natural DNA and RNA structures by Watson–Crick base pairing. However, other base-pair patterns are possible and recently Hoogsteen base pairing has been found

Figure 4.22. Torsion angles for polynucleotides.

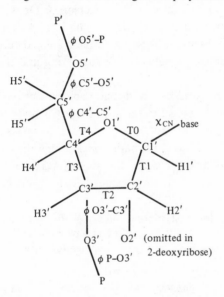

Table 4.1. *Definition of polynucleotide torsion angles*

Fragment	Torsion angle	Atom nos.				
Ribose	T0	C4'	O1'	C1'	C2'	
ring	T1	O1'	C1'	C2'	C3'	
	T2	C1'	C2'	C3'	C4'	
	T3	C2'	C3'	C4'	O1'	
	T4	C3'	C4'	O1'	C1'	
Chain	ϕ C4'–C5'	C3'	C4'	C5'	O5'	
	ϕ C5'–O5'	C4'	C5'	O5'	P	
	ϕ O5'–P	C5'	O5'	P	O3'	
	ϕ P–O3'	C5'	P	O3'	C3'	
	ϕ O3'–C3'	P	O3'	C3'	C4'	
Sugar	χ CN	O1'	C1'	N1	C6	Pyrimidines
base	χ CN	O1'	C1'	N9	C8	Purines

in a nucleic acid crystal complex with the drug triostin A (Wang *et al.*, 1984). As we saw in section 3.4, hydrogen bonding, although directional, is not geometrically rigid. This flexibility inherent in hydrogen bonds has led to different conformations of base planes in a base pair; the planes may be twisted in a propeller fashion or bent. The orientation of the base planes can then be expressed by rotation round three coordinate axes as various angles of twist, tilt and roll defined in figure 4.23. In crystal structures of DNA, a considerable variation in these parameters is found and this is related to whether the base sequence is purine followed by a pyrimidine or *vice versa* in reading in the direction 5′ → 3′ (Dickerson & Drew, 1981). Two factors seem to determine the orientation of the base plane: steric effects from neighbouring bases and the glycosyl torsion angle determined by the local conformational energy between the sugar ring and the base (Dickerson, 1983).

The sugar ring is non-planar and has five torsion angles. The pucker of the sugar ring is described with respect to the atoms C5′, C3′ and C2′; it is *endo* if the atom lies on the same side as C5′ and *exo* if the atom lies on the opposite side to C5′. The atom showing the major puckering is then used as the description, eg C3′-*endo*, or C3′-*exo* has the major puckering for C3′ respectively on the same side or the opposite side to C5′. Another

Figure 4.23. The axes for twist, tilt and roll across the base planes of nucleic acid helices; axes for tilt and roll lie in a plane normal to the helix axis. Watson–Crick base pairing is illustrated for purine and pyrimidine bases.

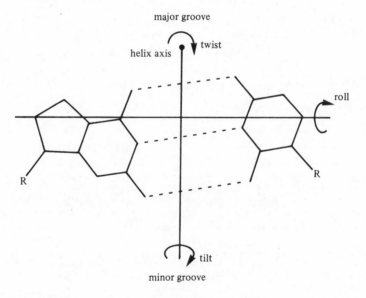

convention is to use the major pucker followed by the minor pucker, eg C3'-*endo*-C2'-*exo* or C3'-*exo*-C2'-*endo* (see figure 4.24). The most commonly encountered ring puckers are C3'-*endo* or C2'-*endo*. The puckering of the sugar ring is very important in the conformation of the polynucleotide chain and is responsible for major differences in helix structure of polynucleotides. In the sugar ring there are five torsion angles and ten possible descriptions of the five constituent atoms as the *exo* or *endo* conformation. The conformational energies for these positions are shown in figure 4.25 for *D*-ribose and *D*-2,deoxyribose. There are two energy minima for each sugar, and for both molecules they are found at C3'-*endo* and C2'-*endo*.

The chain of atoms containing the phosphate between the two sugar rings is the other major conformational pathway affecting the structure of the backbone. There are five torsion angles in this region with a large number of possible conformations. One of the difficulties for calculations of the conformation by *ab initio* methods is the molecular size of the whole unit. In general the bases have to be omitted and a semi-empirical calculation performed on the sugar–phosphate sugar fragment. The phosphate group poses a problem having a second row element with its consequent representation in a molecular orbital calculation. PCILO calculations suggest that the pair of torsion angles ϕ P–O3, ϕ O5'–P contain the most flexible link with two stable regions at ($+60°$, $+60°$) and

Figure 4.24. Conventions to describe the conformation of the *D*-ribose ring.

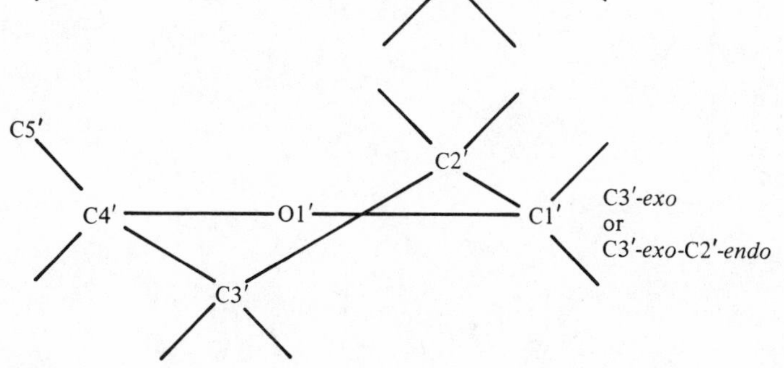

($-60°$, $-60°$). The other torsion angles have the following minima: $\phi\,O3'–C3'$, $\phi\,P–O3$ (180°, $-90°$) and (180°, 90°); $\phi\,O5'–P$, $\phi\,C5'–O5'$ ($-60°$, 180°) and (60°, 180°); $\phi\,C5'–O5'$, $\phi\,C4'–C5'$ (180°, 60°) and (180°, 180°).

The glycosidic links between the sugar ring and the nucleotide bases determine the conformational relationship of the backbone atoms to the plane of the base. This torsion angle χ is found in base-paired nucleic acids to have roughly three different values. With χ near 0° the conformation is termed *anti*, at about 90° it is termed high *anti* and at $-90°$ it is termed *syn*. In both *anti* conformations the atoms of the base are generally pointing away from the sugar ring, whereas in the *syn* conformation the atoms of the two moieties are much closer.

The secondary structure of double-stranded polynucleotides is helical. The artificial synthesis of small oligonucleotides and their crystallization has provided an impetus to recent structure determination. New types of helix have been discovered and in many cases the helical parameters are determined by the salt concentration of the solution from which the polynucleotide was crystallized. In x-ray fibre studies, double helical DNA is right-handed and forms two structural types, A and B; A-DNA is characterized by a helical corkscrew-like shape. The base-pair twist is 33° and the rise per base pair is 2.6 Å; this results in a narrow major groove and

Figure 4.25. Conformational energies for different ring conformations of *D*-ribose.

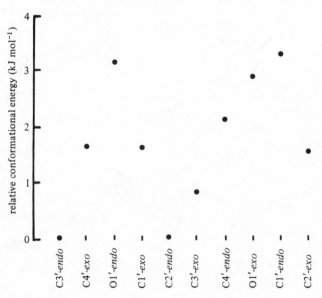

a wide minor groove. In contrast, B-DNA looks more like a twist drill with a wide major groove and a narrow minor groove: the base-pair twist is 36° and the rise is 3.3 Å. The broad principles of DNA conformation in the two forms A and B have been substantiated by x-ray diffraction studies of crystallized oligonucleotides. These recent studies, however, show that the helical conformation is not constant but varies with the local sequence. When the crystal parameters are averaged, the structural coordinates are similar to those found in the fibre studies. A remarkably different new form of DNA, termed Z-DNA, has been found for very short oligomers. Z-DNA is left-handed and the repeating unit is two base pairs, not one which is found in A- and B-DNA. With Z-DNA the rotation per base pair is $-60°/2$ and a rise per base pair along the helix is 3.8 Å. The glycosyl torsion angle for A- and B-DNA is *anti*; with Z-DNA, χ is *anti* for cytosine and *syn* for guanine (Dickerson *et al.*, 1982).

The structural studies on conformations of DNA are still at a very early stage and are not as well defined as analogous studies on protein structures. Environmental effects on the oligonucleotides appear to be the factors that lead to the three different helical forms. In low salt conditions A-DNA is generated, and in high salt the B-DNA conformation predominates. Z-DNA is also formed in high salt solutions and can be catalyzed to form by trivalent cations. No conformational studies have yet been carried out to investigate the energy changes along the pathway leading to the three different forms.

4.6 Prediction of polypeptide structure from amino acid sequence

The prediction of polypeptide tertiary structure from amino acid sequence is one of the most challenging problems in structural molecular biology today. The simple fact that many scientists are now trying to predict the structures indicates the confidence, or perhaps the blind optimism, that has been engendered by earlier conformational studies with small molecules. Structure prediction would be an enormous advantage to the drug designer since there are many more protein sequences available than crystal structures. The bottleneck in rational drug design at the moment is the small number of resolved structures. Predicted structures would also provide massive short cuts in the determination of crystal structure by speeding up the refinement procedure.

At the moment no single method of structure prediction has emerged pre-eminent. Secondary structure prediction is an obvious first step before structural folding can be considered to give a tertiary structure. Most prediction methods are based on a statistical evaluation of secondary structural features from known protein structures and their sequences.

Probability estimates for sequences of di, tri and tetrapeptides can be made for particular local conformations such as α helices, β strands and U-turns. The statistical method of Chou & Fasman (1977) can give a successful prediction of α helices and β strands for 77 % of all examined cases. Another method, advocated by Lim (1974a, b), uses a set of stereochemical rules and crude hydrophobicity parameters for the residues to predict α-helical, β-strand and coil regions to a reported accuracy of 81 %. Thus the results of secondary structure prediction are very encouraging and with time it should be possible to improve the rule base by more rational principles and perhaps make use of energetic criteria as well.

A tertiary structure is made up from a folded secondary structure and the topology determines the shape of the protein. A topological classification for proteins has been given earlier in section 4.5. However, the actual mechanism of folding is not understood. Since hydrophobic residues are found predominantly within the tertiary structure it would seem logical for folding on both sides of the strand to proceed from some central hydrophobic portion. It is unlikely, on probability grounds, that strands could weave within themselves; a winding process is therefore favoured. This folding of the strands around themselves will have some combinatorial element and the probability of each turn will, in the natural process, be determined energetically. In myoglobin the combinatorial problem can be illustrated with reference to table 4.2. There are 10^{32} possible conformations, a number which is too large to be handled by brute-force computer methods. This number of possible conformations has to be whittled down to a handful by the successful application of constraints. In hydrophobic regions the free energy of docking of strands will be related to the change in solvent accessibility. Solvent accessibility can easily be calculated along the secondary structure and this has been used as a basis for studying the packing of α helices by Richmond & Richards (1978), and Cohen, Richmond & Richards (1979). In a study of

Table 4.2. *Reduction in the combinatorial folding problem for myoglobin*

Number of conformations	Restraints
10^{32}	None
10^{24}	Globular folding
10^{8}	Helix–helix interactions
20	Chain connectivity and restriction of contacts
2	Accommodate the haeme group

myoglobin, six α helices are packed, 3×10^8 folds are possible but the number could be narrowed down to 20 by imposing two constraints: a distance constraint between the ends of the helices and a small number of close contacts. Further distance constraints imposed by the haeme group generated only two possibilities for the myoglobin folding problem. One structure gave the correct pairing of α helices and an rms error for the Cα atoms of only 4.5 Å compared with an rms Cα value of 16 Å for a random structure (Cohen & Sternberg, 1980). This is an encouragingly small rms value, particularly if we bear in mind that for a molecular dynamics calculation, the rms for Cα movement is about 1.1 Å from the crystal structure positions. This is a reasonably good prediction for the local topology of a folded protein considering the simple assumptions used to formulate the method. Similar work has progressed on the packing of α helices against β sheets, and the stacking of β sheets together. An obvious progression in this research field would be to use the predicted structure as a starting point and see whether the coordinates could be refined further by a molecular dynamics calculation.

The acid test for this brave new world of structure prediction is to forecast the tertiary structure of a protein which has not been experimentally determined. Sternberg & Cohen (1982) suggested that the structure of interferon will be like that shown in figure 4.26. Time will tell. If

Figure 4.26. The tertiary structure for interferon predicted by Sternberg & Cohen. Cylinders represent helices and the broad arrows are β-sheets. (Re-drawn from Sternberg & Cohen, 1982.)

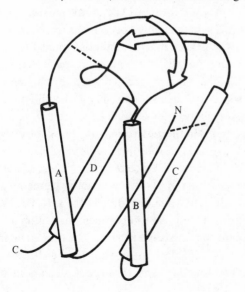

these prediction methods work adequately, then we will have the facility for going directly from a gene sequence (held in a data bank) to polypeptide tertiary structure in one package of computer programs! An attempt is given in section 6.4.2.

4.7   Dynamic conformational changes in proteins

Conformational calculations for small peptide segments show clearly that with most residue pairs there is no precisely fixed conformation. The torsion angles have an energetic leeway with a consequent probability for freedom of movement. Thus there are dynamic positional relationships between one residue and its neighbour. The theoretical method of molecular dynamics, outlined in section 3.2, enables one to study dynamic conformational changes in a quantitative manner. Results from this type of study are very important for the drug designer; static dimensions for a drug binding site are too simple a description. The atoms do not just vibrate with small amplitude around some fixed mean position, as they tend to in crystal structures of small molecules. There is a much greater flexibility in the protein and cooperative movements can occur along segments of the protein chain with hydrogen bonds breaking and new bonds forming. These cooperative shifts in conformation are termed trajectories.

The elegant studies of Levitt (1983b) on bovine pancreatic trypsin inhibitor (BPTI) will be used as an example of what can be gleaned from molecular dynamics about protein flexibility. This work is highly relevant to anyone studying molecular interactions with protein; the moral being – first understand the structure of your protein! The study includes 527 atoms yielding a problem of 1581-dimensional space to be analyzed for 66 000 successive time intervals. Distance matrix methods, applied to the moving atoms over 111 ps, reveal four distinct regions of space that the trajectories occupy (figure 4.27). These regions are delineated by clustering in the rms deviations for successive determinations of the difference distance matrix. The total rms deviation for a small time interval is not large since all 527 atoms are included in the matrix and may mask localized large shifts in position by the averaging procedure. A better illustration would be to use a partitioned difference distance matrix, or an animated version, to give the scale of movements in various positional segments of the polypeptide chain. Nevertheless, four regions, designated I, II, III, IV, can be discerned in the trajectory. The transition between one region and another is rapid, taking less than 2 ps, and is irreversible. The time spent in each region is: I 22 ps; II 39 ps; III 13 ps in moving in the direction III → IV, 21 ps in the backwards direction; and IV 44 ps. Calculations of the various energy contributions in each region show that the torsion angle energy, the van der Waals energy, and the hydrogen-bond energies shift in concert. Each is involved in a complex play-off to minimize the total energy within a particular region. These results give a fascinating insight into intramolecular interactions described formally by the equation

$$V_{\text{TOT}} = \sum V_{\text{TOR}} + \sum V_{\text{VDW}} + \sum V_{\text{HB}} \qquad (4.12)$$

The summation within each term can be adjusted to give a different value but, at the same time, the terms can add to give an equivalent minimum energy V_{TOT}.

An analysis of the torsion angles shows that the conformational flexibility is large and an rms deviation of $34°$ for ϕ, $38°$ for ψ and $51°$ for χ is found. Some pairs of torsion angles exhibit greater flexibility during particular time intervals and correlate with the transition from one conformational region to another (figure 4.28). Note that this illustration shows that ψ_{15} and ϕ_{16} have torsion angles that are anticorrelated. This observation is a frequent feature for ψ_i and ϕ_{i+1} torsion angles where large changes are observed and helps to confine the conformational swing to a localized region rather than propagate massive movements down the chain. Side-chain torsion angles (χ) can show dramatic movements and can

Figure 4.27. Four definable clusters in which the rms deviation Δd_{ij} for all pairs of Cα atoms are found. The full line indicates the trajectory of the protein moving through the conformational states; p and q are 2-dimensional projections of 1581-dimensional space. (Redrawn from Levitt, 1983c.)

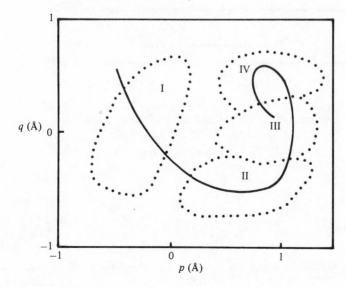

flip between g^-, g^+ and t conformations; an example is shown in table 4.3 for Lys 15. Residues that are close to each other can exhibit synchronous changes in their side-chain conformation.

Hydrogen bonds can break or form during the dynamic movement of the protein. In total, 49 hydrogen bonds are encountered but the number existing at different time intervals can vary; this contrasts with 32 hydrogen bonds present in the x-ray structure. Not surprisingly, the hydrogen bonds vary in their stability; 27 bonds are stable and exist for 90 % of the time, 14 bonds are weak and are present for only 50 % of the time and eight bonds are unstable and exist for less than 10 % of the time. Weak hydrogen bonds are associated with temporary links between side-chains. Cooperative

Figure 4.28. Variation between pairs of torsion angles ψ_9 ϕ_{10} and ψ_{15} ϕ_{16}. Anticorrelations can be observed over 70 ps. (Re-drawn from Levitt, 1983c.)

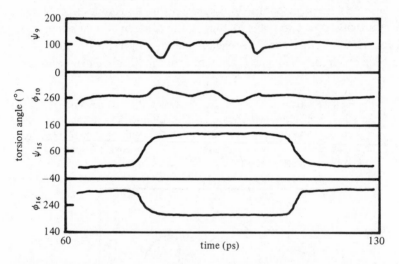

Table 4.3. *Changes in side-chain torsion angles for Lys 15 in bovine pancreatic trypsin inhibitor during a molecular dynamics simulation*

Torsion angle	Elapsed time (ps)					
	8	32	56	80	104	128
χ_2	g^+	g^+	g^-	t	t	t
χ_3	t	t	t	t	g^+	g^+
χ_4	g^+	g^-	g^-	t	t	t

Torsion angles are defined as: $g^+ = 0 \to 120°$; $t = -120 \to 120°$; $g^- = 0 \to -120°$.

breaking of six hydrogen bonds at the two ends of the β-hairpin facilitates the sharp transition in trajectory from region III to region IV.

In conclusion we may visualize the protein as having its atoms in continual motion. However, for most of the time this motion is restricted and the atoms vibrate about a localized mean position. When the intramolecular energies are favourable, the conformation can flip into another state and remain there for a significant time before a reverse transition takes place, or the structure moves on to another region of the trajectory. The transitional jumps along the trajectory appear to involve cooperative movements of significant numbers of atoms with breaking or reforming of hydrogen bonds.

4.7.1 Atom mobility in proteins

The studies of Levitt (1983a, b) consider atom movement in a molecule by a dynamic calculation method. Atom mobility can, however, be studied by examining the behaviour of each atom in the crystal from the associated temperature factors. Temperature factors were briefly discussed in chapter 2. The temperature factor B is related to the harmonic displacement, $\bar{U}^2$ by the equation

$$B = 8\pi^2 \bar{U}^2 \tag{4.13}$$

For isotropic displacement, B has a single value, but anisotropic displacement can be modelled by splitting B into the six parameters that define the ellipsoid of displacement if the crystal structure is well refined (1.5 Å resolution) (Artymiuk et al., 1979). Thus $\bar{U}^2$ can be plotted for any atom against its position, r, usually the molecular centroid, in the molecule. The scattergram in figure 4.29 illustrates the variation of both parameters for atoms in human lysozyme. Main-chain atoms show less variability in mobility than atoms in the side-chain. Atoms far from the centroid have displacements which are more broadly scattered. All side-chain atoms are displaced more than their local main-chain atoms. This incremental difference between side-chain and main-chain atoms is quite small for hydrophobic amino acid residues but increases markedly in length for hydrophilic side-chains. The active site cleft of human lysozyme undergoes a conformational change on ligand binding and it is found that atoms in the region of the cleft show the greatest displacements.

A useful method for illustrating atom mobility is to plot the temperature factor (colour coded) on the molecular and accessible surface (Tainer et al., 1984). Regions of high mobility appear as 'hot' spots on the surface (figure 4.30). Tainer et al. searched the surfaces of 22 proteins and discovered that regions of low mobility were found very frequently at subunit interfaces for oligomeric proteins, and at the surfaces between adjacent molecules in the crystal lattice. Hot spots tend to be related to lipophilic regions; lipophilic regions are also correlated with the antigenicity of a molecular surface. Therefore, it is not surprising to find that regions of high atom mobility react with antibodies made against these specific regions. Tainer *et al.* conclude that regions of high atom mobility may be an integral part of protein–protein recognition mechanisms. It is probable that the atom mobility shows some leeway in conformational re-adjustment as the antigenic and antibody molecular surfaces come together, perhaps by an induced fit, before a precise binding lattice is established of intermolecular hydrogen bonds.

Figure 4.29. Atom mobility in human lysozyme. Limits on the scatter of the mean square harmonic displacement $\bar{U}^2$ plotted against position r^2 from the molecular centroid of each atom; (*a*) main-chain atoms, (*b*) side-chain atoms.

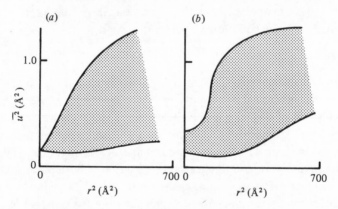

Figure 4.30. Atom mobility at the accessible surface of myohaemerythrin. Regions of high mobility (small dots), intermediate mobility (medium-sized dots), and low mobility (large dots).

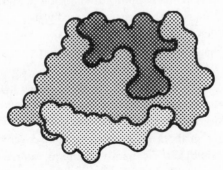

4.8 Software systems for shape display

The display of molecular shape and surfaces has developed rapidly since 1980. This swift growth of activity has been caused by two developments in computing technology, the construction of good raster graphics displays and dedicated hardware for graphics handling. In parallel with these developments, molecular graphics has become a specialized discipline introducing its own journal in 1983. A surprising amount of molecular graphics output is connected, in one way or another, with drug design. The great advantage of these graphical methods has been to remove the tedium of molecular model building and at the same time to make it possible to examine docking manoeuvre simulations between a drug molecule and its binding site. Animated graphics, linked to molecular interaction energy calculations, provide a very clear insight into how drugs might work at the molecular level.

4.8.1 Molecular structure

The traditional representation of the chemical formula of a molecule is by a series of bonds connecting points labelled for particular atoms. Perspective symbols for bond direction can readily convert the bond description into a three-dimensional representation. Bond drawings form the simplest representation of molecular structure. These drawings can easily be adapted to computer graphics to give a wire frame description showing the molecular skeleton of bonds. Wire frame representations are useful because they require little computation and they can be rotated with ease at a molecular graphics terminal. Colour coding of the lines can be used to represent different atoms visually. The disadvantage of the wire frame model is that it does not give any idea of the space-filling properties that can be represented by van der Waals surfaces. However, the wire frame model is very useful for watching conformational changes during molecular dynamics simulations and gives a very good impression of relative movements.

Space-filling graphic models of molecular structure are mostly variant methods of mimicking the CPK (Cory, Pauling and Koltoun) molecular models frequently used in chemistry. Each atom is treated as a sphere with an appropriate van der Waals radius; the conglomerate of atoms can then be treated as one of unions of spheres. The intersections of the spheres appear as elliptical arcs; hidden lines need to be suppressed by using a standard hidden line algorithm on the curves approximated to straight line segments. Each circular region can be colour coded to represent the atom. The display can be improved if a representation of three-dimensionality is given to each region. This can be achieved either by shading, which would be a natural representation, or by drawing the great circles on the spherical surfaces for rotation round the local atomic x- or y-axes. The great circles method has the advantage that it can be used in conjunction with some other property mapped onto the molecular surface. The mathematics of shading spheres to give depth, and light reflectance to give surface highlights, has developed so well that it is now not possible to distinguish between a photograph of a physical CPK model and a raster graphics generated picture of a CPK simulation. A complete solution to the problem of shading molecular surfaces is given by Brickman (1983).

4.8.2   Accessible surface representation

Earlier in this chapter a method was outlined for computing the accessible surface areas on the molecular surface. These surfaces were divided into contact surfaces, re-entrant surfaces and the surface accessible to the centre of the probe rolling all over the van der Waals surface. In practice the accessible surface indicates the surface of nearest approach of the nuclei of an interacting molecule. Consequently, in drug–receptor interaction, accessible surfaces are most useful for defining possible docking regions. Often these binding sites are formed as clefts between the folds of secondary structures and so a method is needed to penetrate beneath the visible external surfaces. An elegant method of accessible surface representation has been given by Langridge et al. (1981). This is the dot representation of the accessible surface; the surface is examined only in the picture segment containing the feature. The probe is moved over the surface and the centre represented by a dot; these positions are spaced to give a density of 2–5 dots per $Å^2$. Dots may be colour coded for local molecular properties, eg atom type or charge density. If the intensity of the dot is altered by depth cueing then a three-dimensional effect for the surface can be obtained. Furthermore, the picture segment can be rotated in real time to obtain a good visual examination of the structure. Molecular skeletons can be superimposed with the wire frame model to relate accessible surface shape to molecular structure. Sections through the surface can be achieved by front and back clipping if an examination between a cleft and a fitted molecule is needed.

Contact and re-entrant surfaces are very important maps of the contact points on a molecular van der Waals surface. They show, respectively the regions which can and cannot be touched by the probe. For a seriously convoluted surface, such as that of a nucleic acid helix or an external surface on a protein, there is a surprisingly small amount of contact surface available. The surface folds can create substantial areas of re-entrant surface. These observations are important for the drug designer looking for directional constraints as only certain contact patches may be usable. These limitations are not appreciated when we are only able to visualize an ordinary space filling model.

A more recent development of Connolly's (1985) is to triangulate the accessible surface as a polyhedron. Thus the surface is shown as a network of triangles. Some molecular property can then be mapped onto the surface triangle to give surface contouring representation.

4.8.3 Graphical display of quantum-mechanical properties on molecular surfaces

A natural extension of identifying atoms in a diagram by colour coding is to project onto the molecular surface some positionally dependent chemical property. The magnitude of the property can then be colour coded to give a limited four-dimensional representation; we view the three spatial dimensions with a fourth property mapped onto them. This brings the application of quantum mechanics to drug design one step nearer. Not only can we visualize the spatial relationship between atoms, we can now examine detailed chemical properties as well. One property that has achieved much attention has been the electrostatic potential mapped onto the accessible surface; this gives an idea of regions of strong electrostatic binding and can be used to assess complementarity between a drug molecule and its binding site. A map of the electrostatic potential at the contact interface for one subunit of lactate dehydrogenase is shown in figure 3.12. Regions of positive and negative electrostatic potential on the surface can be located. Structural and electrostatic complementarity is present. Other quantum-mechanical parameters such as electron density and superdelocalizability can be displayed by similar procedures (Quarendon, Naylor & Richards, 1984).

5

Ligand-binding sites

5.1 Identification of receptors

The task of identifying and defining ligand-binding sites has been one of the major achievements of pharmacologically related disciplines since the 1940s. Perhaps the most worked example has been the binding site for acetylcholine. In general, there are two classes of binding site for acetylcholine, characterized by preferences for particular ligands. The nicotinic site is found predominantly at the skeletal neuromuscular junction and in sympathetic autonomic ganglia; the muscarinic site is located at the post-synaptic junctions between parasympathetic nerves and the effector system. So far, greater progress has been made in isolating and characterizing the nicotinic acetylcholine receptor.

Most attempts to isolate and purify the cholinergic nicotinic receptor have been performed using the electric organs from *Torpedo*, *Narcine* and *Electrophorus*. These species provide large concentrations of receptor in the electric organ. The receptor is extracted from detergent extracts of these organs and purified by affinity chromatography. Marker ligands for purification have been choline derivatives or small protein α-neurotoxins extracted from particular snake venoms; the α-neurotoxins bind to the post-synaptic receptor. Radio-labelled ligands with high binding affinity, which are bound to the receptor, are then used as markers for subsequent purification.

Extraction of the high affinity α-neurotoxin bound to polypeptide yielded a protein tightly associated into four subunits termed $\alpha, \beta, \gamma, \delta$. Further experiments demonstrated two identical α subunits. Sequence analysis of the subunit polypeptides suggested that there was some degree of homology between the subunits but there are also distinctive differences. Great use has been made of these differences in polypeptide sequences as labels for searching the genes that code for each subunit. Gene cloning has been used to elucidate the complete gene sequence and thus the amino acid sequence for each subunit. The α subunit which binds acetylcholine as well as α-neurotoxins consists of 461 amino acids and includes a pre-peptide of 24 amino acids (Noda *et al.*, 1982). The amino terminal 24-amino-acid sequence is rich in hydrophobic residues and may indicate that this region of the α subunit is a signal peptide that might lead to insertion in the cell membrane. Seven cysteine residues are present in the sequence and those at positions 128 and 142 are capable of forming a sulphur bridge on the polypeptide surface. These two cysteine residues are thought to be close to the acetylcholine binding site. Sequence analysis, for the prediction of secondary structure, suggests that these two residues lie in a set of four β sheets possibly aligned in the antiparallel mode. Cloning of the genes for the β and δ subunits show that the subunits consist of 493 and 522 amino acids respectively (Noda *et al.*, 1983). Each is preceded by a pre-peptide portion rich in hydrophobic residues containing 24 (β) and 21 (δ) amino acids. The amino acid sequence of the γ subunit has been determined by Claudio *et al.* (1983); there are 489 amino acids including a 17-amino-acid signal peptide.

Statistical analysis of the gene sequences coding for the amino acids in each subunit show that there are regions of substantial homology. About 40% of the residues are homologous in α, β and δ subunits but regions of homology are not randomly distributed; the amino terminal two-thirds contain most homology with the four proteins. Three blocks of amino residues, termed a, b, c, are found in the γ subunit also. These observations of sequence homology point to a common ancestor gene and suggest the possibility that regions of homology may have a similar structural function, possibly in the association of subunits to form the oligomeric protein of the receptor (Marx, 1983). The elucidation of the sequences for nicotinic acetylcholine receptors from other species should now be possible with the library clones available. We should be able to examine special differences, long suspected from pharmacological experiments, and perhaps genetic malfunctions which may be the cause of numerous neuromuscular disorders, eg myasthenia gravis.

The tertiary structure of the subunits is not yet known. So far all that can be deduced is that the subunits aggregate to form a rosette structure with a central open channel. A scanning transmission electron microscope picture with image analysis averaging over many structures is shown in figure 5.1. Contour values associated with Young's fringes suggest 3–5 discrete regions surrounding an annulus (Zingsheim *et al.*, 1980). There is speculation, but no confirmation yet, that the central channel is the ion pore associated with the cholinergic receptor.

Messenger RNA extracted from *Torpedo marmorata* can be inserted into *Xenopus* oocytes (Barnard, Miledi & Sumikawa, 1982). Culture of these treated cells leads to the formation of nicotinic receptors in the cell membrane. Normal untreated oocytes possess only small amounts of muscarinic receptors and not nicotinic receptors. Cells treated with AchR-mRNA are capable of binding ^{125}Iα-bungarotoxin whereas untreated cells do not bind the neurotoxin. Thus significant amounts of '*Torpedo*' acetylcholine receptors are synthesized after the injection of specifically coded mRNA. Electrophysiological recordings of the transmembrane potential in treated oocytes, with atropine present to abolish muscarinic actions, show that acetylcholine applied iontophoretically alters the potential in a rapid response. The Na$^+$ and K$^+$ permeability is changed by acetylcholine in the mRNA treated cells. Control cells were unaffected by acetylcholine. Furthermore, the induced receptors were unaffected by atropine but inhibited by (+)-tubocurarine and α-bungarotoxin; this is strong evidence that they are nicotinic receptors coded for by *Torpedo* mRNA. Application of acetylcholine to the outer membrane surface and iontophoretically into the cell interior showed that the recognition site was oriented on the external surface since intracellular application did not evoke a response. These results of Barnard *et al.* show clearly that receptors

Figure 5.1. Rosette structure of the subunit arrangement of the acetylcholine receptor. A central channel is flanked by the subunits. The contours represent surface structure revealed by scanning electron microscopy. (Re-drawn from Zingsheim *et al.*, 1980.)

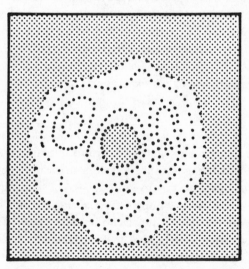

synthesized *de novo* from specifically cloned mRNA behave like nicotinic receptors for acetylcholine in all respects.

The cloning experiments for different subunits of AchR open up the possibility of synthesizing the subunit proteins in sufficient quantity for purification and hopefully crystallization. When pure crystals are obtained, structure elucidation for each subunit should be possible crystallographically. We shall have to await this development before we can identify the acetylcholine binding site unambiguously and learn how the drug interacts with the receptor to control ion movement. The acetylcholine receptor is discussed further in section 9.1.3.

5.2 Identification of binding sites for drug molecules

In the previous section we considered the question of isolating and identifying the receptor macromolecule in a pure form from other cellular constituents. The next stage in working out the mechanism of drug–receptor interaction is to characterize the receptor and locate the region for drug binding. With the acetylcholine receptor this step in the research is nearly ready to be undertaken and the results of crystallographic studies may be expected in the 1990s. For the moment we have to be content with examining another drug-binding site, the enzyme dihydrofolate reductase (DHFR). In many respects DHFR has become an archetype for the study of drug binding to protein. The pace of research on DHFR accelerated during the 1970s due to the important therapeutic value of inhibitors of the enzyme in three fields: cancer chemotherapy with methotrexate as inhibitor, antibacterial action with trimethoprim and antimalarial effects with pyrimethamine. DHFR is currently the most investigated enzyme in medicinal chemistry. Here we shall confine the study of drug-binding sites to the DHFR model to use this as an example for drug binding to proteins. Drug binding to nucleic acids will be reserved for section 5.4.

In this section we shall examine the structural details of the folic acid binding site in DHFR and make comparisons between the molecular architecture of different species for which the DHFR structure has been elucidated. These structural differences can be related to the amino acid sequence and local secondary structure of the binding site. The physical properties of possible site-points on different DHFR enzymes can explain species differences for selective antifolate compounds.

5.2.1 DHFR

The enzyme catalyzes the reduction of dihydrofolic acid (DHF) to tetrahydrofolic acid (THF); the pteridine ring is reduced (figure 5.2). THF is the coenzyme for the transfer of single carbon fragments used in intermediary metabolism. This particular reaction is very important in nucleic acid synthesis by thymidylate synthetase and the coupled cycle is shown in figure 5.3. A methyl group is incorporated into thymine by thymidylate synthetase. DHFR is essential for all living cells but it can be inhibited by analogues bearing a structural resemblance to folic acid. The three inhibitors in widespread usage are methotrexate, trimethoprim and pyrimethamine; their structures are illustrated in figure 5.4. However, the inhibitors are selective for different types of DHFR and their IC_{50} (concentration of inhibitor producing 50% inhibition) values are given in table 5.1. Methotrexate is not absorbed through the bacterial cell wall and so is of no practical use as an antibacterial agent; its great use is in the treatment of leukaemia. On the other hand, trimethoprim is about 50 000 times more active on *E. coli* DHFR than rat liver cells.

The only vertebrate crystal structure for DHFR is from chicken liver (Volz *et al.*, 1982). This structure can be compared with bacterial DHFR from which there are crystal structures for the enzyme extracted from *E. coli* and *L. Casei* (Filman *et al.*, 1982; Bolin *et al.*, 1982). Primary structures have been obtained for a number of vertebrate DHFRs and strong sequence homology is found between chicken liver, bovine liver and porcine liver enzymes; approximately 72% of the residues are in common.

Figure 5.2. The molecular structures of: (*a*) DHF, and (*b*) THF; H_S and H_R show the hydrogen atom configurations. The group labelled R in THF is identical to the side-chain at position 9 in DHF.

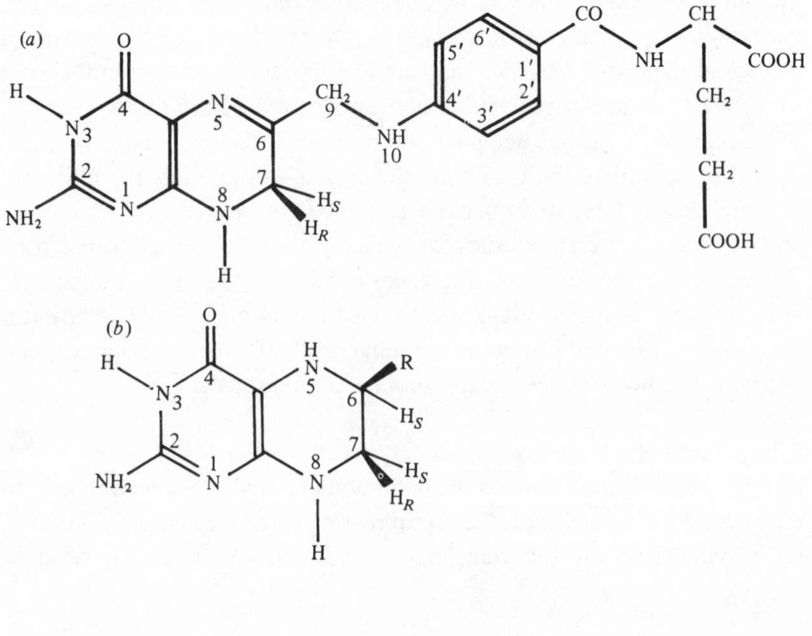

Figure 5.3. The thymidylate synthetase–serine *trans*hydroxymethylase cycle coupled to DHFR.

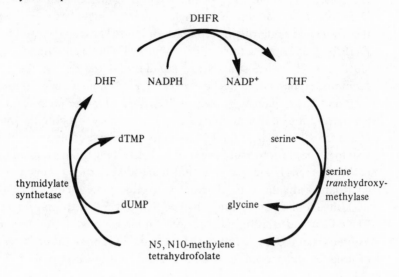

Figure 5.4. Folic acid inhibitors: (*a*) methotrexate, (*b*) trimethoprim, (*c*) pyrimethamine.

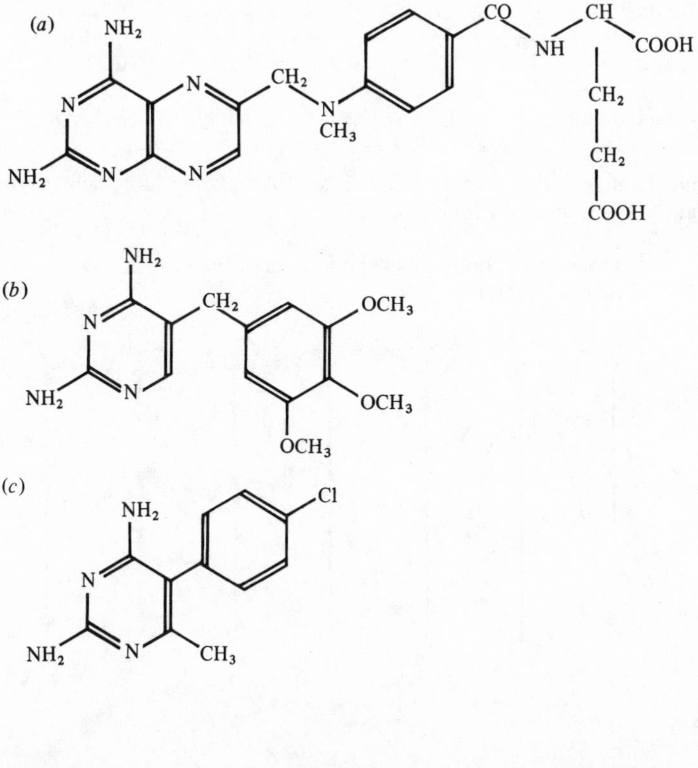

This contrasts strongly with the poorer sequence homology for bacterial enzymes. Only about 25% of the residues of mammalian and bacterial DHFRs are in common. Nevertheless, despite the difference in sequence there is a close similarity in secondary structure between the two groups of DHFRs. Each enzyme molecule is composed of an eight-stranded β sheet labelled A–H; seven strands are parallel and the one at the carboxy terminus (H) lies antiparallel between F and G. Four α helices, labelled B, C, E and F, lie against the β sheet, the other amino residues forming the connecting loops. The topology of the secondary structure is shown in figure 5.5. Chicken liver DHFR contains 30 more residues than the *E. coli* enzyme. The extra residues occur in nine insertion regions and all but one of these regions is a loop connection. This odd insertion occurs in β-G and six residues are added to the chicken liver enzyme. Figure 5.6 shows the backbone folding comparing chicken liver DHFR with that for *L. casei*. The β-G strand is far from the region of inhibitor binding and so it may be assumed that it does not affect the catalytic reaction of the enzymes. A list of residues and secondary structural features is given in table 5.2.

Table 5.1. *The inhibition of DHFR taken from various species*

Inhibitor	E. Coli	S. Aureus	P. Berghei	T. Equi-perdum	Rat liver
Methotrexate	0.1	0.2	0.07	0.02	0.21
Trimethoprim	0.5	1.5	7.0	100	26 000
Pyrimethamine	250	300	0.05	20	70

Numerical values are expressed as $IC_{50} \times 10^8$ mol dm^{-3} for each enzyme. Data taken from Beddell (1984).

Figure 5.5. Topology of the folding of the protein secondary structure of *E. coli* DHFR.

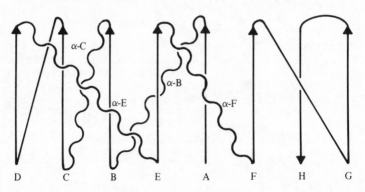

D   C   B   E   A   F   H   G

5.2.2 *The active site*

The active site for dihydrofolate binding has not been determined directly. However, there are crystal complexes for the enzyme and NADPH (reduced nicotinamide adenine dinucleotide phosphate) as well as for the ternary complex with inhibitor. X-ray crystallography can therefore be

Figure 5.6. Folding of *L. Casei* DHFR; dotted lines show insertions in chicken liver DHFR. The substrate binding site is shaded.

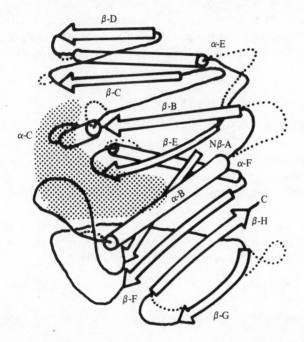

Table 5.2. *Residue numbers and the secondary structure of DHFR from* E. Coli *and* L. Casei

α helices	*L. Casei*	*E. Coli*	β sheets	*L. Casei*	*E. Coli*
			β-A	1–7	1–8
α-B	23–34	24–35	β-B	36–42	39–43
α-C	42–49	43–50	β-C	56–63	58–63
			β-D	74–76	73–75
α-E	79–89	77–86	β-E	93–98	89–95
α-F	99–107	96–104	β-F	111–119	108–116
			β-G	135–144	132–141
			β-H	152–161	150–159

Data from Bolin *et al.* (1982).

used to identify the active site; it is found in a cleft about 15 Å deep with the walls composed of β-A, β-B, β-C, β-E and β-F and portions of α-B, α-C and α-F. This active site binds the cofactor and inhibitor methotrexate. Bacterial DHFRs show a similar folding of polypeptide chains at the binding site. Structural comparisons between the backbone atoms of the helices and sheets show that bacterial DHFR structures are geometrically very similar. Comparison between bacterial and chicken DHFR indicates a general similarity although there are some distinct positional differences, 1–3 Å, in alignment of elements of the secondary structures. Thus we may conclude that, although the amino acid sequence does not show large regions of homology, there is still a great deal of resemblance between the folding of secondary structural features in the bacterial and vertebrate DHFR.

5.2.3   Site-points for binding NADPH

The molecular structure of NADPH together with its reduced form is given in figure 5.7. When bound to the enzyme the cofactor is in an extended state spanning about 17 Å; this is in contrast to the conformation observed in solution where the cofactor folds considerably. The binding region for NADPH is across the cleft of the active site. The geometrical

Figure 5.7. Molecular structure of (a) NADPH and (b) NADP$^+$. The R group on NADP$^+$ is identical to that in NADPH.

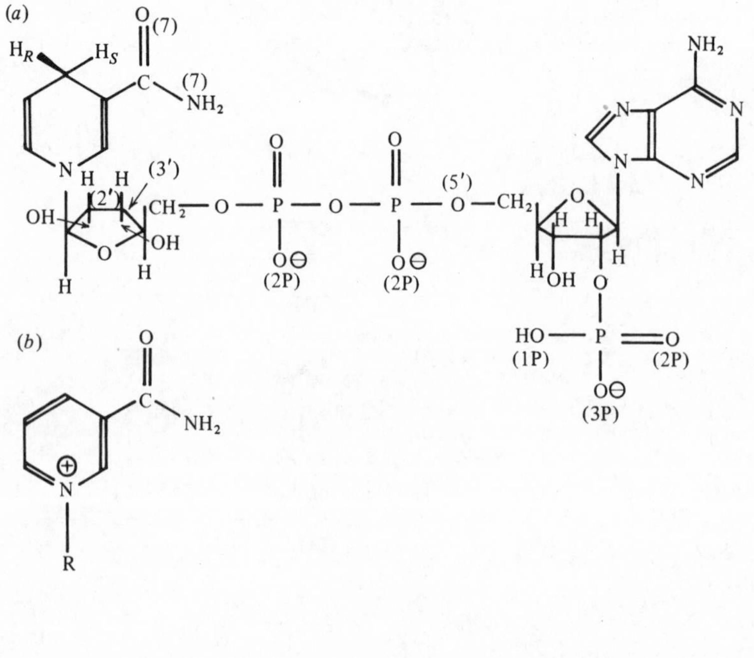

arrangement of site-points for hydrogen bonding to NADPH is illustrated in figure 5.8. Table 5.3 lists the site-points and structural features corresponding to NADP atoms; there are 17 hydrogen-bonding and ionic site-points. These points all have geometrical directionality. Other non-directional binding points also exist and 26 further van der Waals contacts are known.

The adenine moiety is not bound to site-points by hydrogen bonding, it simply fits into a hydrophobic pocket. Similarly the adenine mononucleotide ribose does not make direct hydrogen bonding to the enzyme although there is one indirect hydrogen bond to a water molecule bound to a residue. In contrast the 2-phosphate attached to the adenine ribose binds to four residues and provides a tight binding region. This binding to five site-points may explain why NADPH binds 100 times more strongly to DHFR than NADH.

The site-points for the pyrophosphate group show the interesting feature of being clustered exclusively round the N-terminal end of α-C and α-F helices. Five of the seven site-points are N peptide backbone atoms. These particular nitrogen atoms lie at the positive end of the two helix dipoles and provide favourable electric fields to be associated with binding an anion. Phosphate binding to the N-terminus of α helices is a well-known

Figure 5.8. A stereoscopic drawing of NADPH hydrogen bonded to DHFR. (Reproduced from Filman *et al.*, 1982.)

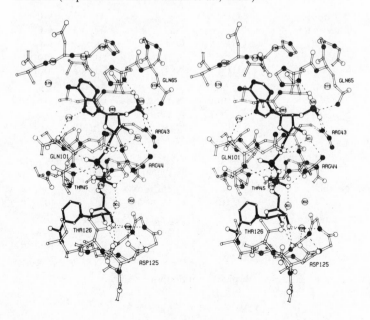

phenomenon of protein crystallography and has been reviewed by Hol & Wierenga (1984). Each peptide unit has a dipole moment of about 3.5 D and when the peptides are arranged in an α helix the dipoles summate. Thus for the α-C helix with seven residues the dipole is about 25 D and for the α-F helix, 28 D. The electric field of the N-terminal end should therefore be considerable and be attractive towards negatively charged groups.

The ribose ring of nicotinamide mononucleotide is bonded to His 18 by O2′ and O3′. Nicotinamide is hydrogen bonded only by the amide portion to Ala 6 and Ile 13. The conformation of the nicotinamide ring is then fixed by these three hydrogen bonds and is not able to rotate. This conformational restriction places the B side of the ring against a hydrophobic pocket and the A side in contact with the pteridine binding site. Hydride transfer is then limited to the A side (figure 5.9). The two possible catalytic sides for the nicotinamide to participate in catalysis are derived from the prochirality of the H atom at position 4. The hydrogen atom donated can be from the *R* or *S* positions (*R* = *rectus* and is in a clockwise position relative to the $CONH_2$ group; *S* = *sinister* and lies in the anticlockwise position). Side A contains H_R, whilst H_S is situated on side B.

Table 5.3. *Hydrogen bonds or ionic bonds from NADPH to site-points on* L. Casei *DHFR*

NADPH moiety	Atom	Site-point	Residue	Secondary structure
AMN-ribose-	O1P	Nε	Arg 43	α-C
2-phosphate	O3P	Nη2	Arg 43	α-C
	O1P	Oγ1	Thr 63	β-C
	O2P	N pep	His 64	loop β-C to β-D
	O1P	Nε2	Gln 65	loop β-C to β-D
AMN 5′-phosphate	O5′(a)	N pep	Arg 44	α-C
	O1P(a)	N pep	Ile 102	α-F
	O2P(a)	N pep	Thr 45	α-C
	O2P(a)	Oγ1	Thr 45	α-C
	O2P(a)	N pep	Gly 99	α-F
NMN 5′-phosphate	O2P(n)	Nγ1	Arg 44	α-C
	O1P(n)	N pep	Gln 101	α-F
NMN ribose	O2′(n)	O carbonyl	His 18	loop β-A to α-B
	O3′(n)	O carbonyl	His 18	loop β-A to α-B
Nicotinamide	O7	N pep	Ala 6	β-A
	N7	O carbonyl	Ala 6	β-A
	N7	O carbonyl	Ile 13	loop β-A to α-B

(a) = adenine moiety, (n) = nicotinamide moiety.
Data from Filman *et al.* (1982).

5.2.4   Site-points for methotrexate

The crystallographic structure of methotrexate in a ternary complex with NADPH and DHFR from *L. casei* has been resolved by Bolin *et al.* (1982). In most respects the ternary complex shows identical features with the binary complex between methotrexate and *E. coli* DHFR. An illustration of the inhibitor binding site is shown in figure 5.10 and the site-points are listed in table 5.4. In addition to these six hydrogen bonds there are two further bonds to water molecules from the 2 amino nitrogen and N8 of the pteridine; these add to the stability of inhibitor binding to DHFR. The *p*-aminobenzamide is largely bound to the enzyme by hydrophobic interactions and van der Waals contacts. Glutamyl carboxylates anchor the end of the molecule to DHFR.

Figure 5.9. The nicotinamide ring showing the hydride transfer sites A and B.

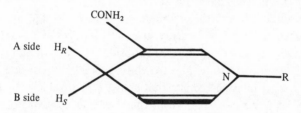

Figure 5.10. A stereoscopic drawing of methotrexate bound to DHFR. (Reproduced from Bolin *et al.*, 1982.)

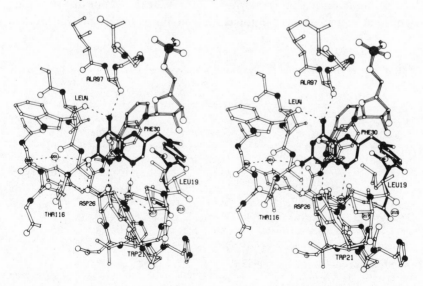

5.2.5   Substrate binding

There are no crystal structures for substrate complexes with DHFR. However, the orientation of the pteridine ring in the methotrexate ternary complex can act as a pointer to likely orientations of substrate at the active site. NADPH is aligned so that hydrogen transfer takes place from H_R on the A side of the nicotinamide ring. If the pteridine ring of dihydrofolate is bound exactly as in methotrexate then the product will have the *R* configuration at C6. But this is not the observed configuration of the product tetrahydrofolate which is in the *S* form. Thus one would expect the substrate pteridine ring to be rotated 80° about the C6–C9 bond. A hydrogen-bond scheme suggested by Bolin *et al.* (1982) is shown in figure 5.11 to illustrate the two modes of binding of pteridine rings in substrate and methotrexate complexes. Methotrexate would, in this scheme, have one further hydrogen bond with Ala 97 to stabilize its binding.

5.2.6   A structural mechanism for catalysis

A mechanism for the reduction of dihydrofolate has been proposed by Filman *et al.* (1982). Only structural features of the active site and substrate coenzyme orientations are considered; these limitations say little about the transition state. Nevertheless, geometrical considerations are an obvious part in a mechanism that shows structural specificity and there is much that we can learn from their scheme. Dihydrofolate has hydrogen atoms at positions 7 and 8; 5,6,7,8-tetrahydrofolate is reduced at the N5—C6 bond (figure 5.2). The stereochemistry of hydride transfer is from the A side of nicotinamide to the C7-si face of the pteridine ring. In the binding mechanism of the coenzyme and substrate there appears to be little significant re-arrangement of the enzyme conformation. However, in the

Table 5.4. *Hydrogen bonds from methotrexate to site-points on* L. Casei *DHFR*

Methotrexate moiety	Atom	Site-point	Residue	Secondary structure
Pteridine	N1	O carboxyl	Asp 26	α-B
	N2 amino	O carboxyl	Asp 26	α-B
	N4 amino	O carbonyl	Leu 4	β-A
	N4 amino	O carbonyl	Ala 97	β-E
L-glutamate	Oα carboxyl	N guanidinium	Arg 57	β-C
	Oγ carboxyl	N imidazole	His 28	α-B

Data taken from Bolin *et al.* (1982).

hydride transfer reaction, two hydrogen atoms are transferred but only one comes from the nicotinamide ring of NADPH and is believed to attach to C6 of the pteridine ring. The second hydrogen atom is assumed to come from bound solvent to attach to N5.

Figure 5.11. Hydrogen-bond networks linking amino acid residues in DHFR to the pteridine rings; (a) a possible network for 7,8-dihydrofolate, (b) orientation of the pteridine ring of methotrexate in the enzyme crystal complex. (R is the p-aminobenzoylglutamate residue.) (Re-drawn from Bolin et al., 1982.)

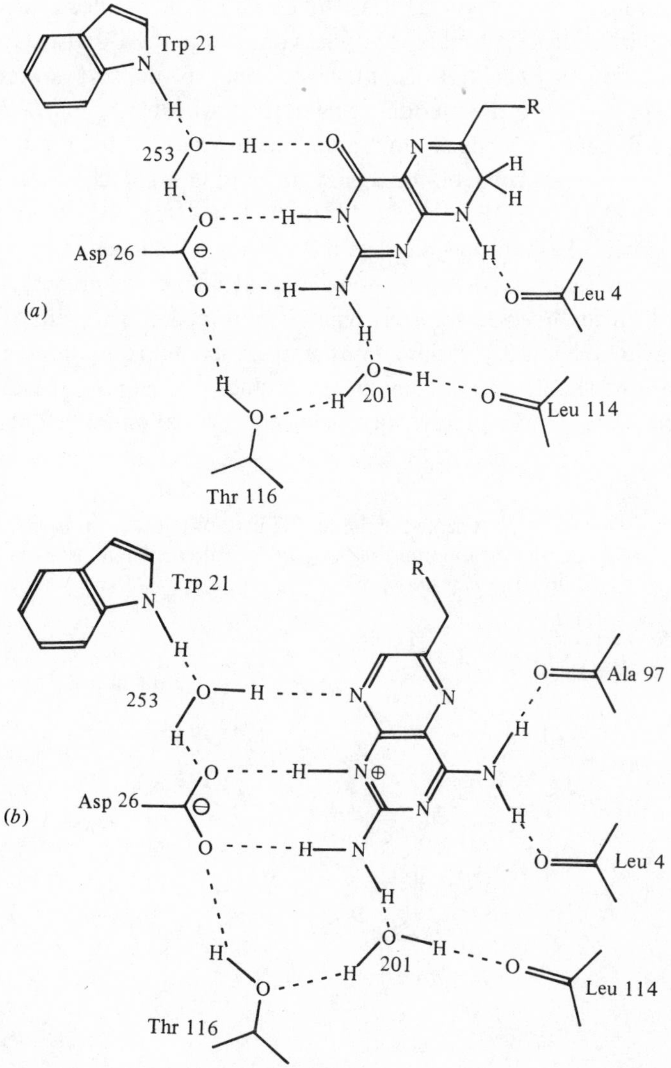

The orientation of the nicotinamide ring is held by three O . . . H—C contacts from Ala 97, Thr 45 and Ile 13. Ala 97 makes a contact with H_S (figure 5.12). The carboxamide group of nicotinamide is also held planar to the ring by hydrogen bonding to Ala 6 as well as Ile 13. A water molecule lies above the A side of the pyridine ring. Contacts between the three ring C—H groups and carbonyl oxygen atoms may help to stabilize a resonating carbonium ion structure. The H_R atom of nicotinamide lies in free space and in a direction pointing towards C6 of the pteridine ring. Little further can be gleaned from structural assessments about hydrogen transfer from NADPH. Reduction of N5 is possible by a water bridging relay (see figure 5.11). H_2O (217) is held to Leu 23 and bonded also to H_2O (253) which in turn is probably hydrogen bonded to the 4-oxo group of the pteridine ring. H_2O (217) is in contact with bulk solvent and can therefore relay a proton onto the pteridine ring surface with N5 being the nearest point and, again, hydrogen atom attachment occurs at the C7-si face.

The equilibrium constant for coenzyme binding is reduced from 1×10^8 $(\text{mol}^{-1} \text{ dm}^3)$ for NADPH to 6.1×10^4 $(\text{mol}^{-1} \text{ dm}^3)$ for NADP$^+$ and so we could expect the oxidized nicotinamide cation to diffuse away from its binding site. The geometrical model outlined here is only part of the mechanism for hydride transfer; what is not yet known is the reaction pathway to form the transition-state complex. *Ab initio* quantum-chemical calculations should provide energetic schemes for defining the catalytic mechanism in a more precise way, using this crystallographic data as an

Figure 5.12. A scheme for the mechanism of hydride transfer from nicotinamide to the pteridine ring of 7,8-dihydrofolate. (Re-drawn from Filman *et al.*, 1982.)

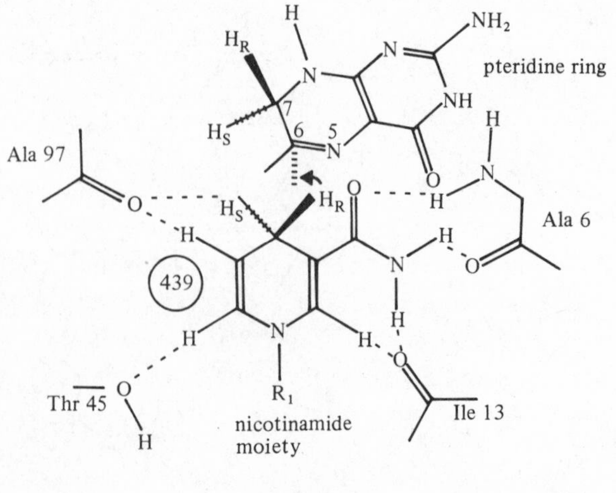

initial input for scanning the geometrical re-arrangements which may take place in the transition state.

These crystallographic studies of the active site of DHFR and the identification of site-points corresponding to the ligand-points mark a great step forward in our understanding of drug–receptor interaction. The information derived from these studies enables the medicinal chemist to study QSARs in a more complete manner than if he has to infer the structure of the site. Drug design by rational principles should be possible and it can be assumed that a large variety of new inhibitor compounds will be discovered that have an increased selectivity for DHFR extracted from different sources.

5.3   Geometrical searching for binding sites

If the macromolecular structure of the receptor is known, together with a real or putative structure for the ligand, we then have the problem: where, in the surface topography, does the ligand fit on the macromolecule? Crystal complexes of ligand and macromolecule often provide an unambiguous assignment of the binding site. But we may not always be so fortunate to have this crystal complex, and methods of geometrical searching and fitting have to be developed to generate candidate binding sites. This type of problem is very dependent on the structural requirements for the fit. If the ligand has a large number of specific attachment points, often gleaned from QSAR studies, then the structure constraints for a candidate site are large and only a few potential sites will be probable. On the other hand, if the ligand has only a few attachment points there may be many potential binding sites on the whole of the accessible macromolecular surface. Another part of the problem is that we may not know the orientation of the ligand that is needed for docking to the binding site. We thus have a problem containing six degrees of freedom. Three translational movements are needed to bring the centre of mass of the ligand to its docked position, and three rigid body rotational movements are needed to orient the ligand correctly. In reality the docking of a drug molecule to its binding site is determined by the interaction energy between the two moieties. With six degrees of freedom, an energy search calculation is not feasible for scanning a large surface and an approximation has to be used to give a measure of the goodness of fit between the partners. Since the interaction energy at any position is a summation of attractive and repulsive forces it may be possible to model both terms as geometrical parameters which can easily be manipulated by analytic geometry. Kuntz *et al.* (1982) have used this type of geometrical simplification as a rapid and universal method of searching for candidate binding sites. Repulsion is considered as a hard sphere interaction. In this interaction no full atomic hard sphere overlap between ligand and sphere is allowed, although small overlaps are permitted, and a proportionality constant of 0.42 converts the value of $E_{overlap}$, computed by equation (5.1), to $kJ\,mol^{-1}$.

$$E_{overlap} = 0.42 \sum (r_i + r_k - d_{ik}) \qquad (5.1)$$

where r_i and r_k are the van der Waals radii of atoms in the ligand and the receptor and d_{ik} is the interatomic distance between them. This very approximate equation is only useful for distances where $E_{overlap}$ is positive. Attraction is considered only as hydrogen bonding, and possible hydrogen-bond positions between acceptor and donor atoms have a ligand–macromolecule separation less than 3.5 Å.

The algorithms of Kuntz *et al.* (1982) are a major development in identifying ligand-binding sites by rational search methods and we shall examine them in some detail. Their methods are preferable to visual identification because the limitations, and visual bias, of human observation are removed. The receptor and ligand are considered as rigid bodies composed of atomic spheres which define respectively a 'lock' and 'key'. Having defined the two structures, a geometrical match between the 'key' and all possible 'key-holes' has to be searched for. Candidate 'key-holes' are then examined in greater detail by optimizing the ligand position in the putative binding site. Thus a number of candidate binding sites may be found and ranked by their associated fitting parameters for the best ligand orientation.

In fitting a key into a key-hole two surfaces have to be characterized, the external surface of the key and the void surface of the key-hole. These surfaces are said to fit perfectly when their geometrical surface properties, in the limit, are identical. How do we characterize the two surfaces for geometrical matching? The procedure adopted by Kuntz *et al.* was to represent each surface by a collection of imaginary spheres of different radii to fill the hole and to fill the key, ie binding site and ligand surface respectively. A fit is said to occur when clusters of the two sets of spherical geometries match with a specified tolerance. The assignment of spheres to fit the surface is defined by three rules. Each receptor imaginary sphere lies on the void surface. Similarly each ligand imaginary sphere lies within the ligand surface. The imaginary sphere touches the molecular surface at two points (i, j) (see figure 5.13) so that its centre is normal to the point i. For n atoms in the receptor cleft, a maximum of $n - j$ imaginary spheres associated with each atom can be constructed touching the other atoms, but the radius of the imaginary sphere and its centre position must not allow it to cut atomic spheres lining the cleft. We thus have to calculate the smallest sphere between the surface points i and j together with angle θ. Spheres with $\theta > 90°$ or $r > 5$ Å are rejected since they are more likely to span on invagination rather than fit into a cleft. A further rule applies to receptor clefts: if the atoms of the cleft span four consecutive amino acid residues they are ignored as possible binding sites since these atoms are likely to form a groove of an α helix. These rules define the receptor surface as a collection of imaginary spheres, $n - 1$, per atom. This number can be reduced to one imaginary sphere per atom by taking the sphere with the largest radius from the $n - 1$ collection; each selected imaginary sphere is characterized by three parameters, the coordinate position of its centre, its radius and θ. Imaginary spheres belonging to the same putative binding site are limited by overlap (figure 5.14). An analogous procedure is used to characterize the ligand surface.

The macromolecular external surface and the ligand internal surface are now expressed by collections of spheres. The geometries of the assemblies of spheres then have to be matched by searching the macromolecular surface against the set, or a subset, of those spheres describing the ligand. Systematic searches are the bottleneck in matching, since if we have $n = 50$ atoms in the receptor cleft and $m = 25$ atoms for the ligand then there are $n!/(n - m)!$ possible arrangements (2×10^{39}). The two collections of spheres can be represented by their respective DM which needs only to be calculated once. A structural match between the receptor and ligand occurs when the DDM sum is a minimum (see section 4.2), or, if we are content with partial matching, a partitioned DDM would be used. Searching involves re-assignment of the order of elements in one DM and computing the DDM sum between the putative site and ligand. One short cut in the search procedure is to use the fact that four non-planar spheres from each assembly define rigid docking; if four spheres can be matched between receptor and ligand, this match forms a possible candidate solution and overlapping spheres are searched for and compared through the DDM.

Figure 5.13. Fitting an imaginary sphere to ligand surface atoms IJ.

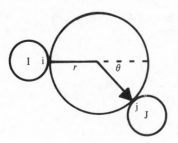

The output from the matching routine generates a number of possible matches which then have to be assessed for feasibility and ranked according to goodness of fit with the ligand.

Structure matching described above is a preliminary filtering method to collect only those matches that are possible within the spherical approximation of the respective surfaces. There may be something in the region of 1000 candidates generated and further optimization of the fit is required firstly to eliminate ambiguities in handedness of the four-point rule defining docking, secondly to eliminate candidates showing steric repulsion and thirdly to select candidates with maximum hydrogen-bond pairing. Each candidate contains a list of matched spheres with the ligand spheres. The ligand spheres are oriented by a least squares procedure to match the receptor candidate spheres; if the error of the least squares match is large this could signify a handedness problem and the candidate is eliminated. From the least squares fitting of the spheres, the centroids of the receptor void and the ligand shape are computed together with the translation and rotation matrices. Steric repulsion is then treated by computing the overlap of atomic spheres from the relationship of equation (5.1). $E_{overlap}$ has to be minimized by moving the ligand relative to the receptor and re-orienting it. Hydrogen-bond formation is handled by attempting to get the most matches between donor and acceptor atom

Figure 5.14. Consider five atoms to form the binding site cleft. Contact and re-entrant surfaces line the void into which two imaginary spheres fit; these two spheres are used to characterize the cleft.

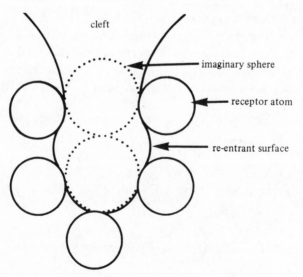

types on the ligand and receptor structures that lie within 3.5 Å for d_{ik}. Therefore this part of the optimizing routine searches for the best hydrogen-bonding pattern by ligand displacement that is not violated by atom overlap. Various possible orientations of the ligand within a candidate binding site can be ranked by the overlap error. Furthermore, there may be groups of candidate sites that can be classified according to location and rms areas of ligand fit.

The algorithms have been extensively tested with two known ligand–receptor interactions; haeme docking into metmyoglobin and thyroxine fitting into pre-albumin (Kuntz *et al.*, 1982). The myoglobin–haeme binding study was a full search of the macromolecule surface, whereas with thyroxine the study was confined to the known binding site region. The surface of metmyoglobin without the haeme contains 2289 contact surface regions for which 204 spheres could be constructed. Three large clusters of spheres are found at the surface corresponding to three possible binding sites. The largest cluster, containing 54 spheres, is the only one able to accommodate 43 spheres of the haeme ligand. A cross-section through this site is shown in figure 5.15 and it can be seen that the best fit orientation of the haeme fits well. A search for the site took 40 minutes on a PDP 11/70. With the site identified the next problem was to fit the haeme into the pocket by matching with the distance matrix; 573 feasible arrangements were generated, at 30 s CPU per fit. A refinement procedure using the overlap error equation showed that four classes of orientation of the haeme were possible. The class with smallest overlap error, ~0.2 Å, coincided with the orientation found by x-ray crystallography. A second class with a slightly larger $E_{overlap}$ value was generated by a 180° rotation of the haeme group. Curiously this corresponds to the rotation of the haeme group in some insect haemoglobins. Classes (3) and (4) corresponded to ±90° rotation of the haeme and are poor docking candidates because of the large $E_{overlap}$.

This study demonstrates the practicality of searching a macromolecular surface for putative ligand-binding sites. The algorithms work well for a rigid ligand neatly docked into a structural cleft in the binding site. However, where docking is only partial and much of the ligand surface in the bound state is accessible to solvent, these algorithms would not be efficient. Nevertheless this work is a landmark in the identification of binding sites.

5.4   Identification of latent binding sites

In section 5.2 we looked at methods for locating binding sites from crystal structures of the receptor and identifying the region where a drug molecule bound; as an example we used substrate or inhibitor binding to an enzyme. In this case the binding site was exposed and identified by positional association with the ligand. Similarly in section 5.3 the binding site on a macromolecular surface was identified by moving a ligand template to search systematically for a positional match. In both these

Figure 5.15. A scheme showing the fitting of a haeme group into its pocket in metmyoglobin; (*a*) section through the molecular surfaces in the plane of the ring, (*b*) a section perpendicular to the haeme plane passing through the cleft opening.

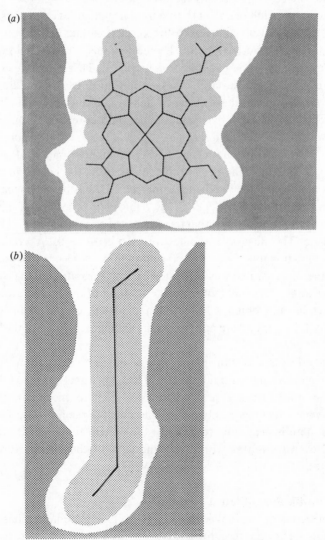

methods the site is freely available for identification. However, there are examples where the binding site is not so easily identifiable because it is hidden. An excellent example of this is the case of intercalative binding to nucleic acids. Here the crystallized oligonucleotides, either A, B or Z forms, show no intercalative pockets in the absence of drug molecules. The putative binding sites are hidden because the nucleic acid adopts the minimum energy conformation along the sugar phosphate backbone. This results in tight and regular stacking of adjacent base pairs. However, in solution the local conformational energy barriers permit some shift in the structure and a cleft can open between the base pairs, although it is still not clear whether the drug molecule enters passively into a pre-formed open site or whether the drug molecule interacts by forcing an entry.

Biochemical and biophysical experimental pointers that have indicated the intercalative site to lie between the base pairs of nucleic acids have been adequately summarized by Waring (1981). Studies of drug binding to DNA showed that the helix was extended as drug binding occurred and that this extension was accompanied by unwinding of the helix. Extension of the helix could be described by the equation

$$L/L_0 = 1 + 2r \tag{5.2}$$

where L/L_0 is the fractional increase in length and r is the number of molecules bound per nucleotide. The increment in length is about 3.4 Å per drug molecule for single intercalated insertion. For a bifunctional intercalator the equation is

$$L/L_0 = 1 + 4r \tag{5.3}$$

and the extension is about 6.8 Å for each bis-intercalator. Early x-ray fibre studies of ethidium binding to DNA revealed that the regular pattern of x-ray reflexions, due to regular stacking of base pairs in pure DNA fibres, was lost. It is possible to distinguish ten different intercalative sites in DNA because the base pairs adjacent to each other have a structural directionality: a purine followed by a pyrimidine in reading from 5′ → 3′ is not the same geometrical arrangement as the reverse order of bases. Ethidium shows a preference for pyrimidine (3′,5′)-purine nucleotide sequences (Krugh, Wittlin & Cramer, 1975). This preliminary experimental work led to a model for intercalation where the adjacent base pairs at a site become separated and the local helix unwinds by about 20°. Monofunctional intercalative drug molecules, like ethidium, then slide between the base planes from the narrow groove side of the helix. The experimentally derived model appears to be substantially correct and a series of crystal structures of monofunctional intercalators with dinucleotides show a drug–nucleotide complex remarkably similar to the earlier experimental predictions. The positioning of the phenanthridinium ring of ethidium with respect to the base pairs in (dC-dG)·(dC-dG) is shown in figure 5.16.

The triumph of the early model-building studies, with monofunctional intercalators, in predicting how the drug molecule binds to the dinucleotide unit must be set against the crystal structure for a bis intercalator, triostin A, with a double helical hexanucleotide. In this crystal complex the binding site is modified markedly (Wang *et al.*, 1984). Triostin A is a cyclic octadepsipeptide antibiotic containing two quinoxaline rings. The quinoxaline rings lie parallel to each other and are separated by the cyclic peptide (figure 5.17). The early model studies for intercalation suggested that the quinoxaline rings were inserted into the DNA helix by intercalation but that two base pairs were sandwiched between the quinoxaline rings, giving rise to a helix extension of about 6.1 Å and an unwinding angle of 47° (Lee & Waring, 1978). Both assertions proved to be approximately correct in the crystal structure of the complex. What was not expected was the extensive re-working of the DNA to mould itself round the antibiotic. The cyclic peptide backbone appears relatively rigid and the movement required to fit the two quinoxaline rings into the DNA is generated by a combination of torsion angle changes in the deoxyribose ring, considerable altering of the alignment of Watson–Crick base pairs and re-arrangement of the hydrogen bonding of A-T base pairs from a Watson–Crick form to Hoogsteen base pairing (figure 5.18).

Figure 5.16. An intercalated complex between ethidium (dashed lines) and (dC–dG)·(dC–dG); atom positions are those derived from the crystal structure. The sugar phosphate backbone is omitted for clarity.

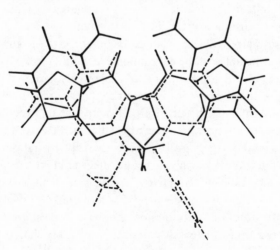

In the intercalated complex with the hexanucleotide, the quinoxaline rings of the antibiotic are held approximately at right angles to the cyclic peptide chain and parallel to each other. The cyclic peptide is nearly rectangular and the quinoxaline rings joined to *D*-serine through a planar *trans*-NHCO link are projected away from the peptide. The N-methyl group of N-methyl valine is found at the other corner of the rectangle projecting towards the nucleic acid surface (figures 5.17 and 5.19). All the peptide groups are in the *trans* configuration. In the narrow groove region where intercalation takes place there are no water molecules sandwiched between drug molecule and receptor. The alanine residue forms a hydrogen bond from its backbone NH to N3 of G2 and by the other alanine to N3 of G12. Conformational asymmetry allows a further hydrogen bond between the carbonyl group of alanine and a neighbouring N2 of G12 but no hydrogen bond exists between the other alanine carbonyl and N2 of G2.

The nucleic acid conformation is remarkably modified to accommodate the intercalated drug molecule compared with the normal B-DNA structure. Watson–Crick base pairing is maintained between the C–G pairs; these are the bases of the dinucleotide portion sandwiched between the two parallel quinoxaline rings of triostin A. Nevertheless, the guanine bases are tilted out of the plane by 20° whereas the cytosine bases show only

Figure 5.17. Molecular structure of triostin A, hydrogen atoms omitted.

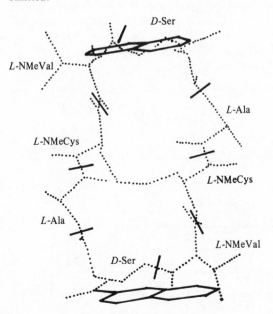

Figure 5.18. Base-pair arrangements for the accommodation of triostin A at the binding site. (a) A–T bases in a Watson–Crick arrangement with the adenosine base in the *anti* conformation; this is the conformation found normally. (b) A–T bases in a Hoogsteen pairing arrangement with the adenosine base rotated round the glycosidic bond in the *syn* conformation; this is the conformation found in the triostin A complex.

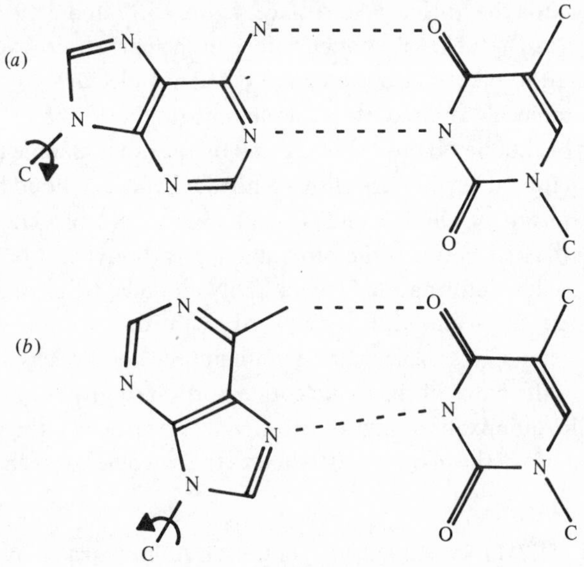

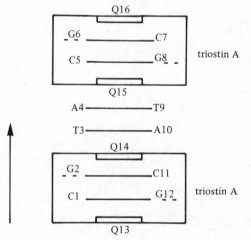

Figure 5.19. Schematic representation of the binding of two triostin A molecules to a section of DNA. The quinoxaline moieties are intercalated between adjacent G–C, A–T pairs. Hydrogen bonding between N3 of guanine and NH of the alanine residue of the cyclic peptide stabilizes the complex.

4° deviation from the plane. The two central A-T base pairs that lie together between the two intercalated drug molecules show the most marked variation; Watson–Crick base pairing has been lost and replaced by Hoogsteen pairing. The adenine bases are rotated 180° from the normal *anti* conformation to the *syn* conformation and the N6, N1 hydrogen bonds are replaced by N7 and N6 of adenine. One structural consequence of this is that the C1 atoms of A10 and T3 are 2 Å closer together; presumably this is energetically a more favourable structure. What is not understood is the mechanism whereby this base pair conformational change has taken place. It has to be assumed that the normal A–T Watson–Crick pair has been broken and that the backbone has undergone a conformational change to swing the adenine base out into one groove to allow the half rotation to occur before the base is flipped back in its turned over state for Hoogsteen pairing. Van der Waals attraction between the valine carbonyl group and the sugar of A10 appears likely together with attraction between the serine α and β carbons and the sugar of T3. In all there are 19 van der Waals contacts between the peptide and nucleic acid moieties. Since there are these marked changes in base pairing and shortening across the helix in the A–T region, it is not unexpected that the backbone torsion angles are very different from the normal angles encountered in B-DNA. Furthermore, the sugar ring pucker is extensively altered. In B-DNA the deoxyribose ring has the C2-*endo* conformation, but in the triostin A complex there is: O1'-*exo*(C1); C2'-*endo*(G2); C3'-*exo*(T3); O1'-*endo*(A4); C3'-*endo*(C5); and C3'-*endo*(G6). The unwinding of the base pairs is not constant; C1 G2 are unwound by 27°, G2 T3 which intercalate Q14 are unwound by 12° and T3 A4 show minimal unwinding of 3°. Thus the net unwinding is about 54° for a complete intercalation of one triostin molecule and is close to that measured experimentally. The large differences in sugar ring pucker contribute to the altered unwinding of the base pairs as well as to other changes in torsion angles along the sugar phosphate backbone.

Thus one may conclude from the work of Wang *et al.* (1984) that with nucleic acids the binding site for a ligand is very flexible. This torsional tolerance in the nucleic acid backbone is greater than the conformational freedom of the cyclic depsipeptide of the antibiotic triostin A. Therefore the binding site actively moulds itself around the intercalated molecule. This drastic modification of the binding site induced by the drug had not been predicted in previous model studies. The results are therefore a warning against taking a crystal structure for a putative binding site without accounting for widespread conformational flexibility and its effect on the position of possible site-points. Apparently stable hydrogen-bonding structures, such as Watson–Crick base pairs in DNA, are not sacrosanct and may be substituted for other arrangements if the overall energy is more favourable.

5.5   Subsets of site-points

In section 5.2 the active site region of DHFR was examined from crystallographic data. The cofactor, NADPH and an inhibitor, methotrexate, could each be assigned a set of site-points for attachment of the moieties by hydrogen bonds. These allocated site-points are a subset of the whole set of available site-points in the region of the active site. Other local moieties in the binding site, such as hydrophobic regions, may enhance or diminish ligand binding but their effect is regional rather than directional; in other words, their effect is not specifically orientational as is the case with a hydrogen bond. The inhibitor methotrexate was speculated to bind at the pteridine binding site to a set of site-points shared with the pteridine ring of the substrate although the ligand-points were not identical, and methotrexate could well have a further extra hydrogen bond to Ala 97 (figure 5.11). An inhibitor of substrate binding may be defined as one with a set of site-points that intersects with the set of substrate site-points (figure 5.20(*a*)). Each set of site-points which is a subset of the whole set is geometrically determined. This would give a straightforward 1:1 competition. If cofactor is needed for the biochemical reaction at the active

Figure 5.20. Diagrammatic representation of possible sets of site-points. Each circle contains a set of geometrically defined site-points for substrate (S), inhibitor (I), or cofactor (C). (*a*) A simple intersection of two sets of site-points for substrate and inhibitor; common site-points are found in the intersection. (*b*) An added non-intersecting set of site-points for cofactor. (*c*) Two intersections linking three sets of site-points. (*d*) Different sets of site-points for an inhibitor in two conformational states I1 and I2.

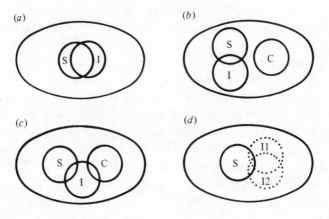

site a further possibility exists; the cofactor site-points may be a target for inhibition as well (figure 5.20(b)). Alternatively, they may form on the cofactor surface a new partial set, or supplementary set, of site-points for ligand binding which could enhance binding cooperatively (figure 5.20(c)). Furthermore, protein structures have a dynamic component and the geometry of site-points can change for different conformational states (figure 5.20(d)). Thus there is the added problem of ligand binding to fluctuating sets of site-points. In this section we shall examine some of the problems just outlined to illustrate how they can be tackled and what bearing the problems have on established QSARs.

5.5.1 Coenzyme binding to DHFR

So far, there are very few crystallographic complexes of DHFR with coenzyme analogues and drug molecules. An unambiguous structural assignment of subsets of site-points is therefore not yet possible. However, experimental studies are strongly indicative of different modes of binding of ligands and coenzyme analogues. Birdsall, Burgen & Roberts (1980a, b) determined fluorimetrically the binding of numerous analogues of NADP$^+$ and NADPH to DHFR L. casei. Table 5.5 lists the binding constants for the binary complexes with the enzyme. The reduced nicotinamide ring binds more tightly to the enzyme, the binding constant ratios ranging from 14 for thionicotinamide analogues to 1600 for NADPH. Weaker binding is associated with modification of the nicotinamide ring.

A glance at the site-points for NADPH (table 5.3) shows that of the 17 site-points only three are to nicotinamide; thus, 14 site-points bind the other moiety. It is assumed that the 14 non-nicotinamide site-points are involved in binding most of the analogues shown in table 5.5. Furthermore, the binding constants of TNADP$^+$, which would be expected to make a weak S7 hydrogen bond, has a similar binding constant to NADP$^+$; also PADPR—Ome, which has no nicotinamide moiety, binds as strongly as NADP$^+$. Although a detailed structural picture of coenzyme binding is not yet possible, high-resolution NMR can provide information on local proton environments from shifts in proton resonance signals. Thus NMR shifts have been used to examine the adenine and nicotinamide moieties at the binding site (Hyde et al., 1980a, b; Roberts, 1983). NHDP$^+$ and NADP$^+$ bind in a similar way; the adenine ring does not bind to any site-points (table 5.3), but TNADP$^+$ and APADP$^+$ bind in a different way to NADP$^+$. The nicotinamide ring of NADP$^+$ binds in the anti conformation round the glycosidic bond and TNADP$^+$ is found as a mixture of syn and anti forms. In the ternary crystal structure of NADPH, methotrexate and DHFR, the NADPH binds in the anti conformation. The NMR differences between bound $NADP^+$ and NADPH do suggest differences in the environment of the two molecules in the bound state. Perhaps changes are related to a modified geometry in the hydrophobic pocket that lines the nicotinamide binding site, since in the crystal structure there are nine van der Waals contacts to the nicotinamide moiety (Filman et al., 1982).

DHFR binding of coenzyme is modified with the ternary complex including methotrexate, trimethoprim or folic acid (table 5.6). The coenzymes bind more strongly in the ternary complexes with the exception of PADPR—Ome. In the complex with methotrexate, coenzymes bind more tightly if the nicotinamide ring is present. There is a clear difference in binding affinity between the three ternary complex forms. Reduced coenzyme analogues bind very much more tightly in the ternary complex (table 5.7) and stronger binding is observed with methotrexate. Therefore, in all cases the binding of coenzyme exhibits a marked cooperativity in the ternary complex; in moving from $NADP^+$ in the binary complex to NADPH with methotrexate, binding is enhanced by a factor of 89 000.

Table 5.5. *Equilibrium constants for the binding of coenzyme* $NADP^+$, *NADPH and some analogues to DHFR*

Coenzyme	Binding constant $(mol^{-1} dm^3)$
$NADP^+$	6.1×10^4
$NHDP^+$	9.4×10^3
$\varepsilon NADP^+$	4.9×10^4
$TNADP^+$	1.4×10^4
$APADP^+$	8.9×10^3
PADPR—Ome	5.2×10^4
NADPH	1.0×10^8
NHDPH	3.1×10^7
$\varepsilon NADPH$	3.1×10^7
TNADPH	2.0×10^5
APADPH	2.1×10^6

$NADP^+$ = nicotinamide adenine dinucleotide phosphate, $NHDP^+$ = nicotinamide hypoxanthine dinucleotide phosphate, $\varepsilon NADP^+$ = nicotinamide 1,N6 etheno-adenine dinucleotide phosphate, $TNADP^+$ = thionicotinamide adenine dinucleotide phosphate, $APADP^+$ = 3-acetylpyridine adenine dinucleotide phosphate, PADPR—Ome = a methyl β-riboside derivative of 2'-phosphoadenosine 5'-diphosphoribose, NHDPH = reduced form of $NADP^+$ (etc.).
Data from Birdsall, Burgen & Roberts (1980a).

5.5.2 Inhibitor binding to DHFR

Methotrexate binds to *L. casei* DHFR by six direct hydrogen-bonding interactions (table 5.4), of which four are made to the pteridine ring. Birdsall *et al.* (1980b) studied inhibitor binding to DHFR by taking fragments of the molecule separately and measuring the binding fluorimetrically in the presence of coenzyme. The fragments taken were *p*-aminobenzoyl glutamate and 2,4-diaminopyrimidine as an analogue of the pteridine ring of methotrexate. The pyrimidine ring is not an exact replacement for the pteridine ring but it could also function as an analogue of trimethoprim. The possible equilibria in the binding reaction are shown in figure 5.21, although not all of them are independent equilibria. Binding constants with a variety of oxidized and reduced coenzymes are shown in table 5.8 and table 5.9 respectively. With all coenzyme analogues, the binding constant for the ternary complex between enzyme–coenzyme and

Table 5.6. *Relative changes in the binding constants of oxidized coenzymes to DHFR in the presence of methotrexate, trimethoprim or folic acid compared with binding of coenzyme to the enzyme alone*

Coenzyme	Methotrexate	Trimethoprim	Folic acid
$NADP^+$	12.5	2.0	20.0
$NHDP^+$	11.7	2.9	14.9
$\varepsilon NADP^+$	10.2	2.5	36.7
$TNADP^+$	4.6	4.7	3.0
$APADP^+$	3.7	6.1	22.5
PADPR—Ome	4.8	4.8	0.84

Data taken from Birdsall, Burgen & Roberts (1980a).

Table 5.7. *Relative changes in the binding constants of reduced coenzymes to DHFR in the presence of methotrexate or trimethoprim compared with binding of coenzyme to the enzyme alone*

Coenzyme	Methotrexate	Trimethoprim
NADPH	680	135
NHDPH	322	
TNADPH	1500	870
APADPH	2190	195

Data taken from Birdsall, Burgen & Roberts (1980a).

2,4-diaminopyrimidine is larger than the corresponding binding constant for the *p*-aminobenzoyl glutamate moiety. This finding is in line with expectations from the larger number of site-points for 2,4-diaminopyrimidine. There is a marked effect of the structure of the coenzyme analogue on the binding of the inhibitor moiety. For example, the binding of *p*-aminobenzoyl glutamate to ECD varies over a 36-fold range. Negative cooperativity is observed in the binding of ED, P compared with ECD, P with NADP$^+$ as coenzyme. Binding of *p*-aminobenzoyl glutamate to form the ternary complex EC, P is only marginally affected by the oxidation state of the coenzyme with differences in the range of two-fold. In contrast, the oxidation state of the coenzyme markedly affects the binding of 2,4-diaminopyrimidine in the ternary complex. However, PADPR—Ome, which has the nicotinamide ring replaced by a methoxy group, has little effect on the binding of *p*-aminobenzoyl glutamate and 2,4-diaminopyrimidine. This observation provides strong evidence that any cooperative effect in binding of the inhibitor moieties is due to the nicotinamide ring of the coenzymes. This hypothesis is strengthened by the fact that NADP$^+$, NHDP$^+$ and ε NADP$^+$, which have the nicotinamide ring in common but an altered adenine portion, have very similar effects on separate binding of the two inhibitor fragments.

Figure 5.21. A scheme to represent the equilibrium constants for coenzyme (C) binding, 2,4-diaminopyrimidine (D) and *p*-aminobenzoyl-*L*-glutamate (P) to DHFR (E). There are eight possible enzyme-bound components in the equilibrium. (Re-drawn from Birdsall, Burgen & Roberts, 1980b.)

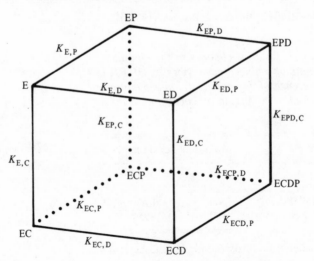

Table 5.8. *Binding constants ($mol^{-1} dm^3$) of oxidized coenzymes, p-aminobenzoyl-L-glutamate and 2,4-diaminopyrimidine to DHFR*

Coenzyme	$K_{E,C}$	$K_{EP,C}$	$K_{ED,C}$	$K_{EDP,C}$	$K_{EC,P}$	$K_{EC,D}$	$K_{ECP,D}$	$K_{ECD,P}$
None	6.1×10^4				8.1×10^2	1.3×10^3	7.25×10^4	4.7×10^4
NADP$^+$	9.4×10^3	2.5×10^5	1.0×10^6	1.3×10^5	3.3×10^3	2.2×10^4	3.8×10^4	6.0×10^3
NHDP$^+$	4.9×10^4	2.9×10^4	1.4×10^5	3.7×10^4	2.5×10^3	1.8×10^4	9.2×10^4	1.3×10^4
εNADP$^+$	1.4×10^4	1.8×10^5	7.1×10^5	1.5×10^5	3.0×10^3	1.8×10^4	5.8×10^4	9.6×10^3
TNADP$^+$	8.9×10^3	8.9×10^3	2.7×10^4	1.3×10^5	5.2×10^2	2.4×10^3	1.0×10^6	2.2×10^5
APADP$^+$	8.9×10^3	3.5×10^4	5.0×10^4	9.7×10^4	3.3×10^3	7.0×10^3	2.0×10^5	9.2×10^4
PADPR—Ome	5.2×10^4	5.1×10^4	8.6×10^4	8.9×10^4	8.0×10^2	2.1×10^3	1.3×10^5	4.9×10^4

Data taken from Birdsall, Burgen & Roberts (1980b).

The measured equilibrium constants for analogue binding to enzyme and enzyme–coenzyme complexes given in tables 5.8 and 5.9 indicate clearly that cooperativity operates in the binding mechanism. Cooperativity can be explained by assuming that the enzyme exists in two conformational states and that the binding of various ligands alters the proportions of the two conformations. Evidence for different conformational states of the small protein BPTI has been presented in section 4.7. Molecular dynamics calculations showed that BPTI exists in four interconverting conformational states. Gronenborn et al. (1981) have shown directly, using ^1H and ^{31}P NMR, that two conformations of trimethoprim–NADP$^+$–DHFR ternary complex exist in solution. Further characterization of the two conformations has been made using four trimethoprim analogues (Birdsall et al., 1984). The fractional populations of the two conformations for different analogues in the ternary complex are given in table 5.10; it would appear that the structure of the analogue

Table 5.9. *Binding constants* $(mol^{-1}\ dm^3)$ *of reduced coenzymes,* p-*aminobenzoyl-*L-*glutamate and 2,4-diaminopyrimidine to DHFR*

Coenzyme	$K_{E,C}$	$K_{EC,P}$	$K_{EC,D}$	$K_{ECP,D}$	$K_{ECD,P}$
None		8.1×10^2	1.3×10^3	7.3×10^4	4.7×10^4
NADPH	1.0×10^8	3.8×10^3	1.1×10^4	2.7×10^5	9.1×10^4
NHDPH	3.1×10^7	1.4×10^3	8.6×10^3	4.7×10^5	7.6×10^4
εNADPH	3.1×10^7	1.4×10^3	1.0×10^4	4.1×10^5	5.6×10^4
TNADPH	2.0×10^5	7.0×10^2	1.6×10^4	1.3×10^5	5.8×10^3
APADPH	2.1×10^6	2.4×10^3	2.9×10^4	8.8×10^5	7.2×10^4

Data taken from Birdsall, Burgen & Roberts (1980b).

Table 5.10. *Populations of two conformations for the ternary complex:* DHFR-NADP'-*trimethoprim analogue*

	Fractional populations	
Complex	Conformational state I	Conformational state II
E–NADP'–trimethoprim	0.55	0.45
E–NADP'–6-aminotrimethoprim	0.70	0.30
E–NADP'–6-methyltrimethoprim	0.55	0.45
E–NADP'–4-fluorotrimethoprim	0.30	0.70
E–NADP'–Ro16-3034	0.40	0.60

E is the enzyme DHFR.
Data taken from Birdsall et al. (1984).

determines the proportion of each conformational state. In the region of the active site there are three histidine residues, His 64, His 77 and His 18, but only His 64 and His 18 appear to be in contact with $NADP^+$ in the crystal complex; His 77 forms a hydrogen bond with water which is then attached to adenine of the $NADP^+$ ring. The environment round His 64 does not appear to be changed between the two conformations. His 18 forms two ligand-point interactions with NMN ribose O2' and O3' (table 5.3). The NMR data is consistent with the hypothesis that conformation I of the $NADP^+$–trimethoprim complex is very similar to the DHFR–$NADP^+$–methotrexate complex. In conformation II the adenosine 2'-phosphate binding is similar to conformation I, the ligand-point O2P and site-point Npep His 64 interactions being similar. The difference between the conformations appears to lie in the binding of the nicotinamide end of the coenzyme. In conformation II, rotation round the nicotinamide glycosidic bond in the ternary complex is similar to that for $NADP^+$ in free solution *and* rotation of the carboxamide is also free (compare with *L. casei* ternary crystal complex where carboxamide is anchored to Ala 6, table 5.3). Therefore, in conformation II the adenosine is anchored in a similar way to conformation I but the dihedral angles in the pyrophosphate part of the chain in $NADP^+$ are different and the C5'–O bond of the nicotinamide–ribose changes by 50° when going from conformation I to conformation II. Thus in conformation II $NADP^+$ is only loosely bonded and anchored by the adenosine portion with the nicotinamide ribose in a different conformation; the nicotinamide ring is more or less free to rotate unhindered by interactions with the site-points that anchor it in conformation I. This difference in binding the coenzyme has effects on binding trimethoprim in the ternary complex. In conformation I the trimethoprim molecule appears to be in van der Waals contact with the nicotinamide ring of the coenzyme and retains the same conformation in the ternary complex as in the binary complex. So far, no direct comparison can be made between the binding of the inhibitor in the ternary complex for both conformational states.

Cooperative binding of the moieties exists in the ternary complex; how does this come about? $NADP^+$ causes the inhibitor trimethoprim to bind twice as tightly to DHFR than it does in the absence of coenzyme. Thus $NADP^+$ and $APADP^+$ produce a ternary complex with 90% in conformation II with a 4.7- and 6.1-fold increase respectively in binding of inhibitor; PADPR—Ome, in which there is no nicotinamide ring, shows a 4.8-fold cooperativity with trimethoprim. Thus if the nicotinamide ring does not participate in inhibitor binding in conformation II because it is turned away from the nicotinamide binding site-points, and it is missing altogether in PADPR—Ome, then it has to be assumed that some other interaction, such as an induced conformational change in DHFR, is responsible for cooperativity. Binding of the adenosine and of the coenzyme appears to be very similar for conformations I and II. Thus Birdsall *et al.* (1984) were left with the hypothesis that the nicotinamide ring induces an energetically unfavourable conformational change in the enzyme for inhibitor binding. Possible candidate residues in DHFR responsible for this conformational change were revealed by model building studies. An early crystallographic study of *E. coli* DHFR in the binary complex with trimethoprim (Baker *et al.*, 1981) showed that there were two enzyme inhibitor complexes in the unit cell and that the trimethoprim molecules bound to each enzyme molecule in the same location but with a slightly different orientation. However, this crystal structure was not sufficiently well refined to allocate specific differences in site geometry. Ser 48 is close to the 4-methoxy group and Leu 19 makes contact with the nicotinamide ring (Bolin *et al.*, 1982; Filman *et al.*, 1982). Thus it is possible that these residues could interfere with trimethoprim binding in conformation I by virtue of the fact that they are bound to the nicotinamide moiety. In conformation II, the nicotinamide moiety is not present in this region and conformational freedom of the two residues is not constrained. Therefore the orientation of trimethoprim is less restricted and a more favourable binding mode may be assumed to produce the cooperative binding of inhibitor in the ternary complex. More detailed studies are needed to assess the validity of this hypothesis.

The work of Birdsall and her colleagues is a landmark in studies of drug–receptor interaction. They have shown that ligand binding to the site region is a very complex phenomenon and encompasses many of the possible schemes outlined in the Venn diagrams of figure 5.20. An important consequence arising from this work concerns the problem of structure–activity relationships at the molecular level. By analogy with the Fischer lock-and-key hypothesis, the binding site-points and ligand-points are usually speculated to form fixed geometric arrays and the energy of binding can be understood in terms of a summation of each separate pair of site-point, ligand-point interactions. However, if the lock is deformable and the key is flexible, to continue the analogy, this simple relationship based on geometrical juxtaposition can no longer be held (Roberts, 1983). QSAR studies are quite simply statistical correlations. But a correlation, however good it might be, does not necessarily imply a mechanistic equivalence. For example, in chapter 4 we mentioned two methods for studying the similarity between arrays of geometrical points: the distance matrix method and the superposition method. Although there is a good statistical correlation between the results of the two methods, they are not equivalent and the relationship does not necessarily hold for a small sample of points. We shall return to the QSAR problem in chapter 8.

5.6 Topographical mapping of ligand-binding sites by inference methods

In section 5.3 we considered a method for identifying the binding site region on a defined macromolecular surface. That method relied heavily on the ligand-binding site being a structural cleft for the search procedures to function adequately. If the binding region is not so well defined in terms of a structural concavity then other methods need to be employed to improve the efficiency of the search procedure. We need to infer a tentative geometrical structure for the site before searching. These inference methods are based on QSARs for the molecular geometry of the ligand within a congeneric series. The basic assumption of the method is that within a related series of molecules there exist intramolecular points on the ligand that enhance or diminish the affinity of drug binding. These are known as ligand-points and, for an enhanced affinity, more points need to be occupied by a molecular group complementary to the corresponding attachment point (site-point) in the receptor. Thus, searching for points on a representative ligand is a preliminary step needed before the topography of the receptor can be mapped by inference. Once the site-points have been identified, their geometrical relationships can be established and matching between site-points and actual physical points on the receptor can be explored.

The approach just outlined has the implicit assumption that members of a congeneric series bind in the same orientation to an identical portion of the total binding site available. This assumption may not always be valid and a variable binding mode may have to be considered. In addition, it is assumed that a fixed conformation is needed in the binding mode. The binding surface of the ligand can be discerned by adding bulky groups to the representative ligand and observing the effect of these groups on the energy of ligand binding. If the binding energy is decreased by an added moiety then a possible ligand site has been identified. Thus the interacting face on the ligand can be deduced in a stepwise manner. Distance matrix methods can then be used to evaluate the geometry of the ligand-points with the energy of binding. This method has been pioneered by Crippen (1981), but it is fiendishly complicated and only an outline of the salient steps is given here. A reader needing a more practical understanding of the method should return to the original monograph where the mathematical development is eloquently explained.

It is essential to have samples of pure receptor macromolecule for *in vitro* measurements of binding energies, ΔG_{bind} for each ligand of the congeneric series. For example, the binding energy may be measured from the IC_{50}. In this case ΔG_{bind} is given by

$$\Delta G_{bind} = RT \ln \{K_M [IC_{50}]/(K_M + [S])\} \qquad (5.4)$$

where K_M is the Michaelis constant for the enzyme, $[S]$ is the substrate concentration and R, T have their usual thermodynamic meaning. ΔG_{bind} is an association free energy and can include solvation, enthalpy and entropy. Therefore, by making systematic chemical modifications to a possible ligand-point, we can have a measure of the change in interactive energy produced by a particular group. At the same time we can tentatively infer a possible type of site-point, for example a hydrophobic pocket or a hydrogen-bond group, to match each ligand-point with a three-dimensional geometry of site-points. The broad principle of the method as stated is a simple concept. The practical solution, however, is not so easy.

We could represent the ligand by a set of ligand-points on each atomic nucleus. This is usually avoided, to save computing time, and a smaller set is used because these contain the positions to which groups will be added to test statistically for the required ligand-points. Aryl rings might be represented by a single ligand-point. If the number of ligand-points initially chosen is too small to describe the data adequately, then a larger set has to be used. Assume that we have selected a set of initial trial ligand-points, a DM is needed to describe the set of points. This matrix is usually constructed to account for upper and lower bounds of ligand-point distances to include conformational changes. The upper bound matrix, U, stores the longest set of interpoint distances and the lower bound matrix, L, stores the shortest. These values are stored in a single matrix B, L in the lower and U in the upper triangle. The matrix B contains a representation of the conformational flexibility. The rows and columns are labelled with their ligand-point number and their ligand-point type, identical groups having the same type. A B matrix is constructed for each ligand in the congeneric series. Next we have to find a set of common structure features within the series from the distance matrices called the base-group, and a set of substituent groups that describe the differences between members. It may well be that no single base-group is common to all ligands; in this case we may need to partition the ligands into separate base-groups having common features within each group. Determination of the base-group is by an exhaustive tree-search of pairs of points between two ligands for matched distances between U and L within some allowable range; distances outside the range are deemed not to match. The optimum match is called the intersection. The intersection is then compared with the DM of the third ligand and a new intersection found. This procedure is repeated until all ligands are considered. The base-group contains the ligand-points in common. The search for substituent groups is similar but first we remove the base-group from each ligand and we are left with substituents. An iterative tree-search for substituents and their intersections is performed until every substituent group on every ligand has been considered. We should then know which substituents are geometrically distinct.

For each ligand we have the measured value of the free energy of binding ΔG_{bind}. This value now has to be decomposed into a set of additive interaction energies for each ligand-point, site-point and substituent. Site-points are allocated to each point in the base-group and to the substituent point. The fitting of interaction energies between each point in the base-group and site-point is done by a least squares method so that the object function, F, has to be minimized

$$F \equiv \sum_{\text{ligands}} \left(\Delta G_{\text{obs}} - \sum_{\text{contact}} \varepsilon_{ij} \right)^2 \tag{5.5}$$

where ε_{ij} is the interaction energy between ligand base-group point and site-point.

The end result of these calculations is to provide an acceptable DM for the site-points. From that matrix, the geometry in Cartesian coordinates can be generated with the appropriate chirality for all connected points and each point has a chemical signature attached to it, such as hydrophobic pocket, hydrogen bond (acceptor/donor), polar or ionic region. Crippen's method is a triumph in deductive geometry and makes it possible to produce a topographical map of site-points. If there is a known structure for the binding region, then the inferred map of site-points can be matched with the actual geometry to identify the site atoms.

6

The importance of solvent in drug–receptor interactions

One of the serious shortcomings of the application of quantum-mechanical methods, or empirical potential studies, to drug–receptor interaction is that the calculations are nearly always performed as though the interaction takes place *in vacuo*. These limitations have led many to be sceptical of this type of computation. The primary reason for the exclusion of water from the calculation is, of course, understandable; the medium is extremely difficult to parameterize because, firstly, its structure varies over very short time-scales and, secondly, it exhibits marked structural changes in local environments close to solute molecular surfaces. How can these variations in local solvent structure be studied and what bearing may they have on the problem of drug–receptor interaction? Perhaps the time has come to grasp the nettle and face up to the questions posed by attempting to include aqueous solvent in molecular interactions. A single unifying approach to the inclusion of solvent in chemical computations is not yet feasible; therefore, this chapter will dissect the problem into constituent parts. The first section will survey the structure of water in the liquid state. Secondly, the organization of water will be considered at molecular surfaces. Thirdly, different computational techniques will be scanned for the applicability to simulate an environment of solvent. Fourthly, recent methods for examining hydrophobicity and hydrophilicity will be mentioned. Finally, we shall attempt to incorporate the effect of solvent into electrostatic computations. This subdivision of the chapter is designed to illustrate contemporary research methods from chemical physics which could shed most light on the influence of solvent in drug–receptor interactions. Many of the examples used are classic pilot studies from widely differing research projects but, nevertheless, are pertinent for molecular pharmacology.

6.1　The structure of water

The structure of liquid water has always presented a puzzle to scientists. The interaction of water with solute molecules has pronounced effects on the behaviour of the solute; large macromolecular structures may interact differently with water at various regions on the macromolecular surface. This interaction is not passive and cannot be ignored for, in many ways, it is responsible for maintaining the molecular architecture as certain groups seek water (hydrophilic moieties), and others avoid a local aqueous environment (hydrophobic groups). It is widely believed, and with good reasons, that these interactions between a macromolecule and aqueous solvent determine the folding of many proteins into their characteristic tertiary structures. Water is distinguished from other media by a high capacity to form hydrogen bonds from one solvent molecule to another as well as between solute molecules and solvent. The hydrogen-bonding capacity of water is responsible for many properties that are implicated in biomolecular interactions.

6.1.1　Liquid water

A molecule of water is able to donate two hydrogen atoms to neighbouring water molecules and, at the same time, its oxygen atom can function as an acceptor for two hydrogen atoms. Thus four hydrogen bonds are possible in a tetrahedral geometry with adjacent water molecules (figure 6.1). Many properties that are peculiar to the liquid state are believed to be related to this hydrogen-bonding capacity. In ice, the hydrogen bonds are specifically directed throughout the lattice to form the regular solid arrays of molecules in the crystal. However, when ice melts at 273 K only about 20% of the hydrogen bonds break. The fraction of hydrogen bonds decreases slowly as the temperature is raised at a rate of about 1% for every 2.5 K. The hydrogen bonds therefore appear to link the

Figure 6.1. The tetrahedral structure of water, two lone-pairs and two hydrogen atoms at the vertices and the oxygen atom at the centre.

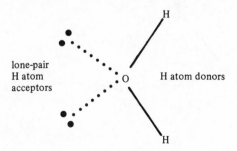

water molecules into a network that has the characteristics of a gel. For water at 310 K, about 65 % of all hydrogen bonds are intact at any instant. The local network of hydrogen bonds determines the connectivity between the water molecules and the microenvironment. There are two major regions in liquid water: high density regions where the water molecules have a low connectivity because the hydrogen bonds are broken, and low density patches where the water molecule has four intact bonds in a tetrahedral arrangement.

Evidence in support of the gel model for liquid water structure has recently been provided by molecular dynamics simulations (Geiger & Stanley, 1982). The simulation allows 216 water molecules, placed in a cube of length 18.6 Å to give a density of 1 g cm^{-3}, to be moved about and interact through an ST2 potential (Stillinger & Rahman, 1974). This potential describes accurately the electronic features of the water molecule (figure 6.2). Snapshots taken during the period of simulation reveal that a connectivity of four-bonded molecules occurs frequently and spontaneously; moreover, in this local network the density of water is shifted significantly lower from the global value. The size of the small clusters of water molecules is comparable with that found by x-ray

Figure 6.2. A structural scheme for hydrogen bonding in water used in the calculations of ST2 potentials; $l_1 = 0.8$ Å, $l_2 = 1.0$ Å, $\theta = 109.47°$, $q = 0.2357e$, $\sigma = 2.852$ Å (optimum), $V = -28.62$ kJ mol^{-1} (minimum).

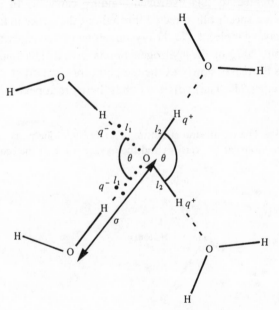

scattering. Further studies on hydrogen-bonded networks by molecular dynamics simulations (Stanley, Blumberg & Geiger, 1983) show that low density patches, and fluctuations in the size of hydrogen-bonded networks, are consistent with the gel model; furthermore polygonal closures, reminiscent of those found in the ice lattice, can exist temporarily. One of the difficulties of the molecular dynamics method is defining when a bond occurs. The hydrogen bond in water does not have a single value but can vary depending on the length of the bond. The most favourable hydrogen-bond energy, ~ 28 kJ mol^{-1}, occurs when the hydrogen bond has a length of 2.85 Å and the hydrogen atom lies along the O–O internuclear axis. This variation in bond length makes it more difficult to compute the average number of hydrogen bonds formed round a molecule at any instant, thus a range of bond energies needs to be considered; a cut-off value in hydrogen-bond lengths of 3.5 Å is commonly used. Nevertheless, accepting those provisos, an average number of hydrogen bonds $\langle n_{HB} \rangle$, an average network size S_n and cluster size S_c, can be calculated for different temperatures with a constant bond length. For weaker hydrogen bonds (length ~ 3.5 Å) at 290 K, $\langle n_{HB} \rangle = 2.5$, $S_c = 5$, and $S_n = 6$ (Geiger et al., 1984). The average lifetime of a hydrogen bond for the whole block of water molecules is 0.02 ps; this short lifetime may be contrasted with the much longer life of the hydrogen bond in dynamic protein structures where the lifetime is 10–40 ps (chapter 4).

Molecular dynamics simulations can shed light on macromolecular properties as well as provide details of hydrogen bonding. For example, the change in density of water at different pressures can be modelled. Calculations of the total dipole moment show temperature changes in line with experimental data and the bulk dielectric constant is of the order of 100. Furthermore, the specific heat at constant volume is in agreement with the experimental values. Therefore, we may conclude that the molecular dynamics method provides a good simulation of bulk water properties and these results are strongly dependent on the hydrogen-bonding capacity of the molecules.

6.1.2 Water structure at a macromolecular surface

The behaviour of water molecules at a macromolecular surface has been studied extensively and the work has been summarized in reviews by Rupley, Gratton & Careri (1983) and Finney (1984). How is water at the macromolecular surface organized as the solute molecule is progressively hydrated, and how is the function of the macromolecule gradually altered? At what stage in incremental hydration does the macromolecule behave as though it were in dilute solution? What are the energetics of hydration?

These questions have been studied by biophysical measurements using IR spectroscopy, ESR relaxation, NMR spectroscopy, neutron diffraction, amide hydrogen exchange, thermal capacity, and dielectric relaxation on the protein lysozyme although the results are probably applicable to other globular proteins. We shall not consider here details of the measurements but take the results aggregated together. Incremental water hydration appears to occur in three well-defined stages.

(a) At the earliest stage, where the hydration is increased from 0 to 60 moles of water per mole of lysozyme, water binds to charged groups, one water molecule per ionized site. At about 60 moles of water the pKs of the ionizable groups are normalized. Water mobility is much reduced, about 1/100th that of bulk water. The enzyme is inactive. Water condenses only on the charged groups and not on the peptide backbone, thus the first event in hydration is ionization of the side-chains. Bound ligands have a mobility constant of about 4×10^{-9} s which is associated with the movement of domains in the protein. The structure of the protein molecule is not detectably different from that found in solution.

(b) Between 60 and 220 moles of water per mole of lysozyme the water molecules at low hydration are located principally at charged sites but spread to polar atoms on the macromolecular surface with increased hydration. At 220 moles water per mole protein the water forms clusters at ionic sites. The sizes of water clusters exhibit spontaneous fluctuations reminiscent of cluster size changes in solution. Local hydrogen-bonded networks in these charged regions are established and the physical properties of water are very similar to those found in solution. These effects are only detectable in regions associated with a full charge and are not observed along the peptide backbone. Hydrogen exchange increases a 1000-fold from 60 to 220 moles of water per mole of protein and at the greater hydration value the exchange is close to that found in bulk solution.

(c) The third stage of incremental hydration is from 220 to 300 moles of water per mole of lysozyme. Water condenses substantially onto polar regions of the enzyme surface. Enzyme activity is detectable and increases in parallel with mobility of the surface water. Surface water is still clustered, but the clusters are widespread to encompass all polar regions. Changes in backbone conformation are noticed as the hydration spreads. At a ratio of 300:1, hydration is nearly complete and a monolayer of water covers all polar regions. Water is ordered into hydration shells; clusters form instantaneously and are similar to those found in bulk solution. Water mobility of the surface is close to the bulk solution value and the mobility constant for a ligand is about 7×10^{-10} s. Enzymatic activity is 1/10th of that found in solution. Most internal motions of the protein are observed. Interdomain water bridges are present to allow thermal movements of the protein chains. Towards high hydration values in this third stage, the substrate cleft is fully activated. Each water molecule is allocated about 20 Å^2 of protein surface which is roughly twice the area capable of being covered by a single water molecule; hydrophobic regions still do not have water covering them. Monolayer patches are confined to the polar regions and water is efficiently arranged over these surface zones (Rupley *et al.*, 1983).

These studies of incremental hydration of a protein indicate that many properties of the protein become manifest when the level of hydration is surprisingly low. In many respects a monolayer of water is sufficient for group ionization, domain movements, side-chain conformational changes and the presence of substantial enzyme activity. Ligand behaviour in the monolayer shows two-dimensional mobility comparable with that found in solution. This layer of water covering polar regions of the enzyme surface is vital for functional structure as well as for functional activity of the enzyme. A hydration layer at the surface of a macromolecule is still only a partial picture of the macromolecule in solution; hydrogen-bonding networks on a planar surface would be trigonal whereas tetrahedral arrangements extending into the bulk solution would be expected with further hydration. Nevertheless, the monolayer of hydration can account for many of the properties found at the protein–solution interface. These experimental studies are pointers to the importance of water at macromolecular surfaces. We now have to elucidate how the water is ordered at the molecular surface. Two techniques will help us to understand this ordering: x-ray crystallography of water molecules in the crystal cell, and molecular dynamics or Monte Carlo simulations of water behaviour at the solute interface.

6.2   Organized water in crystal structures

The biophysical experiments outlined in the previous section suggest strongly that water molecules may, in certain regions, cover the molecular surface. Monolayers of water can be detected in crystals obtained from aqueous solvent. The precise positioning of these water molecules can be demonstrated directly by x-ray crystallography. However, this study of water in crystal structures is at a rudimentary stage and only a few examples will be given to illustrate the organization of water round a host molecule. At the present time, there are no rules for the arrangement of water molecules in the monolayer of the first hydration shell, although pentagonal arrays are a frequently observed feature. One of the difficulties of direct observation of hydrated structures by crystallography is that the packing forces, and arrangement of the host molecule in the crystal cell, may preclude water binding in regions of tight packing. Water may be squeezed out from between the host molecules and not appear in the crystal structure. Thus only water molecules in free space between host molecules can be observed.

6.2.1   Water surrounding small molecules

Multihydrated host molecules provide excellent material for the study of water arrangements. One of the earliest reports of medicinal interest was on the anthelmintic piperazine (Schwarzenbach, 1968). This is a secondary diamine compound and the molecular structure is shown in figure 6.3. The cyclic drug molecule is in the chair form. Water molecules in the first hydration layer are arranged to surround the ligand by forming a pentagonal array and each pentagon forms a partial dodecahedron structure in a three-dimensional arrangement of pentagons. The nitrogen atoms are directly hydrogen-bonded to the oxygen atoms of two water molecules. Each oxygen atom has four hydrogen bonds to neighbouring atoms. The hydrogen bond N—H . . . $O(H_2O)$ has an angle of 168° and so is slightly distorted; the next contact N . . . O . . . O also has a bond angle that is not tetrahedral. Proton donation from the piperazine ring is in the equatorial direction. The acceptor hydrogen bond onto the nitrogen atom, N . . . H—O, is axial to the piperazine ring and follows the direction of an sp^3 orbital lone-pair; it is nearly a linear hydrogen bond. The complete hydrogen-bonded layer of water molecules is shown in figure 6.4. Therefore

Figure 6.3. Molecular structure of the anthelmintic piperazine.

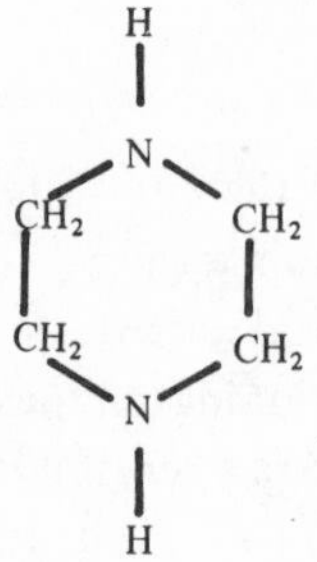

we can conclude that with the drug molecule piperazine, water molecules are hydrogen bonded to the host molecule and at the same time form hydrogen-bonded networks held in a regular arrangement of pentagons. Thus we appear to have a high degree of ordering of solvent from the hydrogen-bonding nitrogen atoms of the drug molecule.

Pinacol (2,3-dimethyl-2,3-butanediol) has the molecular structure shown in figure 6.5. In the anhydrous crystal, three conformational structures are found corresponding to a *trans* arrangement of the two oxygen atoms and two *gauche* conformers. This suggests that these three conformations are energetically almost equivalent. In the hexahydrate, the oxygen atoms of the two hydroxyl groups lie in the *trans* position (Jeffrey & Robinns, 1978). These two oxygen atoms form anchor points for hydrogen bonding to water molecules in a structural process that is remarkably similar to that found with piperazine. The water molecules form pentagons which are arranged in a partial dodecahedral array. The oxygen atoms of pinacol both donate and accept hydrogen atoms to form two hydrogen

Figure 6.4. Pentagonal water networks surrounding piperazine in the crystal structure. Shells of water lie above and below the plane of the drug molecule. (Re-drawn from Schwarzenbach, 1968.)

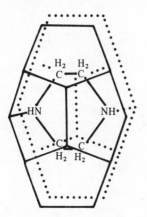

Figure 6.5. Molecular structure of pinacol.

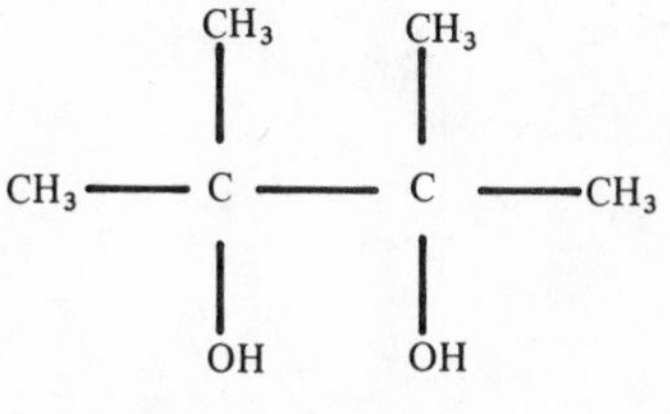

bonds to the structured water system. The pinacol molecule is located in a cage of pentagons (figure 6.6). There are eight pentagons and two hexagons that form the sides of a barrel-like enclosure.

n-Propylamine crystallizes from water to form a clathrate hydrate $4NH_2C_3H_7 \cdot 26H_2O$ (Brickenkamp & Panke, 1973); the structure of the amine is shown in figure 6.7. This structure is interesting because it contains a hydrophilic amine group capable of hydrogen bonding and a hydrophobic chain. Water molecules are largely linked together to form pentagons and the pentagonal faces form polyhedra. These polyhedra may be convex or concave and are all joined to each other in the crystal cell. The polyhedron cages form different structural types but it is sufficient to observe that the hydrophobic portion of the n-propylamine chain lies completely enclosed within a cage of water molecules. The pentagonally arranged water oxygen atoms lie on the accessible surface of the hydrophobic chain, a finding which is confirmed by interatomic distance measurements between $C \ldots O(H_2O)$. One portion of the polyhedral cage

Figure 6.6. Pinacol lying in a cage of water molecules formed from pentagonal and hexagonal arrays. (Re-drawn from Jeffrey & Robinns, 1978.)

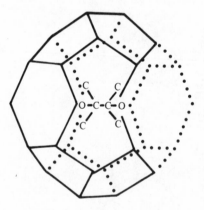

Figure 6.7. Molecular structure of n-propylamine.

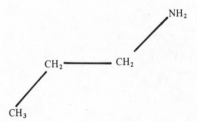

is shown in figure 6.8. The nitrogen atom forms three hydrogen bonds which stabilize the polyhedral clathrate frame round the host molecule.

Therefore, with the three examples for hydration, it can be seen that the hydrogen-bonding atom on the ligand acts as an anchor point for water molecules. Usually, these water molecules make the maximum possible number of hydrogen bonds to the ligand atom and, at the same time, generate a polygonal structure for ordering the water. Pentagonal arrays of water forming a partial dodecahedron are common, although hexagons are also observed. This polyhedral arrangement of water can spread round a small hydrophobic chain. Hydrophobic residues enclosed by water molecules on proteins, and at known drug receptors, are now being searched for. What is not clear so far, is how strong is this polyhedral hydrogen-bonded network? Studies of hydrated drug structures are not common and this research area could be a fruitful one to pursue.

6.2.2 Water at macromolecular surfaces

The arrangement of water molecules on protein macromolecular surfaces has not been studied extensively because of the tremendous complexity in analysing the structural data. However, recently the water organized round a very small polypeptide, crambin, has been studied by x-ray and neutron diffraction to atomic resolution. Crambin is a seed protein with 44 amino acid residues and is of no medicinal interest; nevertheless, we can use the results of Teeter (1984), for crambin, to illustrate how water can be arranged on the surface. Crambin is an approximately T-shaped molecule with only four charged side-chains. The hydrophobic residues lie on the opposite side to the charges and three of the four charges are found very close to each other. Where water molecules are bound to the protein backbone, twice as many hydrogen bonds form to the carbonyl oxygen atom compared with the amide nitrogen atom. Water molecules along the

Figure 6.8. A partial dodecahedron of water molecules surrounding n-propylamine. (Re-drawn from Brickenkamp & Panke, 1973.)

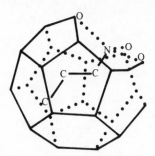

backbone regions show a pentagonal arrangement (figure 6.9). Five pentagonal rings (A–E) form at a hydrophobic surface and are held in place by hydrogen bonds to the backbone atoms as well as to charged groups. Three of these rings, A, C and E, form a cap round the hydrophobic end of the Leu 18 chain, thus in part enclosing the hydrophobic chain in a way that is reminiscent of the clathrate cage of n-propylamine.

With crambin, the positions of the bound water molecules on the macromolecular surface are determined by the intermolecular environment in the crystal cell; only water molecules in the free space are able to bind to the host surface. The effect of local environment on water binding has been well illustrated in the crystal structures of *L. casei* DHFR and *E. coli* DHFR (Bolin *et al.*, 1982). Firstly, these authors observed a tendency for fixed water molecules to be bound to the carbonyl oxygen atoms of an exposed α helix. This water-binding site has a hydrogen-bond length of 2.8 Å, a $C=O \ldots O(H_2O)$ angle of $120°$ and a $C\alpha—C=O \ldots O(H_2O)$ dihedral angle of $33°$. An example of this class of site is shown in figure 6.10 for a length of exposed α helix. Secondly, in the crystallographic cell two molecules of *E. coli* DHFR are arranged asymmetrically and therefore have different surface environments. Only about 50% of bound water molecules occupy corresponding sites on the two macromolecules, although solvent molecules bound to main-chain atoms occupy 60% of the sites. Thus the contact environment between molecules in the unit cell will affect water binding. Therefore, not all the potential water-binding sites will be revealed by crystallographic structure analysis. Water binding in the region of the active site of DHFR was discussed in the previous chapter

Figure 6.9. A pentagonal arrangement of water molecules over the hydrophobic region of leucine 18 in the protein crambin. (Re-drawn from Teeter, 1984.)

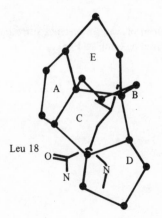

where it was shown that certain water molecules could form a network of hydrogen bonds to orientate the coenzyme and substrate correctly for the catalytic reaction. In DHFR, water is bonded only to polar atoms; no clathrate, or pentagonal, arrangements of water are observed round non-polar groups.

The conformations of nucleic acids in their crystal structures are strongly dependent on the ionic strength of the solution from which they were crystallized. So far, three major structural forms of DNA are known, A, B and Z. Both A and B form right-handed helices, where Z-DNA has a left-handed helix (section 4.5.3). It is not clear what factors force the conformations into these major structural forms during crystallization. Drew & Dickerson (1981) studied hydration of a B-DNA dodecamer where 72 ordered water molecules were observed. An unexpected finding is that the water molecules near the phosphate groups are not tightly bound and are thus not observed bound directly to phosphate oxygen atoms. Water molecules are bound predominantly in the two grooves of B-DNA. In these regions the hydration is ordered. In the major groove hydration sites are: adenine N7, N6; thymine O4; guanine N7, O6; and cytosine N4. Highly ordered and regular arrays of water molecules are found in the minor groove with hydrogen bonding to polar atoms of the DNA.

In the minor groove of B-DNA, water binding is sequence dependent and layers of water are tightly packed into the groove. For the tetranucleotide sequence AATT, the first hydration-shell water molecules

Figure 6.10. A small portion of an α helix from DHFR Pro 25 to Arg 33 showing hydrogen-bonded water molecules to main-chain carbonyl oxygen atoms; carbonyl oxygen atoms not bound to water are omitted for clarity.

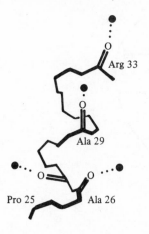

form hydrogen-bonded bridges from thymine O2 to adenine N3 atoms but these bridges are not formed within the base pair. Instead, the bridge links the base of one pair to the opposite complementary base of the next base pair (figure 6.11). This bridging water molecule is then linked to a second-shell water molecule to give a tetrahedral coordination. A zig-zag arrangement of water molecules is thus found in the minor groove. Third- and fourth-shell water molecules then pack the groove often bridging to the phosphate atoms at the edge of the groove. The bulky N2 amino group of guanine disrupts the hydrogen-bond network and fewer water molecules are bound in the CG CG sequence so that this part of the minor groove is drier.

The major groove of the dodecamer shows water binding which is usually monodentate; water in the second hydration shell is not regular. Polar atoms of the base pairs often show water molecules hydrogen-bonded to them. With adenine, water bridges N6 and N7 of single bases. The methyl group of thymine pushes water towards the phosphates by a simple steric effect and prevents a bridge forming between the phosphate bound water molecules and those in the major groove. GC-rich regions have little ordered water. Thus, in the major groove of B-DNA the water is generally monodentate and bound to the polar atoms; there is no cooperative water binding here, in marked contrast to that found in the minor groove.

Hydration of the A-DNA helix has been studied by Dickerson *et al.* (1982) and Dickerson *et al.* (1983), although the work is not as advanced as

Figure 6.11. Hydrogen-bonded solvent molecules in the minor groove of the minihelix A–A–T–T. The spine of water molecules are bonded to each other and to the base pairs.

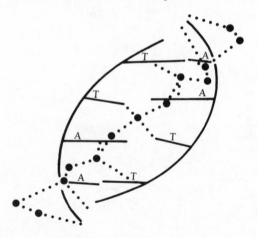

that for the B-DNA dodecamer. The free phosphate oxygen atoms are hydrated to the same extent as in B-DNA. Hydration in the two grooves of A-DNA follows the reverse pattern for B-DNA. Thus the major groove is packed with an ordered lattice of water; in part, this may be due to the major groove being very narrow and deep with many polar atoms lining the cavity. In contrast, the minor groove of A-DNA is almost dry, possibly because the groove is flat and shallow, and does not present the deep U-shaped structure that is encountered in the major groove. There is no regular ordered structure of water in either groove of A-DNA that is comparable with the zig-zag of water found in the minor groove of B-DNA.

Does hydration of DNA govern its overall conformation? Polymers lacking G are always in the B-conformation, whereas polymers containing A, T, C, G may exist in either the B or A forms (Leslie *et al.*, 1980). Polymers containing G may be driven into the A form by altering the ionic concentration. These observations suggest that regular hydration in the minor groove, to give an ordered spine of water molecules, with A–T sequences may provide helix stability and lock the conformation into the B form. G–C base pairs are disruptive for minor groove hydration and may allow B–A transitions. Further experimental work is needed to establish this hypothesis.

Water molecules are definitely hydrogen bonded to the external surface of DNA helices. So far, all helical fragments of DNA studied by x-ray crystallography have the base pairs stacked. If the base pairs were unstacked, could water be intercalated between them? Parthasarathy *et al.* (1982) have examined how water is intercalated between stacked purine and pyrimidine bases as well as between nucleotides. It is found that water molecules act as spacers between stacked pyrimidines to produce a sandwich structure with the pyrimidine base planes unstacked to a separation of 6.4 Å. Hydrogen bonds are not formed between the bases on either side of the sandwiched water molecule. Purine bases appear to intercalate water in an identical manner. Some hydrogen bonding to phosphate oxygen atoms in nucleotides is possible and usually, when this property is found, the hydrogen bonds are often planar with an apparent sp^2 hybridization of the oxygen atom. Therefore, we may conclude that water is capable of intercalation between the base pairs of a nucleic acid, but we do not know whether this is a frequent phenomenon, or how structurally significant it might be.

6.2.3 *Water surrounding drug–receptor complexes*

The study of water organized round a drug–receptor complex is not well advanced. In part this is due to the scarcity of examples of drug–

receptor complexes; also, where they exist together with a macromolecular receptor, there is the added problem of identifying water at atomic resolution. However, two examples will be considered here; the methotrexate and coenzyme binding sites on DHFR and a striking pentagonal arrangement of water at the proflavine intercalation site with nucleic acid.

The molecular structure of the methotrexate binding site has already been considered in detail in section 5.2. Two direct contacts between the pteridine ring and water 201 and 253 are observed (see figure 5.11) (Bolin *et al.*, 1982; Filman *et al.*, 1982). These contacts are further stabilized by another hydrogen-bonded water molecule, 201 to 252, and 253 to 217, in *L. casei* DHFR. There are minor differences between *L. casei* and *E. coli* enzyme at this point in that the water 253 is hydrogen bonded to a different residue which may reflect a slight conformational difference between the two binding sites.

Crystal structures for the binary complex of DHFR (*E. coli*) with methotrexate, and for the ternary complex of DHFR (*L. casei*) with NADPH and methotrexate, make it possible to study the displacement of water molecules from a binding site by cofactor and the re-involvement of water molecules in binding that cofactor. There are of course important provisos in the analysis of this cross correlation; not all amino acid residues are conserved between the two enzymes. Nevertheless, it is possible to build an approximate picture of water displacement by cofactor since the overall structure of the vacant cofactor site in *E. coli* DHFR closely resembles the site in the *L. casei* enzyme. In the binary complex, the vacant cofactor site is filled by bound water molecules; they are listed in table 6.1. Ten water molecules are located with their oxygen atoms at positions that would correspond to hydrogen-bonded atoms on the cofactor if cofactor were present. These solvent molecules provide a ghostly outline of NADPH. Unfortunately, other partial hydrogen-bonded networks at this site, linking the solvent molecules, have not yet been resolved. When cofactor binds, these water molecules are displaced. Cofactor binds to its site in DHFR (*L. casei*) by 17 direct hydrogen bonds; 14 water molecules are linked to the cofactor (table 6.2); seven of these water molecules directly bridge between the cofactor and residues on the enzyme. Tight binding between cofactor and enzyme can therefore be expected. The nicotinamide moiety of NADPH does not have a water bridge to the enzyme and could, therefore, exhibit some rotational freedom.

From these studies of water organized at the two binding sites for inhibitor and cofactor on DHFR, we can surmise that water plays a role in bonding the ligand to the macromolecule. Many of the water molecules

Table 6.1. *Solvent molecules located at vacant NADPH sites on* E. Coli *DHFR*

Corresponding NADPH region	Atom	Water	Residue
Nicotinamide	O7(n)	624	Ala 7
	N7(n)	625	Ala 7
			Ile 94
	C5(n)	614	Ile 94
NMN ribose	O3'(n)	710	Met 16
	O1'(n)	711	
Pyrophosphate	O1P(n)	611	Arg 98
	O1P(a)	731	Gly 97
	O2P(a)	605	Gly 96
			Thr 46
			Thr 46
AMN ribose	O1'(a)	616	Arg 44
			Ile 62
Adenine	N7(a)	687	Gln 102

(a) = adenine moiety, (n) = nicotinamide moiety.
Data taken from Filman *et al.* (1982).

Table 6.2. *Water bound to NADPH and DHFR* (L. Casei)

Molecular moiety	Atom	Water	Residue
NMN ribose	O2'(n)	439	Ser 48
	O3'(n)	208	Asp 125
NMN ribose-5'phosphate	O2P(n)	301	Thr 126
	O2P(n)	302	
	O1P(n)	276	
AMN ribose	O3'(a)	401	Arg 44
	O3'(a)	401	Gln 101
AMN ribose-2-phosphate	O2P	373, 326	
	O3P	326	
Adenine	N7	279	Gln 101
	N1	318	His 77
	N6	266, 579	

(a) = adenine moiety, (n) = nicotinamide moiety.
Data taken from Filman *et al.* (1982).

function as movable site-points and add to the fixed site-points already present on the enzyme in the binding region. What is not yet known is the coordination number for the bound water molecules; are they loosely bound with a low coordination number, or do they form stable tetrahedral hydrogen-bonded networks?

A highly structured network of water molecules in a drug–receptor complex has been observed in crystals of proflavine and d(CpG) (Neidle, Berman & Shieh, 1980; Neidle & Berman, 1982). The geometry of the intercalated complex was established by Shieh *et al.* (1980). Proflavine is found at two positions in the crystal, one molecule is intercalated between the base planes and the second proflavine molecule is stacked above a base pair to sandwich a base pair between the two drug molecules (figure 6.12). The two nucleoside strands form a self-complementary duplex with Watson–Crick pairing. An unusual feature of the binding site is that the nucleoside strands are not completely symmetric; the deoxyribose ring pucker is different. In the intercalation complex, the heterocyclic ring nitrogen atom of proflavine faces the major groove together with the exocyclic amino groups; 25 water molecules are bound to this small asymmetric drug–receptor complex and each water molecule is joined to another in a polygonal network; 15 water molecules are linked into an

Figure 6.12. Hydrogen-bonded water molecules in the d(CpG)–proflavine crystal complex. Five pentagons are arranged over the drug–nucleoside complex. (Re-drawn from Neidle, Berman & Shieh, 1980.)

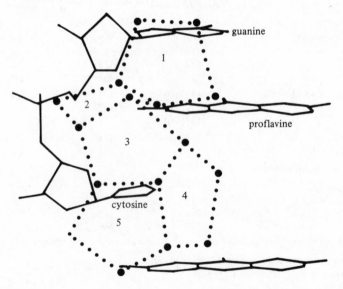

array of five pentagons that run down the edge of one strand in the major groove (figure 6.13). Where oxygen atoms of these pentagons link to the intercalating proflavine or the bases, they do so by a hydrogen bond which lies in the plane of the bases or proflavine. The pentagonal arrangements form a partial dodecahedral motif that is observed round piperazine, pinacol or n-propylamine. In the self-complementary duplex, water molecules fill the major groove by this pentagonal network.

Organized water surrounding drug molecules, their binding sites and drug–receptor complexes appears to be a definite possibility for many drug receptors. Since research on this topic is at such a preliminary stage, it is not possible to predict how common the phenomenon of organized water might be. Studies of water in the crystalline state may place too much emphasis on organization that may not be found in solution. Other methods, that incorporate statistical mechanics, are needed to determine whether the mobility of water molecules, immediately adjacent to the host surface, is sufficient to overcome the organizational tendency reflected in these highly hydrated crystal structures. Some of the methods for simulating mobile water will be examined in the next section.

6.3   Simulation of the solvent environment

One problem associated with transferring knowledge about the structure of molecules from the crystalline state to that in solution is the inclination to think in terms of rigid structures. It is well known from conformational studies that many single bonds are freely rotatable; similarly, the organized water structure found in the hydrated crystal may be disrupted by bond rotation as well as by the kinetic energy imparted to the 'bound' water of hydration by free solvent molecules. Hydrated structures in solution may only have a transient existence; a constant exchange of water molecules between solvent and waters of hydration is probable. Where are hydrated water molecules likely to be found on drug

Figure 6.13. A polygonal disk of water molecules attached to pentagon 2 of water molecules that flank the backbone of the d(CpG)–proflavine crystal complex. (Re-drawn from Neidle, Berman & Shieh, 1980.)

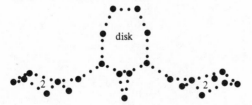

molecules and components of their binding sites? How tightly are water molecules held there? What is the probability of finding a water molecule at a particular position? Can we monitor water molecules skipping over a molecular surface? These questions will be examined systematically in this section.

6.3.1   The supermolecule approach to hydration

This approach is perhaps the most primitive way of assessing the hydration of a molecule. The accessible surface of the host molecule is obtained first. Usually this surface is limited to a planar section. A water molecule is then placed on the accessible surface and the interaction energy is computed between water and the host molecule at different positions (figure 6.14). If the interaction energy at a particular position is less than the sum of the interaction energies at infinite separation, then hydration at that position is favourable. Where possible, supermolecule calculations are performed within a quantum-mechanical framework. Initially, this type of study is carried out by a semi-empirical method, such as PCILO, to find regions of strong hydration and also to determine the optimum orientation

Figure 6.14. Hydration of ethanol by the supermolecule approach. Water molecules are placed on the accessible surface and allowed to relax their orientation to seek the minimum energy position. Two hydration sites are found close to the OH group.

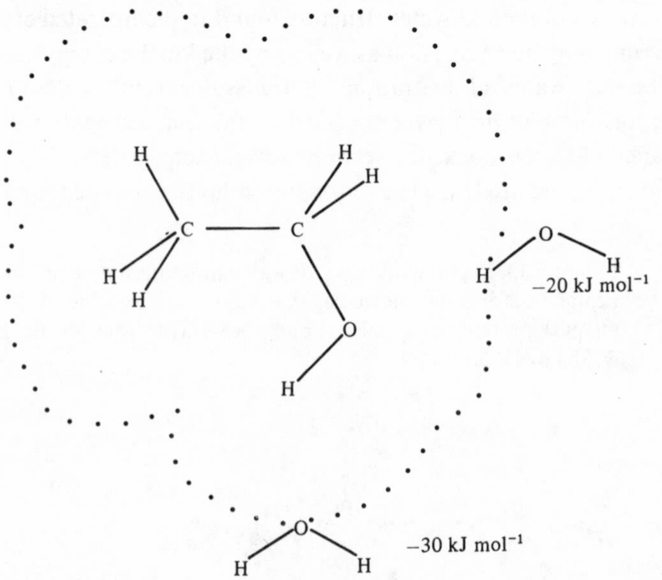

of the water molecule. At each position the water molecule is allowed to relax its orientation by incremental changes in the rigid-body rotation angles to minimize the energy. With a trial position, determined semi-empirically, more rigorous *ab initio* procedures are used to derive better values for the energy of hydration. An example of this type of calculation for ethanol is shown in figure 6.14 (Port & Pullman, 1974). Two hydration sites are observed close to the hydroxyl group; other regions are hydrophobic. Hydration appears to favour the creation of hydrogen bonds, thus the two positions correspond to acceptors or donors in the formation of a linear hydrogen bond with energies of -21 to $-29 \, \text{kJ} \, \text{mol}^{-1}$. Systematic studies with small model molecules have made it possible to predict positions of hydration quite accurately for larger molecules. For example, methanamide can model a peptide bond; water binds by being an acceptor for the NH_2 group in the plane, and a donor to the carbonyl-oxygen atom; methanoic acid is an acceptor in the region of the oxygen lone-pairs of the host and a donor from the hydroxyl hydrogen atom; the amine NH_2 group acts as a good hydrogen donor, with hydrogen-bond hydration energies of above $-25 \, \text{kJ} \, \text{mol}^{-1}$, and a good acceptor; quaternary ammonium groups bind water strongly along the N—H . . . O bond; aromatic hydroxyl groups are more effectively hydrated than aliphatic hydroxyls. These results have provided a useful picture for predicting hydration round many different types of molecule and tentative hydration maps can be built up from the interaction of water with model fragments.

The interaction of water with nucleic acid bases has been studied extensively and reviewed by Pullman & Pullman (1974). The first hydration shell for guanine is illustrated in figure 6.15. Water binds with significant energies, stronger than $-18 \, \text{kJ} \, \text{mol}^{-1}$, to hydrogen-bonding sites on the molecule; the orientation observed is along a linear hydrogen bond for a guanine acceptor atom, or in the direction of an oxygen lone-pair in the water molecule if water is the acceptor. Water bridging is feasible between the carbonyl oxygen atom and N7 of guanine; this water molecule is strongly bound. Similar overall results are observed for cytosine, adenine and thymine. Water binds in the first hydration layer within the plane of the molecular ring. Above and below the plane, water is only weakly bound. These studies thus make it possible to allocate putative hydration sites to the major and minor grooves of helical polymers of nucleic acids. Analogous investigations with amino acids have pointed to favourable hydration sites on proteins.

This early work on hydration by the supermolecule method can be extended to other hydration shells by building further out from the first hydration layer. However, this is a less satisfactory procedure since there is some ambiguity in the choice of initial water positions in the first hydration shell. The supermolecule method also suffers from the fact that it is costly to perform simultaneous relaxations in orientation of many water molecules as the shell is being built, therefore the possibility of hydrogen-bonded networks of water molecules is ignored.

6.3.2   *Water isoenergy contour maps*

The supermolecule approach, using *ab initio* methods to evaluate the interaction energy, is very costly. Attempts to obtain an approximate procedure, using paired potentials, have been investigated exhaustively by Clementi (1980). Effective two-body potentials can be constructed that are nearly as accurate as *ab initio* calculations and, at the same time, are about 1000-fold quicker to compute. This rapid method makes it possible to examine a large space surrounding a host molecule for hydration properties. The principles of the method have been discussed in section 3.1.7. It must be stressed that these contour maps are *not* molecular electrostatic potential maps even though they appear similar. In common with the supermolecule approach, a single water molecule is used to chart the interaction energies by placing its oxygen atom at regular grid points round the host molecule, allowing the water to relax to its preferred

Figure 6.15. Positions of water molecules in the first hydration shell of guanine computed by the supermolecule method; values are given in kJ mol^{-1}. (Re-drawn from Clementi & Corongiu, 1980.)

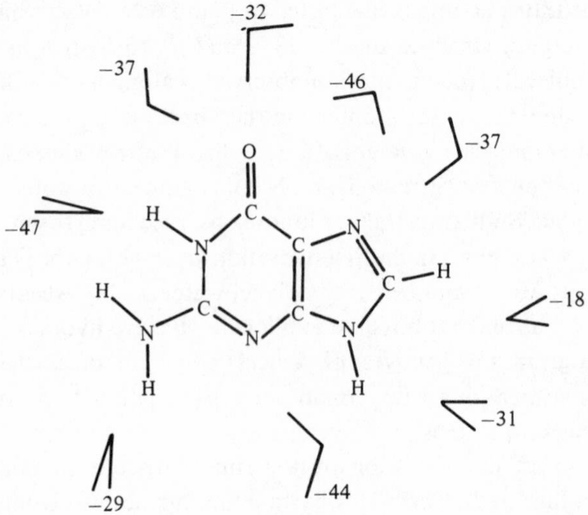

orientation, and then calculating the hydration energy. The grid planes are then contoured at isoenergy levels. Four regions in the contour maps can readily be distinguished. Firstly, there is a region within the accessible surface with a very large positive energy value where water cannot penetrate; this area maps out the accessible surface. Secondly, a region of positive hydration energy may extend out from the accessible surface into solution for 4–5 Å; this would be a hydrophobic zone where no hydration would be found and water molecules tend to be repelled from that region. Thirdly, regions containing an attractive negative energy value are hydrophilic; these have a zero-energy contour boundary separating them from the hydrophobic region and the accessible surface. Fourthly, the region of bulk water extends away from the hydrophilic and hydrophobic regions and is characterized by a small gradient in the interaction energy. An isoenergy map for the nucleic acid base, thymine, is shown in figure 6.16. The hydrophobic region is found close to the 5-methyl group and a very weakly hydrophilic region is located near C6. Strong hydration is expected at all other positions near to the accessible surface.

Figure 6.16. Isoenergy maps for the interaction of thymine with water. The outermost contour is $-5\,kJ\,mol^{-1}$ and the shaded region is $-20\,kJ\,mol^{-1}$. The map is computed in the plane of the base. (Redrawn from Clementi & Corongiu, 1980.)

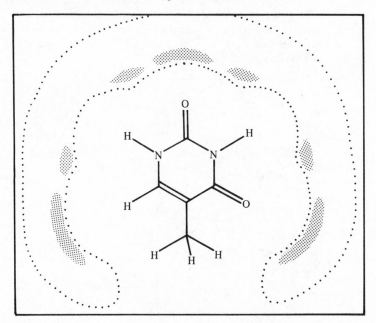

Hydration round a complete turn of B-DNA has been extensively studied by Clementi & Corongiu (1979) using the isoenergy method. Cylindrical isoenergy maps show regions of strong water binding surrounding the phosphate oxygen atoms; these hydrophilic regions follow the turn of the helix (figure 6.17). The hydration energy in the minor groove is about $-42\,\mathrm{kJ\,mol^{-1}}$ and $-25\,\mathrm{kJ\,mol^{-1}}$ in the major groove. This observation would predict that water of hydration would be found predominantly in the narrow groove of the B helix. Crystal structure studies substantiate this hypothesis and it was remarked on in the previous section.

Isoenergy contour maps for hydration clearly indicate the regions for hydrophobicity and this is a great advantage over the simpler supermolecule method. The technique is ideal for partitioning the space surrounding a molecule into hydrophilic and hydrophobic regions. This has made it possible to extend the calculations to polymers, such as polypeptides, to relate these regional hydration properties to molecular mechanisms such as protein folding or antigenicity (see section 6.4.2). The weakness of the method lies in the fact that only two-body interactions are considered; in actuality we have an n-body reaction problem, water

Figure 6.17. Isoenergy contours along a cylinder surrounding a B-DNA helix, radius 11 Å. The narrow groove in the centre of the frame is a region of high isopotential ($-40\,\mathrm{kJ\,mol^{-1}}$); interaction energies in the wide groove are of lower potential ($-25\,\mathrm{kJ\,mol^{-1}}$). The approximately circular regions are prohibited space where molecular collisions between the water and phosphate groups occur. (Re-drawn from Clementi & Corongiu, 1979.)

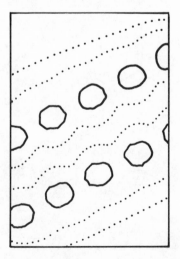

molecules interacting with the host, with the host–hydrate complex, and with themselves.

6.3.3 Monte Carlo simulations of hydration

Molecular interactions between n bodies can be simulated by the Monte Carlo method of Metropolis *et al.* (1953), provided that there exists an adequate pair-potential function for water–water and water–solute interactions. The method will be outlined briefly and is summarized schematically in figure 6.18. The solute molecule is usually placed centrally in a block of space and water molecules are added to the block to give the density of bulk water. Obviously the size of the block has to be sufficient to include the solute molecule, accommodate the waters of hydration and include some bulk water. However, if too many water molecules are present the calculation may be too costly. Some saving in computer time may be made by limiting the number of water molecules to those expected to fill the first and second hydration shell if the object of the problem is to investigate only water of hydration. Suppose that there are N particles in the block of space including the solute molecules, then the potential energy, E, of the system is

$$E = \frac{1}{2} \sum_{\substack{i=1}}^{N} \sum_{\substack{j=1 \\ i \neq j}}^{N} V(d_{ij}) \tag{6.1}$$

where V is the potential between the molecules i and j and d_{ij} is the distance between them. V would be a pair potential, water–water and water–solute, depending on the molecule pair. The water molecules are then moved by a stochastic process to generate a Boltzmann-weighted distribution from the initial distribution. A particle chosen randomly is moved from position

Figure 6.18. A decision-making scheme for Monte Carlo calculations incorporating the Metropolis condition.

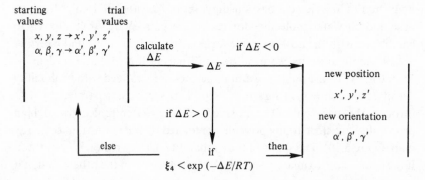

x, y, z to x', y', z', according to the equation

$$x' = x + a\zeta_1$$
$$y' = y + a\zeta_2 \qquad (6.2)$$
$$z' = z + a\zeta_3$$

where a is the maximum allowed displacement and $\zeta_1, \zeta_2, \zeta_3$ are random numbers between -1 and 1. A similar procedure is carried out to adjust the orientation of the molecule with old Eulerian angles α, β, γ to α', β' and γ'. If a particle moves through one face of the block it is deemed to enter the opposite face by the same amount of displacement; this keeps the number, N, of particles in the block constant.

For each move the energy change, ΔE in the system is calculated. If $\Delta E < 0$ the molecule is allowed to move to the new position x', y', z' with an orientation α', β', γ'. On the other hand, if $\Delta E > 0$ the move is allowed if, and only if, $\zeta_4 < \exp(-\Delta E/kT)$ where ζ_4 is a random number between 0 and 1; this second condition is the Metropolis modification of the Monte Carlo method and if that condition is not satisfied the molecule is not moved (Metropolis *et al.*, 1953). This has generated a new configuration. After M configurations, the equilibrium value $\bar{F}$, of any property F, is

$$\bar{F} = \left\{ \int F \exp(-\Delta E/kT) \, dV \right\} \bigg/ \int \exp(-\Delta E/kT) \, dV \qquad (6.3)$$

and

$$\bar{F} = (1/M) \sum_{j=1}^{M} F_j \qquad (6.4)$$

where dV is the volume element of the configuration space. The Monte Carlo method is ergodic; this means that the equilibrium value is not dependent on the starting position. Therefore, after a large number of moves, the molecules reach an equilibrium distribution and a probability density function can be calculated through the block of space.

Clementi & Corongiu (1980) applied the Monte Carlo method to the hydration of nucleic acid bases using a set of 40 water molecules for each base and 50 water molecules for each base pair. Water was allowed to equilibrate with the host molecule for 5×10^5 configurations and a further 5×10^5 configurations were needed to analyse the probability density. Round an A–T base pair the water molecules are arranged with probability density maps shown in figure 6.19 for two sections, separated by 1 Å, through the base pair. The positions of the water molecules of high probability are close to the positions predicted by the hydration isoenergy map (figure 6.20). The shape of the probability density region is important. If the field is approximately spherical it is isotropic; on the other hand, if it

Figure 6.19. A probability density map for water molecules surrounding an A–T base pair computed by a Monte Carlo simulation. The shaded regions are local areas of high probability for finding the oxygen atom of a water molecule; the map is drawn in the plane of the bases. (Re-drawn from Clementi & Corongiu, 1980.)

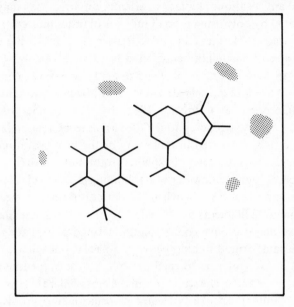

Figure 6.20. Isoenergy contours for water interacting with an A–T base pair drawn in the plane of the bases. (Re-drawn from Clementi & Corongiu, 1980.)

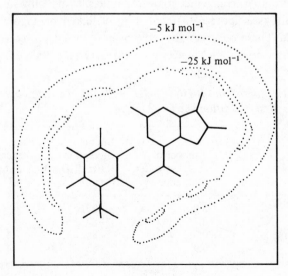

is elongated then there is a difference due to an additive effect from elsewhere in the solute or from the solvent molecules. The actual orientation of water molecules between successive configurations is not correlated with the best orientation since there are numerous equivalent orientations at any location. This lack of orientation correlation may be an artifact generated by only using a small number of water molecules. With the number of water molecules adjusted to the density of bulk water, many water–water interactions might be expected to stabilize the orientations.

Simulations for solvent hydration of nucleic acid have been extended to a complete turn of double helical B-DNA, with and without Na^+ counterions (Corongiu & Clementi, 1981a, b). The number of water molecules, their energy of binding to B-DNA and the interaction between water molecules is summarized in table 6.3, for the first hydration shell. Water is strongly bound to the phosphate group and relatively weakly bound to the bases; significant water–water binding is observed. This latter observation is important because when the pattern of water molecules is examined, structured filaments are revealed (figure 6.21). These filaments span across the major groove connecting interstrand phosphate groups, and along the same strand linking between adjacent phosphates. When Na^+ is added as a counterion to each phosphate group, the interstrand network is disturbed and cyclic structures in the cross-links are introduced. Counterions also alter the packing of water molecules in both grooves of B-DNA.

Solvation of a drug–receptor complex was discovered in the d(CpG)–proflavine asymmetric unit by Neidle, Berman & Shieh (1980). Their work revealed a remarkable arrangement of water molecules in highly structured polygonal forms (see figure 6.13) consisting of linked pentagons and a polygonal disk. These observations have been followed up by a Monte Carlo study of the crystal hydrate complex (Mezei et al., 1983) using three

Table 6.3. Hydration of the B-DNA helix averaged for each constituent moiety for the first hydration shell

Moiety	Number of solvated water molecules	Interaction energy for water–B-DNA $(kJ\,mol^{-1})$	Interaction energy for water–water $(kJ\,mol^{-1})$
PO_4^-	5.9	-102	-6.1
Deoxyribose	0.3	-87	-12.6
Deoxyribose base	0.5	-86	-12.6
Base	0.9	-63	-16.6

starting positions for the water molecules: an arrangement equal to the crystal hydrate positions with 100 water molecules (HC 1), a random arrangement of 100 water molecules (HC 2), and a third arrangement similar to HC 1 but with a further eight water molecules (HC 3). Equilibration was deemed to occur after 5×10^5 configurations and ensemble averages taken during $(2\text{--}4) \times 10^6$ further steps. The ergodic nature of the Monte Carlo procedure was substantiated by the convergence of HC 1 and HC 2. The fraction of calculated positions of water molecules within 0.6 Å of the observed crystal structure positions was 64 %, with good agreement for the polygonal distance. A poor correlation was found for the hydration of guanine. Use of a different potential function to describe the interaction, based on the work of Goodfellow, Finney & Barnes (1982), showed an improved correlation with the pentagon network being reproduced. Thus, from these Monte Carlo calculations, it appears that an ordered network of water molecules found in the crystal structure of a drug–receptor complex can be related to the dynamic situation where water molecules are free to move. Ordered networks of water molecules round drug receptors may well be expected, showing polygonal arrays of water molecules, in solution.

Figure 6.21. Hydrogen-bonded water networks in the wide groove of a B-DNA helix simulated by Monte Carlo calculations. Two types of water filaments appear: transgroove filaments link across the groove between strands and interphosphate filaments link adjacent phosphates on the same strand of the DNA backbone.

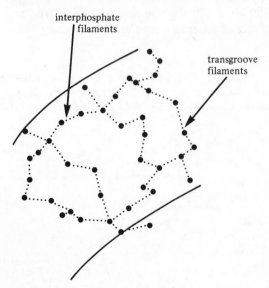

Thus, one may conclude that Monte Carlo methods provide great insight into the arrangement of water molecules round a fixed solute. What happens if the solute molecule also has conformational freedom? How can we incorporate this extra variable into model studies of molecular behaviour in liquid matter?

6.3.4 Simulation of solvation by molecular dynamics

Molecular dynamics calculations applied to a protein molecule have been described in section 4.7. Solvation of a dipeptide in aqueous solution has been extensively studied by Rossky, Karplus & Rahman (1979) and Rossky & Karplus (1979) using the molecular dynamics method. The model they used was an alanine dipeptide placed in a cubic block of water with 195 water molecules to give the required density. The time step for the simulation was 3.7×10^{-16} s for 400 steps giving a total time of 1.5 ps. The alanine dipeptide structure, together with initial torsion angles, is shown in figure 6.22. A full simulation of the dipeptide molecular dynamics included bond length variation, bond angle changes, torsion angle rotation, non-bonded interactions and electrostatic interactions. Water–water interactions were simulated by an ST2 potential and water–peptide interactions modelled by a Lennard–Jones potential incorporating electrostatic terms.

During the molecular dynamics simulation, the average dipeptide structure in water did not change significantly from the structure followed in a parallel *in vacuo* molecular dynamics calculation, although there was a considerable variation in ψ from 20° to 100° in solvent compared with 30° to 80° *in vacuo*. Therefore the analysis of this molecular dynamics study centred on the properties of water in the regions immediately surrounding

Figure 6.22. The alanine dipeptide showing the torsion angles; the preferred conformation occurs when the angles $\phi = -60°$ and $\psi = 60°$; in this C7 conformation an internal hydrogen bond forms to stabilize the structure.

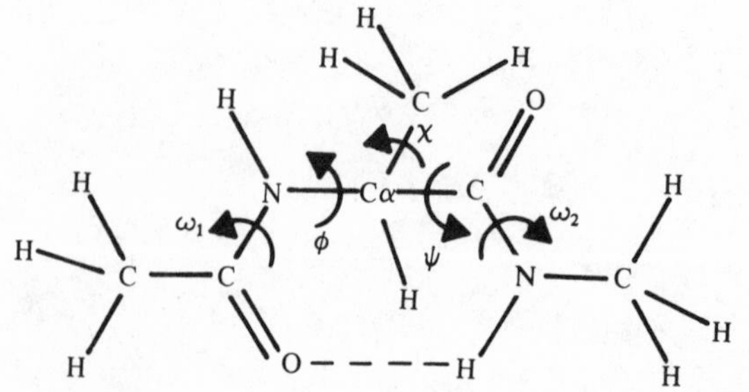

the dipeptide. Three solvation regions are related to the chemical properties of the solute molecule: polar regions near to the amide links, non-polar regions close to the three methyl moieties and bulk regions that extend out into solution (figure 6.23). The polar region lies out from the accessible surface and is contained within a sphere of radius 4 Å centred on the amide hydrogen and carbonyl oxygen atoms; it would contain hydrogen-bonded water molecules attached to the peptide links. Non-polar regions are between the accessible surface and within 5 Å of each methyl carbon atom; parts of the polar region are excluded from this portion. In polar regions 14 molecules are located, 20 in non-polar regions and 161 molecules are designated bulk. Qualitative differences in water molecule behaviour between each region were then searched for. The diffusion constant (D) of water shows a decrease by a factor of 5 for the non-polar region (table 6.4) suggesting that the water is held in this region, a hypothesis which is substantiated by the observation that the diffusion constant of the peptide is identical to this value for non-polar water. Similarly the rotational motion (τ_1) is slower in the non-polar region. What energetic or geometric factors determine these differences in dynamics?

Figure 6.23. Solvation of the alanine dipeptide. Close to the accessible surface two solvation regions can be distinguished; polar regions where the hydration energy is favourable due to hydrogen bonding and non-polar regions where water does not bind significantly by hydrogen bonding. (Re-drawn from Rossky & Karplus, 1979.)

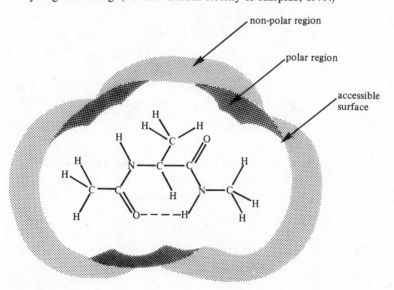

Computed hydrogen-bond energies for pairs of water molecules in the three regions show no significant differences in the mean energies which are in the range -21.76 to $-22.39\,\text{kJ mol}^{-1}$. The average number of hydrogen bonds to water molecules is approximately the same for the three environments and the distribution of the number of bonds with an associated bond energy is approximately equivalent. It appears, therefore, that the energetics of bonding do not explain the difference in water mobility between each region. However, there is a difference in the number of nearest neighbour water molecules between the regions 5.75 (bulk), 4.95 (polar) and 4.70 (non-polar). However, since the average number of water molecules having four hydrogen bonds is the same, the organization between water molecules in the polar environment and the bulk must be markedly different because there is one less immediate neighbour in the non-polar region. Therefore, the geometrical structure of the hydrogen-bonding networks in the three environments must be different. A geometrical limitation on possible networks can be understood with reference to figure 6.24. If we represent the methyl group by a sphere, and the water molecule by a tetrahedral arrangement of bonds, one bond to each hydrogen atom and one 'bond' to each lone-pair, then it can be readily seen that the orientation with $\theta = 0°$ would give the maximum number of unoccluded putative hydrogen bonds. This condition would be satisfied when any of the bonds has $\theta = 0°$. Moreover, this orientation is observed in crystalline structure I clathrates. In the molecular dynamics simulation, the distribution of the orientations in the non-polar region is shown in figure 6.25. A uniform distribution in orientations is found in the bulk solution, but in the non-polar region there are orientational preferences. The probability of finding a water molecule with $\theta = 0°$ is approximately three times that for $\theta = 180°$. Thus the direction of charge on the water molecule, tetrahedrally arranged, points away from the non-polar group. A second peak in the distribution is found at $120°$ which is a similar orientation to that found at $0°$. The orientation preference therefore follows that observed for the clathrate water structure found round polar groups and is

Table 6.4. *Calculated diffusion constants for water and rotational mobility in three regions surrounding a dipeptide*

	Bulk	Non-polar region	Polar region
$D\ (10^{-5}\,\text{cm}^2\,\text{s}^{-1})$	3.5	0.68	2.8
τ_1 (ps)	2.7	8.6	3.7

very similar to hydrate structures described in section 6.2. There is a difference, however, between clathrate networks found in crystals and those expected in solution: in solution the hydrogen-bond networks are transient and very disordered. Nevertheless, the important lesson to draw from the work of Rossky & Karplus (1979) is that the temporary arrangement of water networks in solution round a polar group shows some organization and will consequently affect the molecular properties of polar and non-polar regions. Detailed molecular dynamics calculations

Figure 6.24. Orientation of a water molecule close to the accessible surface of a methyl group. The orientation angle θ is measured with respect to the normal at the accessible surface.

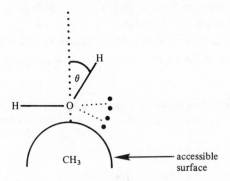

Figure 6.25. The probability distribution for the orientation of water molecules close to the accessible surface of a methyl group.
(*a*) Probability taken randomly from bulk solution. (*b*) Probability taken from the regions close to the methyl groups of the alanine dipeptide.

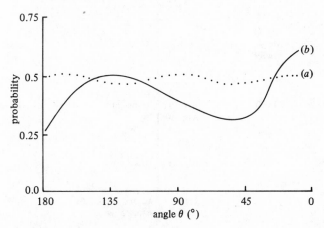

round molecules of biological interest are in their infancy and the work of Rossky & Karplus is a signpost for directing future research work.

The study by Levitt (1983b, c) on BPTI has been discussed in section 4.7, where we placed emphasis on conformational changes along the polypeptide backbone. A further important part of that study was the tracking of bound water molecules during a 62 ps simulation. Although the crystal structure of BPTI reveals 43 external water molecules (Deisenhofer & Steigmann, 1975), only four water molecules are strongly bound. Nevertheless, these water molecules shift their position during the simulation and one of the molecules, water 2, evaporates after 15 ps. Initially water 2 is tetrahedrally bonded but it then skips from an interior position to the macromolecular surface before moving out into solution. Figure 6.26 shows the track of this water molecule. At each step along this pathway, new hydrogen bonds are formed and old ones are broken as the molecule moves from one position to another.

Water potential, Monte Carlo and molecular dynamics calculations are very powerful tools in the study of solvation in biomolecular interactions.

Figure 6.26. The time track of water molecule 2 moving from the interior of BPTI to the outside. Hydrogen-bond connections are documented.

| hydrogen bond to water 2 | | | |
O	H1	H2	comment
Lys 41 Asn 44	water 1	Tyr 10	start
Lys 41 Lys 41		Arg 39	
	Arg 39 Ala 40 Arg 39		transient
	Ala 40	Arg 39 Arg 39 Arg 39	transient
	Ala 40		
Ala 40 Arg 39			
		water 2 evaporates after 15 ps	

Further application of these procedures to the formation of the drug–receptor complex is eagerly awaited. Furthermore, it is possible to link the computed dynamics to raster graphics display units to make cine films of molecular movement, thus giving an excellent insight into how molecular moieties move in a real-time simulation.

6.4 Hydrophilicity and hydrophobicity

Studies of the hydration of drug molecules, or macromolecular binding sites, point to the existence of regions on the accessible surface that show favourable or unfavourable hydration properties. What has now to be assessed is some scale for the hydrophobic or hydrophilic nature of a molecule; this property is not quantal but is a continual gradation between extremes. For example, some amino acids are very hydrophilic because they possess charged moieties, yet others with large non-polar groups are hydrophobic. Can we map the solvation properties of a chain of polypeptides in order to predict structural and functional behaviour of the protein?

6.4.1 Hydropathy

Kyte & Doolittle (1982) have developed a scale for each amino acid, called the hydropathic scale, which measured the relative hydrophilicity and hydrophobicity. The hydropathy index is a hybrid term taken from three measures: $\Delta G^0_{transfer}$, the fraction of side-chain 100% buried and the fraction of side-chain 95% buried. The free energy of transfer was calculated:

$$\Delta G^0_{transfer} = -RT \ln (18.07\gamma/\phi) \tag{6.5}$$

where γ is the partition coefficient and ϕ is the molal volume ($cm^3 \, mol^{-1}$). The fraction of the side-chain buried is given as the fraction of the total number of a given residue that is either 95% or 100% buried within 12 globular proteins. Values for the hydropathic index are given in table 6.5; it can be seen that hydrophobic residues have a positive index value and hydrophilic residues have negative indices. Other methods for measuring indices of hydrophilicity and hydrophobicity are reviewed by Chothia (1984).

Once a hydropathy value has been assigned to a residue, an average value for a length of sequence has to be calculated to give some idea of the overall local environment. This objective is frequently achieved by using a moving average encompassing a number of residues. Hopp & Woods (1981) found that a hexapeptide segment resulted in the best prediction for structural hydrophilic regions. However, Kyte & Doolittle (1982) found that averaging over 7–11 residues gave more consistent results for both hydrophilicity and hydrophobicity where the structural criterion used was whether the region was exposed or buried in the globular protein. For example, in figure 6.27 the hydropathic index for chymotrypsinogen is summated over nine residues and hydrophobic or hydrophilic peaks are correlated with surface or interior regions. Internal burying of residues with large positive hydropathic indices is well correlated in many proteins. Further correlations exist for external surfaces which are hydrophobic but at the same time are membrane bound. For example, glycophorin has a very high hydropathic index in the surface region which spans the membrane bilayer whilst the hydrophilic groups are located at the immediate edge of this region where the aqueous environment begins.

One functional correlation of hydrophilicity that has been pursued extensively is that strong hydrophilic regions form antigenic determinants on a protein surface (Hopp & Woods, 1981). Careful analysis of

Table 6.5. *Values for the hydropathic index of amino acid residues*

Amino acid	Hydropathy index	Amino acid	Hydropathy index
Alanine	1.8	Leucine	3.8
Arginine	−4.5	Lysine	−3.9
Asparagine	−3.5	Methionine	1.9
Aspartate	−3.5	Phenylalanine	2.8
Cysteine/cystine	2.5	Proline	−1.6
Glutamate	−3.5	Serine	−0.8
Glutamine	−3.5	Threonine	−0.7
Glycine	−0.4	Tryptophan	−0.9
Histidine	−3.2	Tyrosine	−1.3
Isoleucine	4.5	Valine	4.2

Data taken from Kyte & Doolittle (1982).

Figure 6.27. Profile for the hydropathic index of chymotrypsinogen. The index is summated over a nonapeptide region along the polypeptide chain. Hydrophilic regions have a negative hydropathic index, hydrophobic regions are positive. (Re-drawn from Kyte & Doolittle, 1982.)

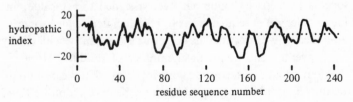

residue sequence number

hydrophilicity values indicates that only regions with the highest hydrophilicity are antigenic and these can be predicted accurately from sequence data. The second and third highest points are not good predictors of antigenicity. Another correlation that has recently come to light is that antigenic regions also have high atomic mobility on protein surfaces (Tainer *et al.*, 1984). This phenomenon was discussed briefly in section 4.7.1. Regions of high atomic mobility would also appear to have a high hydrophilicity as well. This apparent relationship needs to be examined further.

6.4.2 Hydropathy and tertiary structure of proteins

Amino acids can be related to gene sequence by a simple codon translation. Thus it is immediately possible to link the gene sequence to a hydropathic index for the protein amino residue if the codon is known. Furthermore, since there is a relationship between gene sequence and secondary structural features it follows that hydropathy can be related to secondary structure as well. Hydropathy is also related to gross tertiary structural features, such as whether a region is buried or exposed. Therefore, once the gene sequence is established, it should be possible to propose a tentative tertiary structure for any protein. There are of course many assumptions in this simple scheme of logic and each step is not completely clear. Nevertheless, a crude structural model can be formulated.

Noda *et al.* (1984) have determined the amino acid sequence for the Na$^+$ channel from its cDNA sequence and predicted its secondary structure by the method of Chou & Fasman (1978) (see section 4.6). They have also computed the hydropathic index from a nonadecapeptide moving average. The polypeptide chain shows four internal repeats of sequence homology. These repeats are highly significant and could only be attributed to chance with a probability in the range 10^{-9}–1.3×10^{-3}; the repeated homological units are labelled I–IV. Within each homological unit there are five hydrophobic segments (S1, S2, S3, S5, S6) at equivalent positions and one segment (S4) contains a positive charge and is located between S3 and S5. Segments S5 and S6 contain a stretch of 15 highly hydrophobic residues and are predicted to be α helices; these segments have properties consistent with transmembrane proteins. Segments S1, S2 and S3 are essentially hydrophobic but contain some charged residues. The segments appear to be α helices, and if this reasoning is correct then the charges all fall on one side of the helix with the opposite side having hydrophobic residues if the helical structure is amphiphilic. These segments appear to be capable of spanning a membrane. The structure of segment S4 cannot be adequately allocated but it is positively charged and contains a string of residues where every third one is either lysine or arginine. The tentative structure of a single homological unit is given in figure 6.28.

The assignment of secondary structural features is on firm ground. What is not understood, however, is the topology of folding to give the tertiary subunits for each homologous unit. Noda *et al.* (1984) propose a folding across the membrane shown in figure 6.29. Segments S1, S2, S5 and S6 span the membrane whilst S3 and S4 lie in an aqueous environment on the inside of the membrane. The link between segments S5 and S6 contains acidic residues for units I and IV only. In contrast, the link between II and III contains a cluster of acidic residues which are not found between units I and II, and units III and IV. The quaternary arrangement of the four units within the membrane is not understood.

Although the work of Noda and his colleagues outlined here is tentative, it represents a remarkable achievement in the application of hydropathy calculations to problems of protein structure that are of major pharmacological interest. Similar work with subunits of the nicotinic receptor was alluded to in section 5.1.

Figure 6.28. A tentative structure for the homological unit of the Na^+ channel showing the hydropathy of each segment. (Re-drawn from Noda *et al.*, 1984.)

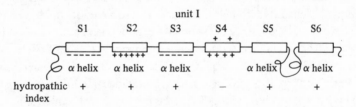

Figure 6.29. A possible topological arrangement for the segments across the cell membrane for a homologous unit of the Na^+ channel. (Re-drawn from Noda *et al.*, 1984.)

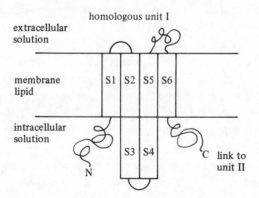

6.4.3 Hydrophobic moments

The Na$^+$ channel receptor protein contained, on each homologous unit, three segments S1, S2, S3 which were assigned to be α helices. The important characteristic of each segment was that the charges were located only on one side of the helix, the other side being hydrophobic. This property of amphiphilicity can be expressed quantitatively as a hydrophobic moment, μ_H, by the equation

$$\mu_H = \left(\left\{ \sum_{n=1}^{N} H_n \sin (\delta n) \right\}^2 + \left\{ \sum_{n=1}^{N} H_n \cos (\delta n) \right\}^2 \right)^{-1/2} \quad (6.6)$$

in a similar way to the familiar dipole moment (Eisenberg, 1984). Each α helix has N residues, and residue n has a hydrophobicity of H_n. The hydrophobicity can be represented by the value of the hydropathic index of n; δn is the angle in radians at which the side-chain, n, emerges. If the hydrophobic moment is small, ie $\langle \mu_H \rangle \sim 0.1$ per residue, there is no appreciable amphiphilicity, whereas if $\langle \mu_H \rangle \sim 0.4$ per residue there is high amphiphilicity. A hydrophobic moment plot for an α helix has a vectorial representation shown in figure 6.30. If H_n is negative, that is for a hydrophilic residue, the direction is effectively reversed.

If a plot is made of hydrophobic moment against hydrophobicity then the position of the helix on the graph indicates the likely environment that is found empirically from statistical studies of many proteins. The deduced environments are globular, surface and transmembrane (figure 6.31). This classification system makes it possible to allocate helices into their probable environments and therefore helps with the logic employed in the prediction of topological folding of polypeptide chains.

Further studies of hydrophobic moments of protein structures contained in the Brookhaven database reveal that β sheets and 3_{10} helices have characteristic periodicities in their hydrophobicities (Eisenberg, Weiss &

Figure 6.30. Hydrophobic moments plotted for residues distributed round a helix (full lines). Dotted lines show the directions of three charged residues; because the hydropathic index is negative the direction of the moment is reversed. The relative lengths of the lines indicate the magnitude of the hydrophobicity.

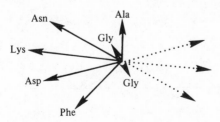

Terwilliger, 1984; Eisenberg *et al.*, 1984). Equation (6.6) can be transformed into the more general case

$$\mu(\delta) = \left| \sum_{n=1}^{N} H_n \exp{(i\delta n)} \right| \qquad (6.7)$$

where $\mu(\delta)$ is the strength of the component having the frequency δ; i is a complex number. In this expression the hydrophobic moment is the modulus of the Fourier transform of the one-dimensional hydrophobicity function. It is found that the characteristic periodicities, in the hydrophobicity, are: for α helices 3.6 residues, for β sheets 2.3 residues and for 3_{10} helices 2.5 residues. These observations indicate that many protein sequences form periodic structures that maximize their amphiphilicity. An application of this approach to the nicotinic acetylcholine receptor has led to the proposal of a model for protein folding, and subunit arrangement, with a central ion-channel spanning across the membrane (Finer-Moore & Stroud, 1984).

A characterization of the environment surrounding a particular region of a molecule as a hydrophobic field, analogous to the electric field surrounding a dipole, has been proposed by Eisenberg *et al.* (1982). This proposal needs further investigation before its possible use can be evaluated. The application of hydrophobic moments to more detailed studies of drug–receptor interaction has not been carried out. This area of research is ripe for exploration.

Figure 6.31. A plot of hydrophobic moment for α helices against hydrophobicity for different local environments. Values are taken from normalized scales. Regions are partitioned from empirical findings from a large number of proteins. (Re-drawn from Eisenberg, 1984.)

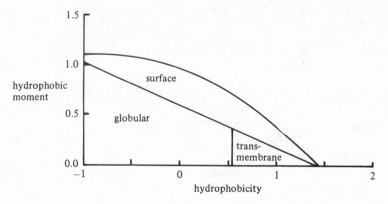

6.5   The effect of water on intermolecular interactions

The interaction between two molecules *in vacuo* is simple compared with their interaction in another molecular medium like a liquid. We know so little about the structure of liquids since the problem in solution is a many-body problem. Nevertheless, this complex problem can be broken down into conceptual parts to calculate the effective force between two interacting molecules. McLachlan (1965) has itemized five effects that are involved in molecular interactions in solution:

  (1) the force depends on the thermodynamic free energy $(U - TS)$, which has to be averaged over the configurations of the solvent;

  (2) local order in the liquid structure is disturbed by the solute molecules and a solvation force is generated between solute and solvent (figure 6.32(*a*));

  (3) the electromagnetic forces that would exist between the two interacting molecules are modified by the dielectric properties of the solvent (figure 6.32(*b*));

Figure 6.32. Factors that modify intermolecular interactions in solution. (*a*) Local solvent networks are disrupted by solute molecules approaching each other. (*b*) Solvent with a dielectric constant ε_s will modify the force between two molecules. (*c*) The water molecules are displaced by Archimedes principle as solute molecules come together. (*d*) Molecular orbitals of water molecules overlap an adjacent solute molecule.

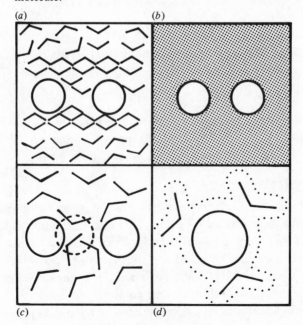

(*a*)   (*b*)

(*c*)   (*d*)

(4) as a dissolved molecule moves in a liquid it has to displace liquid from its path by Archimedes principle (figure 6.32(c));

(5) interaction of the electron orbitals with solvent atoms modifies the electronic properties of the solute molecules (figure 6.32(d)).

At the present time, these five effects cannot be assessed with complete rigour. However, various parts of this problem are being tackled with some success. The thermodynamic free-energy could be obtained from a Monte Carlo simulation of the interaction provided that we have an accurate potential function. At the moment the potential functions are inadequate because effects (2)–(5) are not precisely worked out. Effect (2), local order in the liquid structure, is currently under intense investigation and molecular dynamics studies related to this problem have been discussed earlier in this chapter. The network of hydrogen bonds in water is constantly varying and is modified by polar solutes sometimes to form hydrogen-bonded links to the solute molecule; or with a non-polar molecule, water molecules show a geometrical re-orientation to minimize putative hydrogen-bonding contact with the non-polar region. This results in the tendency of water molecules to form clathrate networks in the immediate vicinity of the solute. Effect (4) is the displacement of solvent by solute along the reaction pathway. Once again this could be resolved by Monte Carlo methods if we had an adequate potential function for the interaction. Effects (3) and (5), the role of dielectric effects and orbital interactions between solute and solvent molecules, will be discussed in the following subsections.

6.5.1   Dielectric theory

The force, F, in a vacuum between two charges, q_1 and q_2 separated by a distance r, is given by

$$F = q_1 q_2 / 4\pi\varepsilon_0 r^2 \tag{6.8}$$

where ε_0 is a fundamental constant, known as the permittivity, with a value of $8.854 \times 10^{-12}\,\mathrm{J^{-1}\,C^2\,m^{-1}}$ (Grant, Sheppard & South, 1978). The electric field, $\mathscr{E}_0$, at a position, $\mathbf{r}$, from a point charge, q_1, is

$$\mathscr{E}_0 = q_1 \mathbf{r} / 4\pi\varepsilon_0 r^3 \tag{6.9}$$

The field at $\mathbf{r}$ is a vector quantity and has direction as well as magnitude. We can imagine these vectors as field lines. By convention they have their origin at, and issue from, a positive charge which acts as a source for the field lines. Field lines terminate at a sink, which may be a point of negative charge. At any point the field line is perpendicular to the direction of the line of isopotential passing through that point.

If the vacuum is replaced by a homogeneous material then the force, F, on the two charges is modified

$$F = q_1 q_2 / 4\pi\varepsilon_0\varepsilon_s r^2 \tag{6.10}$$

and the electric field, $\mathscr{E}_m$ in the medium is

$$\mathscr{E}_m = q_1 \mathbf{r} / 4\pi\varepsilon_0\varepsilon_s r^3 \tag{6.11}$$

Thus both the force and the field strength are reduced by the dimensionless quantity, ε_s, which is the dielectric constant of the material or its relative permittivity. This reduction in the electric field is caused by electric charges and these must be present in the medium producing a field $\mathscr{E}_0(1 - 1/\varepsilon_s)$ in the opposite direction to $\mathscr{E}_0$. This opposing field is generated from molecules in the medium by polarization; that is, the field produced by q_1 tends to align the nuclei of the molecules of the medium away from the charge and at the same time aligns the electrons towards the direction of the field. This alignment in polarization is only partial since very strong forces hold the electrons on the nuclei. Thus the initial charge induces potential charges in the medium which introduce a field in the opposite direction. The polarization charge, q_p, is given by

$$q_p = q_1(1 - 1/\varepsilon_s) \tag{6.12}$$

If a charge is induced in the molecules of the medium by q_1, these molecules will possess an induced dipole moment and the dielectric polarization, $\mathbf{P}$, is then

$$\mathbf{P} = (\varepsilon_s - 1)\varepsilon_0\mathscr{E}_m \tag{6.13}$$

$\mathbf{P}$ is also a vector quantity and in a homogeneous dielectric the vectors $\mathbf{P}$ and $\mathscr{E}_m$ are parallel, but in a crystal ε_s is a tensor.

Where the molecules of the dielectric medium are non-polar, the field generated by q_1 induces a dipole in the molecule by perturbing the electron distribution; but if the molecules of the medium are polar they will have a molecular dipole and will tend to rotate in the external field. The rotation is such that the positive end of the dipole tends to be aligned in the same direction as the field and, although the actual angle of rotation is small, it

Figure 6.33. Factors affecting the orientation of molecular dipoles. (a) In the absence of solvent forces the molecular dipoles lie parallel to the field direction $\mathscr{E}$. (b) Brownian motion of the solvent disturbs the alignment of molecular dipoles.

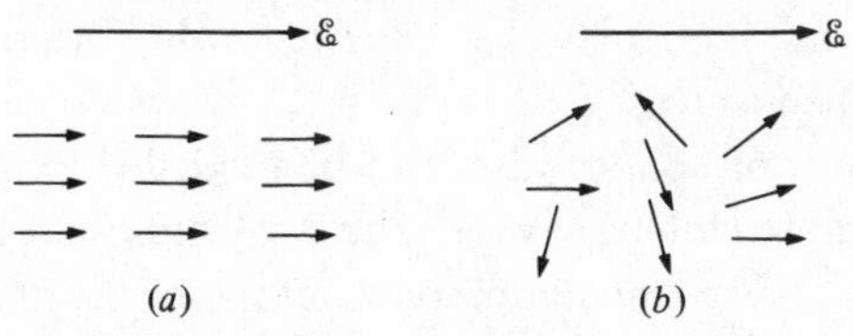

(a)   (b)

can collectively give rise to a large value for ε_s. If the molecules of the medium exhibit Brownian motion, as they would if the medium were a liquid, then ε_s is the contribution from molecules in motion and in different orientations. These two competing effects are illustrated in figure 6.33. Equation (6.13) describes an equilibrium process and **P** would show random fluctuations as the dipoles change; furthermore, the equation is only valid in the dielectric bulk and not in regions smaller than molecular dimensions.

6.5.2   Water as a dielectric medium

The structural properties of water affect its dielectric behaviour in biomolecular interactions. Water molecules situated more than a few molecules diameter away from the molecular surface may be regarded as 'free' molecules and behave in such a way as to give a bulk dielectric effect. However, water molecules hydrogen-bonded to a biomolecule, or temporarily immobilized in a clathrate pentagon network in a non-polar region, will have a different dielectric effect from bulk water since the relaxation time of bound water will be longer. For large proteins the layer of bound water is found experimentally to be 2–3 water molecules thick (Grant, Sheppard & South, 1978).

A dielectric model for liquid structures has been proposed by Frohlich (1958). Molecules with a permanent dipole moment, μ, are considered to be in a medium of permittivity, ε_∞, the equation for the permittivities is

$$(\varepsilon_s - \varepsilon_\infty)(2\varepsilon_s + \varepsilon_\infty) = N_d g \mu^2 / \varepsilon_0 kT \tag{6.14}$$

where N_d is the number density of molecules and g is the Kirkwood correlation factor. The parameter g attempts to handle the orientation of neighbouring dipoles compared with a central dipole

$$g = 1 + \sum_{i=1} z_i \langle \cos \gamma_i \rangle \tag{6.15}$$

where z_i is the number of water molecules in the ith shell, and γ is the angle between a pair of neighbouring molecular dipoles.

An accurate computation of the dielectric constant for bulk water is frustrated by a scarcity of knowledge about the hydrogen-bonding properties. Molecular dynamics calculations give a value of about 100 compared with an experimental value of 80 (Geiger et al., 1984). However, improvements in this method may be expected when the dipolar correlation factor is refined.

The dielectric behaviour of bound water can be modelled by studying water in clathrate hydrates (Johari, 1981). The host molecule lies in a dodecahedral-like cavity and re-orientation of water molecules at the host accessible surface is hindered by the surface network of water molecules. In these dielectric studies the interactions between host and water are ignored. For two clathrate structures the dielectric permittivity computed gave values of $\varepsilon_s = 82$ and 85 compared with measured values of 62 ± 2 and 65 ± 2 respectively. The computed values are about 30 % larger than the measured dielectric constant. We may thus conclude that the methods available for computing dielectric effects in water, although giving values of the right order, are still not sufficiently accurate to be used unambiguously in calculations of molecular interactions between polar molecules.

6.5.3 Ion interactions in water

The hydration of some simple ions of biological interest, Li^+, Na^+, K^+, F^- and Cl^- has been studied by Monte Carlo methods using water–water and water–ion potentials (Mezei & Beveridge, 1981). Coordination numbers for water molecules in the first hydration shell of each cation were found to be approximately 6; for the anions the coordination numbers were 4 for F^-, and 8 for Cl^-, values which are confirmed by x-ray crystallography. The cations interacted with water via ion–dipole interactions whereas the anions interacted mainly through hydrogen bonding. Radial distribution functions for the oxygen atoms of water surrounding the ions indicate a highly structured region of water immediately surrounding the ion. Adjacent to this structured region is a region which is partially destructured in comparison with bulk water lying beyond it. This destructured region of water has different hydrogen-bonding indices compared with the bulk water. The loose layer of water molecules acts as an interface between the network of water in bulk solution and the water of hydration surrounding the ion.

Water molecules appear to be ordered in their arrangement round a central ion. Figure 6.32(*d*) suggests that water molecules adjacent to a solute molecule may exhibit orbital overlap, that is, significant sharing of electrons. A quantum-mechanical investigation of this possibility has been carried out by Pullman, Pullman & Berthod (1978). In this study, water molecules were placed between the anion dihydrogen phosphate ($H_2PO_4^-$) and the cation Na^+. The arrangement of the waters of hydration was studied separately using the supermolecule approach for the subsystems $PO_4H_2^- \ldots (H_2O)_n$, $(H_2O)_n$, $(H_2O)_n \ldots Na^+$, so that the interaction energies and electron distributions can be evaluated for each unit.

An arrangement for the trihydrate geometry of $H_2PO_4^-$ is shown in figure 6.34 for further hydration. There is a negative interaction energy for water binding to $H_2PO_4^-$ up to the fourth hydration shell; the interaction energy is greater for each anion hydrate complex than for the corresponding $(H_2O)_n$ polymer. The interaction energy of the dimer $(H_2O)_2$ is $-21\,kJ\,mol^{-1}$ and the double dimer $(H_2O-H_2O)_2$ is $-36.4\,kJ\,mol^{-1}$. Hydrogen bonding was treated as a linear chain and not as a tetrahedral unit. Electronic charge transfer from $H_2PO_4^-$ to the water molecules is illustrated in table 6.6. A significant fraction, 0.124, of the anionic charge is transferred to the water molecules hydrating the phosphate group. Further hydration in the other shells does not alter appreciably the amount of charge dissipated from $H_2PO_4^-$ but the rearrangement of the charge is affected along the chain of water molecules. Where there is a hydrogen-bonded chain, the charge is shunted to the water molecule in the furthest hydration shell so that when the third and fourth shells are occupied, the intervening water molecules are nearly neutral.

Figure 6.34. A scheme for the hydration of $H_2PO_4^-$. The first hydration shell for the trihydrate unit is enclosed in the box; hydration proceeds to the fourth shell.

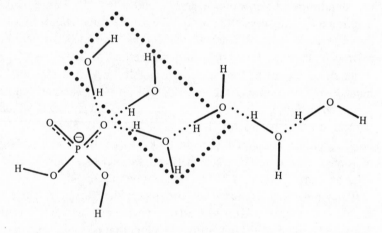

Table 6.6. *Net electronic charges in the hydration of $H_2PO_4^-$*

		Hydration shell			
$H_2PO_4^-$		1st (trihydrate)	2nd	3rd	4th
-0.876	-0.039	-0.040	-0.045		
-0.873	-0.039	-0.040	-0.024	-0.024	
-0.872	-0.039	-0.039	-0.024	-0.003	-0.024
-0.872	-0.039	-0.039	-0.023	-0.002	-0.002 -0.023

Data taken from Pullman, Pullman & Berthod (1978).

This charge shunting to the end of the chain is also mimicked by water-chain polymers (table 6.7). Although the intervening molecules are nearly neutral, in the hydrated anion and the polymer chain, they are strongly polarized. This observation suggests that the hydrogen bond, holding the adjacent water molecules together, is largely electrostatic.

A parallel study of water binding to Na^+ showed that water was strongly bound to the cation up to a chain of $(H_2O)_4$. Electronic charge was transferred away from the cation to the waters of hydration in a very similar manner to that found for $H_2PO_4^-$ ... $(H_2O)_n$ (see table 6.8). Again the intervening water molecules in a linear sequence were found to be nearly neutral.

What role does water play in the binding reaction between $H_2PO_4^-$ and Na^+? If we consider the arrangement of water molecules to be a trihydrate surrounding $H_2PO_4^-$, together with a chain representing the second, third and fourth hydration shells separating Na^+, then the O ... Na^+ distance is 9.78 Å. The interaction energy for this fully hydrated system is about $-197\,kJ\,mol^{-1}$ but if all the water molecules are omitted the interaction energy is $-133.5\,kJ\,mol^{-1}$. Clearly the intervening water molecules contribute significantly, and in a favourable way, to the interaction. This finding suggests that the water molecules play an important role in the

Table 6.7. *Net electronic charges for a linear chain of water molecules* $(H_2O)_n$

	Water molecule number			
n	1	2	3	4
2	+0.019	−0.019		
3	+0.021	−0.001	−0.020	
4	+0.022	−0.002	+0.002	−0.022

Table 6.8. *Net electronic charges in the hydration of* Na^+

	Hydration shell			
Na^+	1st	2nd	3rd	4th
+0.989	+0.011			
+0.988	−0.013	+0.025		
+0.987	−0.015	+0.005	+0.023	
+0.987	−0.015	+0.003	+0.004	+0.021

Table 6.9. *The changes in net electronic charges as Na^+ interacts through water with $H_2PO_4^-$; dehydration proceeds as the cation and anion come together*

		Hydration shell			
$H_2PO_4^-$	1st (trihydrate)	2nd	3rd	4th	Na^+
−0.874	−0.039	−0.022	0.000	+0.001	+0.987
−0.871	−0.039	−0.022	−0.017	−0.015	+0.987 (Na^+)
−0.869	−0.037	−0.043	+0.984 (Na^+)		
−0.974	+0.974 (Na^+)				

early stages of ion interactions in solution. However, what is still not known is the quantitative modification of the interaction energy by the intervening dielectric. There is no suitable way yet for determining dielectric effects through quantum-mechanical procedures.

Charge transfer in the progressive dehydration of the hydrated complex of through-water interactions can be followed (table 6.9). Where the fourth hydration shell is preserved, the phenomenon of charge shunting down the water polymer chain is maintained. Only when all the water from the first hydration shell surrounding the $H_2PO_4^-$ is removed is there a significant gain in ionic character for the phosphate. A small amount of charge transfer from $H_2PO_4^-$ to Na^+ is always present with the intervening water molecules. In contrast, when the O1 ... Na^+ distance is 9.78 Å and water molecules are omitted *no* charge transfer takes place. Therefore, we may conclude that charge transfer through water takes place by a shunting mechanism when there are intervening, and overlapping, molecular orbitals. Overlapping orbitals relay a significant portion of charge away from an anion and facilitate the interaction of a cation at a site which is distant, in this case by 9.78 Å, from the anion. The implications of this study by Pullman, Pullman & Berthod (1978) for drug–receptor interaction have yet to be investigated, but it is sufficient to mention that many pharmacological interactions are of the cation–anion type and water may be expected to play a similar role to that found in the $H_2PO_4^-$–Na^+ pilot study.

7

Ligand docking at a binding site

The previous chapters in this book have outlined the molecular foundations for the major steps in understanding drug–receptor interaction. We must now consider the application of these steps to one important example: intercalative drug binding to DNA. This drug–receptor system is the most widely studied model and provides the ground work for a large amount of basic science aimed at anticancer chemotherapy; it can also serve as a prototype for docking interactions in general.

If we consider the nucleic acid to be a long-chain polymer and the drug molecule to be very small by comparison, the polymer will have a large moment of inertia and will be relatively immobile compared with the free rotational and translational freedom of the small drug molecule. The geometry of the receptor can therefore be regarded as fixed within a coordinate reference frame and we need only consider the trajectories and rotations of the drug molecule moving within the molecular field of the receptor. Collision of the ligand with the binding site, in the correct relative orientation to form a complex, is termed the docking manoeuvre. Molecular recognition is the operation of intermolecular forces which govern a productive collision. A study of molecular recognition in the docking manoeuvre should provide a basis for drug design based on rational and sound molecular principles.

There are many advantages in studying drug binding to DNA. The nucleic acid may be considered as a chain polymer and linear in the region of drug binding. An intercalative site is formed between a dinucleotide of the DNA duplex and the site-points are largely restricted to this locality. Binding sites on proteins are much more complicated because of local conformations of secondary structure, such as α helices or β sheets with further folding into the tertiary structure. Ligand sites on proteins therefore have a more varied structural geometry compared with an intercalative site bounded only by purine and pyrimidine bases. Conformational changes in the dinucleotide are determined by seven torsion angles and puckering of the deoxyribose rings; hydrogen-bonding patterns in a base pair are limited. The molecular geometries of A-, B- and Z-DNA have been established from crystal structures. There are numerous examples of intercalative drugs cocrystallized with dinucleotides whose structures have been determined crystallographically. Therefore, the basic molecular geometries of the isolated receptor unit, drug molecule and drug–receptor complex are well established. Extensive quantum-chemical computations on the molecular moieties of nucleic acids performed over many years have provided a wealth of intramolecular information. Experimental studies of drug binding to DNA have gone hand in hand with computational work. These reasons make the intercalative binding site a judicious choice for modelling drug–receptor interaction.

A partial breakdown of the intercalative process into separate steps can help our understanding of the problem, but it must be borne in mind that such a partitioning is artificially contrived and that the steps really merge into each other along the reaction hypersurface. Initially, the participants are widely separated, so the first step in the interaction is the formation of an intercalation site by unwinding and extension of the helix. This step may not need to be complete; a partial separation of base pairs by the formation of a kinked helix may be sufficient to provide a putative partial binding site. The approach of the drug molecule into the receptor region, where long-range intermolecular forces are sufficient to modify the orientation and trajectory of the drug molecule, marks the second step and a form of molecular recognition becomes apparent. Short-range intermolecular forces operate on closer approach and help to control the docking manoeuvre at the third step. Conformational re-adjustment in the receptor allows the drug to be positioned to form a stable drug–receptor complex. The binding of intercalative drug molecules to a nucleic acid receptor then inhibits the function of the macromolecule, since transcription of the genetic code from DNA is prevented at an intercalated site.

7.1 Mechanical flexibility and the formation of intercalative sites

The docking of a drug molecule into the intercalation site can proceed, in theory, by two possible mechanisms. Firstly, the site may be closed and the drug molecule in some way is able to force an entry to perturb the conformation in the site region by actively altering the relevant torsion angles that govern the spatial characteristics of the binding site.

Evidence in favour of this proposal has not been easy to obtain. Secondly, the intercalation site may exist pre-formed, or partially pre-formed, as a kinked structure occurring naturally on the double helical strand of DNA. That is, the torsion angles are partially pre-set and require only a small further perturbation to complete the intercalative docking by the drug molecule. This second hypothesis has much theoretical and experimental evidence in its favour and has been extensively studied by Sobell *et al.* (1983). We will examine some of the details of this hypothesis.

The B-DNA double helix contains the sugar in the C2'-*endo* configuration; the A-DNA helix has C3'-*endo* sugar residues; mono-intercalated dinucleotides have sugar ring puckers of C3'-*endo*(3'-5')-C2'-*endo*. This change in sugar ring pucker at an intercalation site is as though there is a transition between B and A forms along the helix at the intercalation site. Energy calculations show that this postulated transition point is a low energy conformation and is termed the β-kink; moreover this junction can be moved, like a phase boundary, along the DNA to give the chain a large degree of conformational flexibility and not just a stiff bending mode to the helix.

β-Kinks in DNA can arise naturally through bombardment by solvent molecules with high kinetic energies. Many unharmonic motions are created which are rapidly dissipated but if the solvent molecule has the appropriate momentum and is directed along the dyad axis an acoustic phonon can be generated. These acoustic phonons have a velocity of about 1800 m s^{-1} and can travel about ten base pairs during a time of $\sim 2 \times 10^{-12}$ s. β-Kinks are formed in the region where two phonons meet when travelling towards each other; the lifetime of a β-kink is 10^{-3} s (figure 7.1).

Figure 7.1. The formation of kinks in DNA by bending: (*a*) normal DNA duplex, (*b*) phonons travelling down the duplex in opposite directions, (*c*) β-kinked region where two phonons meet.

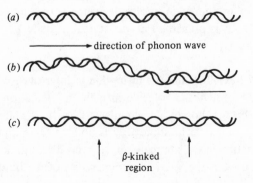

The propagation of a phonon wave in DNA is dependent on the elasticity in neighbouring regions, thus different energy fluctuations would be produced. The mean square energy fluctuation $\overline{\Delta\omega^2}$ can be calculated from the formula

$$\overline{\Delta\omega^2} = Na\pi(kT)^3/6h\mu \qquad (7.1)$$

where N is the number of base pairs, a is the lattice constant (3.4 Å for DNA), k is Boltzmann's constant, T is the absolute temperature, μ is the longitudinal velocity of the phonon and h is Planck's constant; μ can be obtained from $(E/\rho)^{1/2}$ where E is the elasticity and ρ is the linear density of DNA. In a β-kink region the elasticity is about one-tenth that of B-DNA, thus the magnitude of the energy fluctuations at a β-kink is much greater than at other places along the polymer. Therefore, drug intercalation should be enhanced at a β-kink. Furthermore, the energies which introduce stretching of the DNA in a β-kink region may be sufficient to disrupt Watson–Crick base pairing and allow intercalation of larger drug molecules such as porphyrins and triostin A.

An alternative method for studying molecular flexibility, and perhaps a more direct method, is by a molecular dynamics simulation. Levitt (1983a) studied the atomic motion of all atoms in two B-DNA double helices, one of 12 base pairs (CGCGAATTCGCG) and a 24 base pair homopolymer (A–T). The potential function used included bond stretching, bond angle bending, torsional angle changes, van der Waals interactions and hydrogen bonding; an environment of water molecules was excluded from the calculation. The most significant finding from this study was an overall bending and twisting motion for movement of the helix. The bend angle θ_B per Å, varied with time according to the relationship

$$\theta_B(t) = 1.75° - 1.18° \cos(2\pi t/26) \qquad (7.2)$$

so that the period found for the bending motion was 26 ps. Twisting, on the other hand, does not exhibit a simple frequency relationship. These bending and twisting motions are in part due to small variations in torsion angles along the backbone with rms fluctuations in the range 6–11°. Large-scale bending and twisting motions are due to concentrated actions along the backbone torsion angles.

The generalized motions of the helix just described are the predominant motions; however, it is also possible to observe the dynamic formation of kinks. The formation of a kink is rapid, taking only 6 ps, and the torsion angle changes are small, up to 30° (table 7.1) when compared with the x-ray geometry. At a kink the base pairs open up to about 90° which would allow a partial intercalative insertion (figure 7.2). In this particular kink, the helix is bent into the minor groove and the opening of the kink would allow intercalative attack from the major groove. The kink observed here occurs naturally by bending motions of the helix and is not dependent on a high-velocity water molecule to initiate it, since water molecules are excluded from the study. Therefore, kinking of a DNA helix can arise quite naturally from normal motions at an appropriate point in the trajectory. It is not certain how long a kink can last. In Levitt's study the kink persisted for 16 ps but the simulation study ended before the kink had time to disappear. Even so, the kink would appear to be long lived enough, as predicted by Sobell *et al.* (1983), to allow an intercalative attack.

Table 7.1. *Torsion angles at a kinked region of a DNA backbone compared with the angles found in the x-ray structure*

Structure	Torsion angle (°)						
	α	β	γ	δ	ε	ζ	χ
X-ray	313	171	36	156	155	265	142
Kink at 78 ps	298	171	59	94	171	281	93
Kink at 84 ps	303	176	63	106	173	277	94

The angles for the kink are averaged over four nucleotides (8–11) in a 24 base-pair homopolymer (A–T).
Data taken from Levitt (1983a).

Figure 7.2. A schematic illustration of a kink occurring in a molecular dynamics simulation of DNA bending. The local helix axis disrupted by θ_k ($\sim 120°$). As the helix is bent a kink opens in the wide groove and the base planes separate to form a possible intercalative gap.

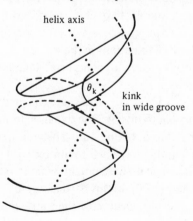

helix axis

θ_k

kink
in wide groove

7.2 Molecular energy changes in the unstacking of adjacent base pairs

The flexibility studies outlined in the previous section dealt with bending and twisting DNA to see whether putative intercalation sites could form spontaneously. Those studies were not designed to assess the specific changes in molecular energy for a dinucleotide unit, in the B-DNA shape, transformed by conformational shifts along the backbone into the open intercalative site configuration. The geometrical pathway for this transformation is not known. However, at the moment it is informative to examine the molecular energies of these two idealized states, namely the open and closed intercalative sites. Unfortunately, the dinucleotide fragment is too large to be treated as a single unit for quantum-mechanical computations so we have to resort to empirical energy calculations. Nevertheless, these less-refined procedures can supply important information about the energetics of particular processes at work in unwinding and unstacking the base pairs.

7.2.1 Interactions between the backbone and base pairs

The base-paired dinucleotide backbone has 14 torsion angles and 12 intermolecular degrees of freedom that determine the opening up of an intercalative cleft. A diagrammatic illustration of dC-dG and the torsion angles is given in figure 7.3. Following the interconversion pathway from the stacked to unstacked dinucleotide is a difficult problem. One way to avoid this difficulty is simply to ignore the conformational route taken and study only the energy of the initial and final states. This method has been pursued successfully by Nuss, Marsh & Kollman (1979) using an empirical energy calculation. The total energy was obtained from a pairwise interaction of all atom pairs using additive terms for electrostatic dispersion, exchange repulsion and the torsional energy. An allocation of torsion angles for the initial stacked dinucleotide was taken from the standard geometry of B-DNA and the geometry for the open dinucleotide, with altered ring pucker, taken from crystal complexes of drug dinucleotides for GpC and CpG. These two starting positions were used as input to an energy minimization routine which located the nearest energy minimum. The structures of the stacked and unstacked dinucleotides are shown in figure 7.4.

A change in the pucker of the sugar rings extends the backbone by 3.4 Å and only small changes are encountered in the torsion angles of the backbone; in fact only two torsion angles change by 40° (ω and ψ), the rest showing changes of less than 20°. Hydrogen-bond lengths between the base pairs are not affected significantly by unstacking the dinucleotides. Energy

Figure 7.3. Torsion angles along a dC-dG dinucleotide.

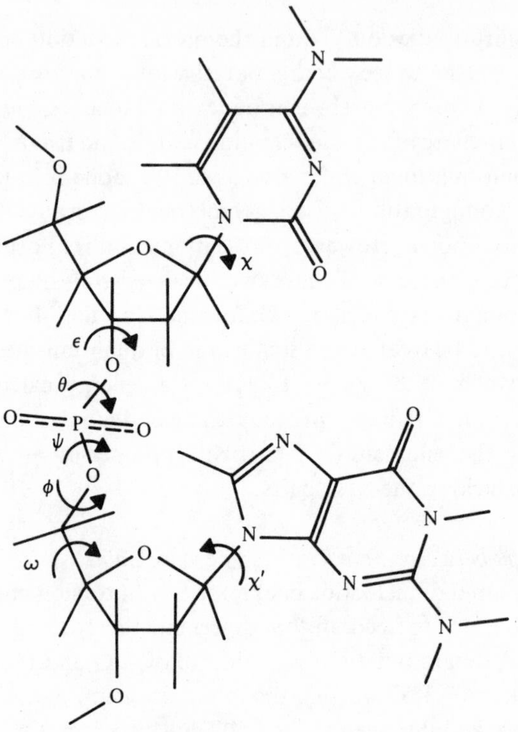

Figure 7.4. Skeletal drawing of a dC-dG dinucleotide duplex, (*a*) in the normal state with base pairs attached, (*b*) with the helix unwound and extended so that the base pairs are unstacked and ethidium is intercalated. (Taken from Nuss, Marsh & Kollman, 1979.)

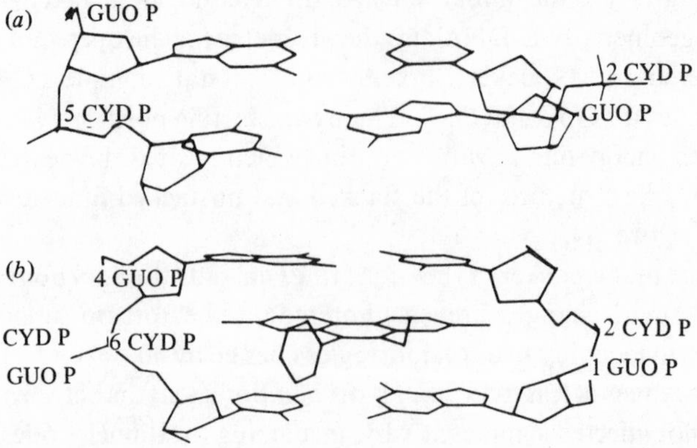

optimization produces a structure for an ethidium–CpG complex with torsion angles differing by an average of only 9° from the crystal structure. The intra- and intermolecular energy components for unstacking B-DNA dinucleotide sequences of GpC and CpG are shown in table 7.2. There are significant differences in energies between crystal structure conformations and the optimized conformations. Unstacking the dinucleotides of either sequence is energetically unfavourable, although the unstacking of CpG is energetically more favourable than GpC. Intramolecular interactions have a larger energy component than intermolecular interactions. Opening an intercalation site for CpG by unstacking requires about $92\,\text{kJ}\,\text{mol}^{-1}$ compared with $121.4\,\text{kJ}\,\text{mol}^{-1}$ for GpC. This ease of unstacking CpG compared with GpC is due entirely to more favourable electrostatic effects in the interaction and amounts to $29.3\,\text{kJ}\,\text{mol}^{-1}$. Changes in the dispersion energy are similar between the two sequences.

7.2.2   Base-pair interactions in an electric field

At physiological pH the nucleic acid phosphate groups are ionized and so produce an electric field radiating from them. This field may affect the stability of the base pair embedded in the field since the pair is only stabilized by hydrogen bonds between the bases. The energy $E_d^{\mathscr{E}}$ of a single dipole, μ, in a field $\mathscr{E}$ is

$$E_d^{\mathscr{E}} = -\mu \cdot \mathscr{E} \tag{7.3}$$

and the dipole will be orientated parallel to $\mathscr{E}$. Now if there are in the base pair two dipoles, μ_1 and μ_2, separated by a vector $\mathbf{R}$ with an angle θ between them, then the energy $E(\text{dd}, \mathscr{E})$ is

$$E(\text{dd}, \mathscr{E}) = -(\mu_1 + \mu_2)\cdot\mathscr{E} = -(2\mu\cos\tfrac{1}{2}\theta)\cdot\mathscr{E} \tag{7.4}$$

Furthermore, the interaction energy E_{dd} between the two dipoles alone, in the absence of a field is

$$E_{\text{dd}} = R^{-3}\left\{\mu_1\cdot\mu_2 - 3\left(\mu_1\cdot\frac{\mathbf{R}}{R}\right)\left(\mu_2\cdot\frac{\mathbf{R}}{R}\right)\right\} \tag{7.5}$$

Thus the total energy of the pair of molecules $E_{\text{dd}}^{\mathscr{E}}$ in the field $\mathscr{E}$ is

$$E_{\text{dd}}^{\mathscr{E}} = E_{\text{dd}} + E(\text{dd}, \mathscr{E}) \tag{7.6}$$

These terms are only electrostatic and neglect any stabilizing effect of hydrogen bonds; nevertheless, the stability of the base pair will be affected by the magnitude and direction of the external electric field. Moreover, if there is an alternative pattern of hydrogen bonding between the base pairs such as Watson–Crick or Hoogsteen pairing then the electric field may be expected from equation (7.4) to have a different, but predictable, effect on the stability of the base pair. This property has been studied extensively by

Table 7.2. *Components for the energy of unstacking the dinucleotides GpC and CpG. Crystal structures are compared with the optimized conformation, energies are kJ mol^{-1}*

	Energy component	B-DNA stacked crystal structure	B-DNA stacked optimized structure	B-DNA unstacked crystal structure	B-DNA unstacked optimized structure
GpC sequence	Intramolecular	348.9	278.5	450.9	375.5
	Intermolecular	−24.8	−33.7	−9.1	−9.0
	Total energy	324.1	244.8	439.4	366.4
CpG sequence	Intramolecular	470.6	364.8	444.1	401.9
	Intermolecular	−95.9	−84.9	−25.3	−30.4
	Total energy	374.7	279.9	418.8	371.5

Data taken from Nuss, Marsh & Kollman (1979).

Langlet, Claverie & Caron (1981). The direction of the individual molecular dipoles, and dimers, with G–C and A–U base pairs are shown in figure 7.5. The effect of a varying electric field on the stability of the complex

Figure 7.5. Monomer and dimer dipole moments for different G–C and A–U base pairs in Watson–Crick or Hoogsteen pairing. (a) G–C Watson–Crick, $\mu GC = 5.9$ D; (b) A–U Watson–Crick, $\mu AU = 1.5$ D, (c) A–U Hoogsteen, $\mu AU = 5.8$ D.

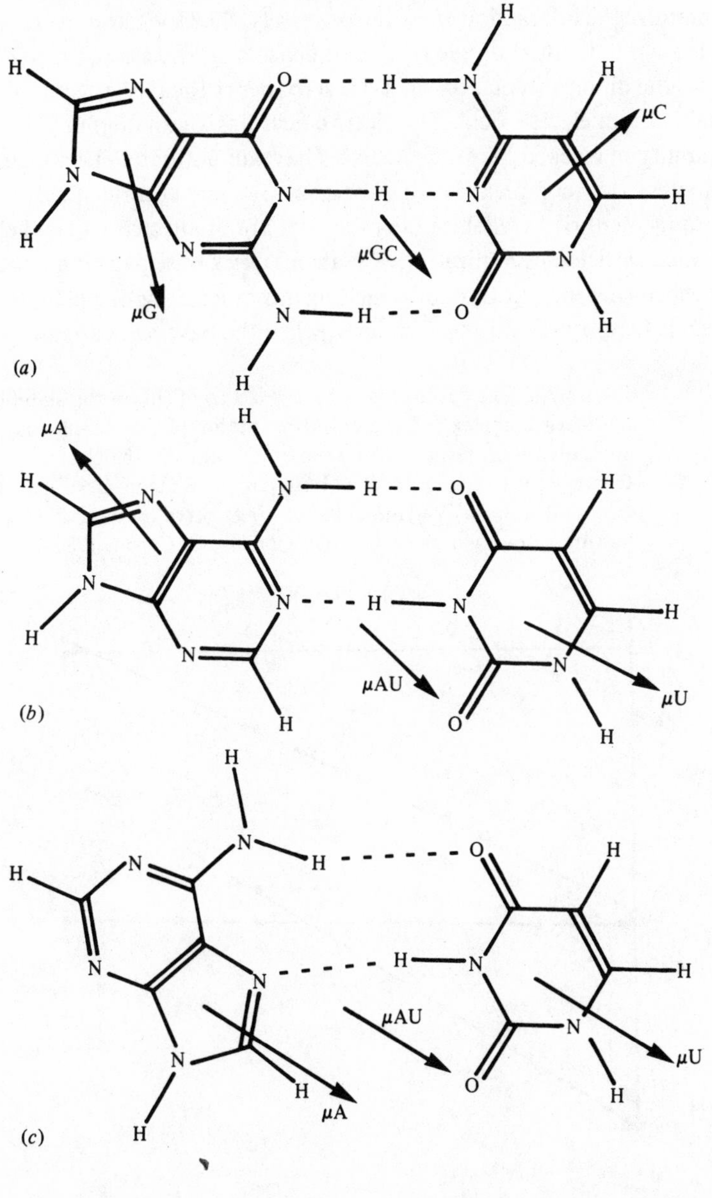

is illustrated in figure 7.6. Watson–Crick base pairs of G–C and A–U are destabilized by an electric field. The Hoogsteen base pair of A–U is relatively unaffected by the electric field. This differential stability with respect to pair pattern has its origin in the direction and magnitude of the resultant dipole of the dimer. For example, the two Watson–Crick structures have a resultant dipole moment which is considerably less than the addition of the separate dipoles due to a large antiparallel alignment. In both these cases an electric field would tend to re-orientate the two bases to minimize the interaction energy. Conversely, the Hoogsteen A–U base pair has each individual dipole lying parallel so that the electric field does not alter the dipole interaction since θ is zero; hence the Hoogsteen A–U pair is stable in an electric field. The electric field has an analogous effect on the stability of stacked bases. If the bases have an antiparallel arrangement of dipoles the complex is destabilized by an electric field, whereas arrangement of parallel dipole moments are unaffected by the field.

Electric field destabilization of Watson–Crick base pairs is an interesting effect since it may lead to local melting at points along the helix where the field is temporarily strong. Similarly, when the base pairs are unwound, as

Figure 7.6. The effect of varying the electric field on the stability of the base-pair complex. The stabilization energy of the complex is computed with respect to the separated bases in the electric field. Curve (*a*) is drawn for a G–C base pair in Watson–Crick pairing; curve (*b*) A–U in Watson–Crick pairing; curve (*c*) A–U in Hoogsteen pairing. (Re-drawn from Langlet, Claverie & Caron, 1981.)

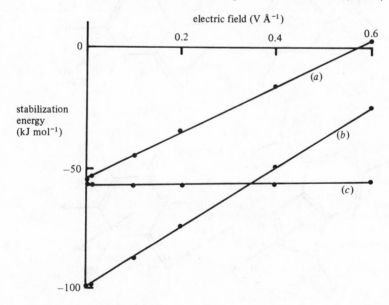

they are in the formation of an intercalation site, the changing direction of the dipole with respect to the field may also alter the stability of the base pair as well as facilitating unstacking. Electric field effects, briefly described here, need to be investigated more fully before they can be assigned a particular role in the flexibility of intercalative binding sites.

We may therefore conclude that, in the formation of an open intercalation site, electrostatic interactions are a very important component of the energy contributions that allow the site to form. It may well be worthwhile re-investigating the work of Nuss, Marsh & Kollman (1979) in the light of subsequent dipole–electric field effects to see whether the electrostatic energy differences between GpC and CpG unstacking can be interpreted further.

7.3   Molecular electrostatic potentials

The computational method for obtaining the molecular electrostatic potential was outlined in section 3.5.1. A pre-requisite for this calculation is the molecular charge distribution.

7.3.1   Calculation of electronic charge distributions

For a small molecule, the electronic charge can often be obtained by a single calculation using *ab initio* or semi-empirical molecular orbital programs. However, the residual charge distribution for a large molecule, like the dinucleotide (dC-dG) $\cdot$ (dC-dG), cannot be obtained by one calculation since the molecule is too large. This problem is commonly circumvented by using the principle of overlapping fragments to derive the electronic population in a piecemeal fashion. The (dC-dG) $\cdot$ (dC-dG) receptor is partitioned into the following moieties: cytosine, guanine, a cytosine–guanine base pair, cytosine deoxyribose, guanine deoxyribose and deoxyribose–phosphate–deoxyribose (Dean & Wakelin, 1979). Hydrogen atoms are added where bonds in the receptor structure are cleaved. A comparison can then be made between each segment and the larger fragment. As may be expected, maximum charge differences are detected where segments overlap. In the base-pair hydrogen-bond region the distribution differs by $0.04e$ compared with atomic charges on the isolated bases. This difference can be reduced ten-fold where atoms are two or three bonds away from the overlapping region. Thus, in the overlap regions, the charges are taken from the larger fragment. This procedure enables one to construct the net atomic charge distribution for the whole dinucleotide after appropriate small adjustments to maintain the total charge at its correct value. If it is desirable, a further partitioning of charge to σ- and π-components can be obtained.

The net atomic charge distribution for dC-dG is shown in figure 7.7 for the stacked and unstacked configurations. In the base pair the charge shows an alternating sign distribution on all atoms except N7 and C5 of guanine where the five-membered ring prevents the phenomenon. Hydrogen atoms attached to nitrogen are more polarized than those linked to carbon, which reflects the relative electronegativities of the heavy atoms.

Figure 7.7. The net atomic charge distribution for dC-dG; (a) with the base pairs stacked; (b) with the base pairs unwound and unstacked to form an intercalation site. (Taken from Dean & Wakelin, 1979.)

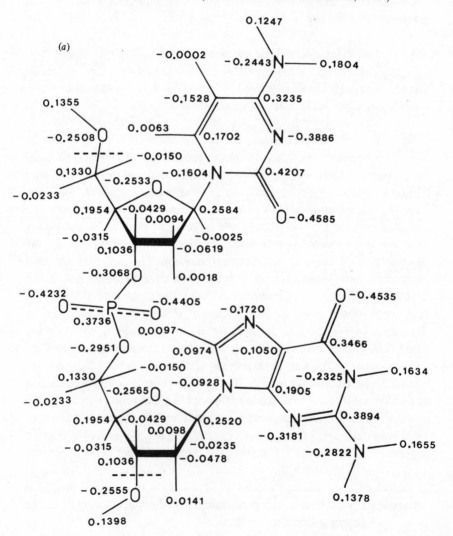

The carbonyl oxygen atoms of the base pairs are strongly electronegative. Both deoxyribose rings have a similar pattern of charges. The furanose oxygen atom has a net charge of $-0.25e$ and could possibly make hydrogen bonds to an intercalated group such as NH_2 of ethidium. In the phosphate group, the single anionic charge is strongly delocalized and the excess charge is held in the p orbitals of the four oxygen atoms. A positive charge on the phosphorous atom results from a loss of electrons from the p and d orbitals. Unstacking the base pairs and extending the helix to

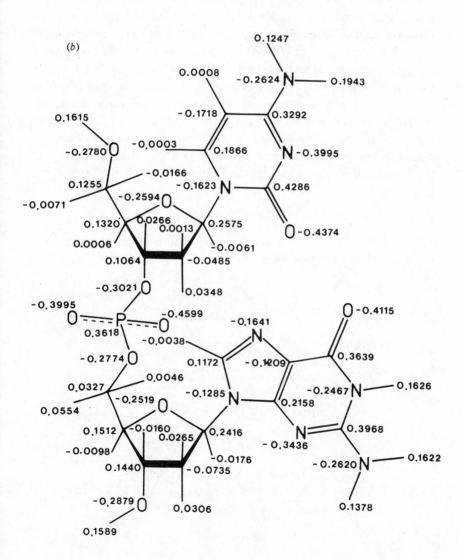

accommodate an intercalated ethidium molecule produces small changes in residual charge along the sugar phosphate backbone atoms (cf. figures 7.7(a) and 7.7(b)). The charge distribution is therefore sensitive to conformational differences in molecular structure.

The charge distribution for the intercalative drug ethidium is given in figure 7.8. The plane of the phenyl ring is at 97° to the phenanthridinium mean plane and the terminal carbon of the ethyl substituent is 84° out of the plane. At position 8, the amino nitrogen atom is sp^3 hybridized and the hydrogen atoms lie on the opposite side of the phenanthridinium plane to the ethyl group; the lone-pair of the nitrogen atom is orthogonal to the plane. The main feature in the charge distribution of ethidium is the extensive delocalization of charge from the quaternary nitrogen atom which has a charge of only $0.081e$. A proportioning of the net residual charge distribution between the functional moieties of the molecule reveals that the ethyl group carries a net charge of $+0.2236e$, the phenyl group $+0.1702e$, the amino groups $+0.1e$ and the phenanthridinium ring $+0.5067e$. A non-quaternized derivative of ethidium has no residual charge on the remaining groups, therefore the majority of the charge is delocalized onto the aminophenanthridinium ring. Furthermore, 26 % of the delocalized charge resides on the hydrogen atoms of the heterocyclic ring.

Figure 7.8. The net atomic charge distribution for ethidium. (Taken from Dean & Wakelin, 1979.)

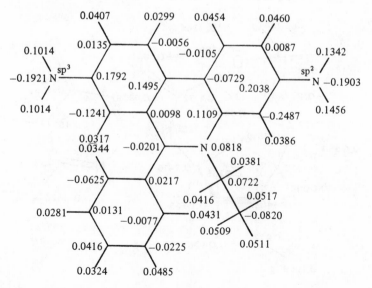

7.3.2   Molecular electrostatic potential surrounding (dC-dG)·(dC-dG)

The molecular electrostatic force is long range and is the first significant intermolecular force encountered when two polar molecules approach each other. The disposition of the potential is not easy to visualize since it has varying magnitude throughout the space surrounding the molecule. Four variables have to be assimilated and any portrayal of the potential surface will present difficulties in perception. The standard way of illustrating the potential is either to confine the examination to a single plane and map out the potential as a set of isoenergy contours, or contour a limited number of potential values in a three-dimensional representation.

In figure 7.9 the dinucleotide duplex is unwound and extended to form an intercalation site between the planes of the base pairs. Isoenergy contours of the molecular electrostatic potential are then drawn at $-50 \, \text{kJ mol}^{-1}$ intervals in a plane midway between adjacent base pairs. The electrostatic potential is dominated by wells close to the phosphate groups of the backbone but regions between the bases are attractive towards a cation. The partial symmetry in the distribution of electrostatic potential reflects the geometrical symmetry of the dinucleotide. However, one very important difference is found in the distribution of potential in the two grooves of the DNA duplex. In the minor groove there is a concentration of more negative potential. This fact could have a profound influence on the direction of drug–receptor interaction if there is a choice of intercalative attack from either groove. This hypothesis is strengthened by examining the Boltzmann probability distribution for a proton placed along the midline in each groove (figure 7.10). The probability becomes significant at about 8 Å from the helix axis in the minor groove, whereas in the major groove there is no parallel distribution. Intercalative attack is therefore more probable from the minor groove. This finding is confirmed by crystal structure studies of the complex between dinucleotides and ethidium. This deduction of the side of intercalative attack is only valid if there is no steric hindrance in the interaction. For a larger molecule which cannot intercalate from the minor groove there is still an attraction for a cation into the major groove side of the minihelix.

An idea of the complexity of the potential distribution surrounding the whole receptor site can be gained by inspecting plate 1 where a single potential of $-334 \, \text{kJ mol}^{-1}$ is drawn in three-dimensional space. A large bulbous field of potential is generated by each phosphate group and the potential then spreads in between the bases to the binding site of the intercalative cavity. Local perturbations round each plane of the bases can be observed.

7.3.3   *Molecular electrostatic potentials surrounding polynucleotides*

Whilst the dinucleotide forms a model for docking in drug–receptor interaction, a more complete understanding of the general problem of recognition can be gleaned from considering a larger polynucleotide. Two repeating sequences, the homopolymer poly(dG)· poly(dC) and the alternating polymer poly(dG-dC)·poly(dG-dC), 12 base pairs long and in the B-DNA helix conformation have been studied (Dean

Figure 7.9. The electrostatic potential map drawn in the plane midway between unstacked and unwound base pairs of (dC-dG)·(dC-dG); contour interval 50 kJ mol^{-1}. This graphic plane is the intercalation plane of the receptor and the minor groove occupies the lower portion of the figure. (Taken from Dean & Wakelin, 1979.)

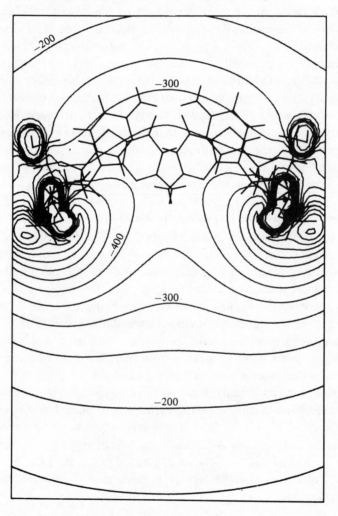

& Wakelin, 1980a). The electrostatic potentials for the fully charged molecules with a dielectric constant set to 1 are shown in plate 2.

Two contours illustrate the molecular electrostatic potential at -1254 and $-2570\,\mathrm{kJ\,mol^{-1}}$; these values are large and would be modified by a dielectric. If water were the solvent, the potential would be reduced by a factor of 80. The lower value of potential shows contour lines that just cut the edge of the block of space about 15 Å from the helix axis; a second contour region lies close to the molecular surface where the electrostatic field diminishes in magnitude until there is a sign change. In the homopolymer, regions of high electrostatic potential, $-2570\,\mathrm{kJ\,mol^{-1}}$, are located within the major and minor grooves and along the sugar phosphate backbone. This observation was later confirmed by Lavery, Pullman & Pullman (1982). The fields associated with the phosphate moiety point only into the wide groove and lie on the side to which the oxygen atoms predominantly point on each antiparallel strand. The region of high potential generated by the carbonyl oxygen atoms of the bases is as large as that derived from the ionized atoms of the phosphate group, an observation which correlates well with a similar net residual charge (figure 7.7). The annular regions of potential in both grooves follow the pitch of the helix and form a continuous band in the homopolymer. Larger values of

Figure 7.10. Boltzmann probability distribution for cation attack in the midplane between the unstacked base pairs of (dC-dG)·(dC-dG). (Taken from Dean & Wakelin, 1979.)

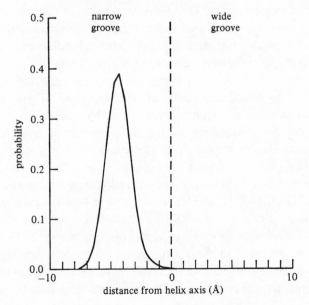

potential are found in the major groove. Electrostatic end effects confine the region of high potential to the central half turn of the helix. With the alternating sequence, the distribution of molecular electrostatic potential is broadly similar, the exception being the pattern of high potential within both grooves. The spiral nature of the fields generated by the homopolymer in the major groove is lost in the alternating polymer. Instead, the high potential follows the alternating geometry associated with the base carbonyl oxygen atoms. From these comparative studies it can be observed that the pattern of potential in both grooves of the B-DNA helix is strongly dependent on the sequence of bases, and that the electrostatic profile can extend a few angstroms away from the molecular surface. This pattern of potential may therefore be important in the recognition mechanism between a drug molecule and its receptor binding site.

7.3.4   *Electrostatic complementarity*

For a stable complex to form between a drug molecule and its receptor binding site, the disposition of intermolecular forces must be so arranged that the total interaction energy is less than the sum of the molecular energies of the isolated components. This principle has a geometrical correlation that has long been recognized in the Fischer lock-and-key hypothesis for the similar problem of enzyme substrate interaction. In other words, two molecules must present geometrically complementary surfaces to each other in the binding orientation and, if the energy of interaction is to be a minimum, local regional intermolecular forces must also be complementary. Thus for polar molecules, the pattern of electrostatic potential at the interfacial surface should be complementary with adjacent areas of positive and negative potential reflecting an attractive electrostatic force between moieties at any position.

A clear example of complementarity between a drug molecule and its receptor is provided by actinomycin intercalation into DNA (Dean & Wakelin, 1980b). The antibiotic inhibits transcription by obstructing the passage of RNA polymerase along the nucleic acid template. An absolute specificity for guanine is shown in natural DNA and there is a strong sequence specificity in its binding behaviour. The phenoxazone chromophore is intercalated between the bases of a self-complementary G-C sequence. Guanine specificity is attributed to hydrogen-bond formation involving the 2-amino group and the N3 atom of both guanines and the amide residues of L-threonine in each cyclic pentapeptide (figure 7.11).

The electrostatic potential in three orthogonal planes surrounding actinomycin is shown in figures 3.9 and 7.12. A stereoscopic illustration of the electrostatic potential contoured through a block of space is drawn in plate 3. Inspection of figures 3.9, 7.12(*a*) and (*b*) reveals that the potential falls into two structurally distinct halves. Most of the chromophore is surrounded by positive potential except for the 3-carbonyl group (figures 3.9 and 7.12(*a*)). In the intercalated state the molecular surface of the chromophore, with its positive potential, is sandwiched between the base pairs with their associated negative potential (figure 3.9). The positive potential extends away from the chromophore towards the two pentapeptide loops to the molecular surface which would be adjacent to the helix groove. Only a small section of negative potential is located on this side of the peptides near to the *L*-threonine oxygen atoms. These negative potential zones impart a twist to the shape of the positive fields seen in figure 7.12(*b*). The twist in the positive fields in this region of the drug molecule matches the shape of the negative potential down the helical grooves (plate 2). The potential generated from the surface on the peptide, at the opposite side to the chromophore, is negative. A structural reason for

Figure 7.11. The molecular structure of actinomycin.

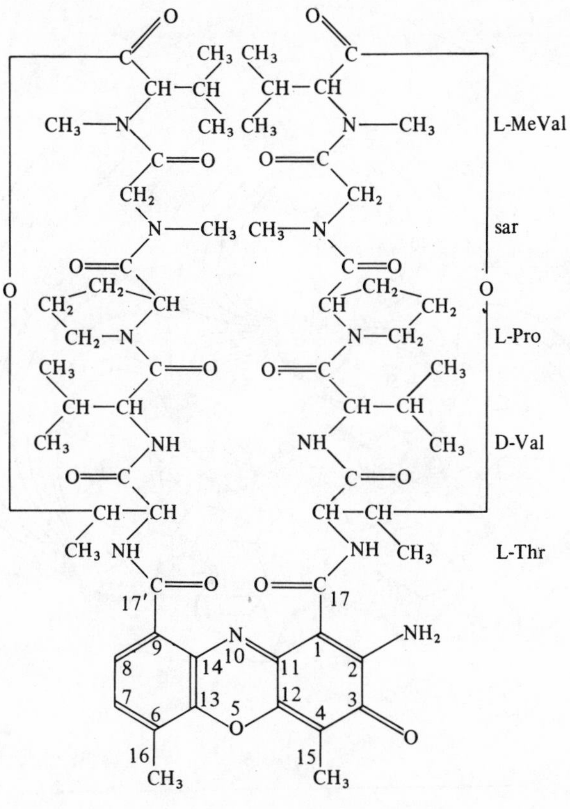

this gross partition of the molecule into positive and negative potential regions can be determined by observation of the pentapeptide geometry. The carbonyl oxygen atoms of L-proline, sarcosine and L-methylvaline point away from the chromophore; they are highly negatively charged and have strong associated fields on the back of the pentapeptides.

Thus the spatial arrangement of the positive field round the chromophore provides a plug for DNA intercalative binding because it is able to fit the negative fields of the binding site. Where the peptide surface lies close to the sugar phosphate backbone, the helically perturbed field matches the shape of the DNA field in the minor groove. The only negative

Figure 7.12. The molecular electrostatic potential for actinomycin drawn in orthogonal planes (*a*) and (*b*). Contour values in kJ mol^{-1}. (Taken from Dean & Wakelin, 1980b.)

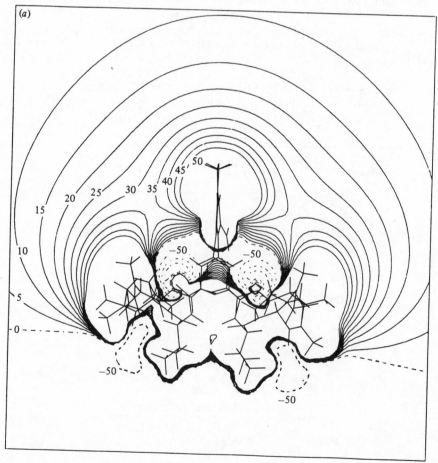

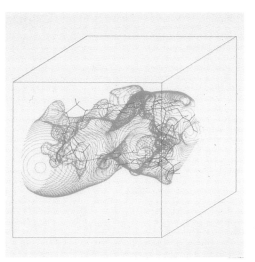

Plate 1 (*left*) Three-dimensional molecular electrostatic potential surface for the (dC-dG)·(dC-dG) receptor. The potential surface is drawn at −334 kJ mol⁻¹ (Taken from Dean & Wakelin, 1980a.)

Plate 2 (*below*) Stereo drawings of the molecular electrostatic potential round B-DNA; (*a*) homopolymer poly (dG)·poly (dC), (*b*) alternating polymer poly (dG-dC)·poly (dG-dC). Contour values −2570 kJ mol⁻¹ (red), −1254 kJ mol⁻¹ (green). (Taken from Dean & Wakelin, 1980a.)

(*a*)

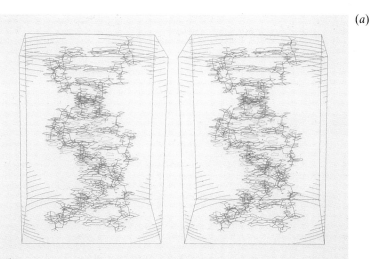

(*b*)

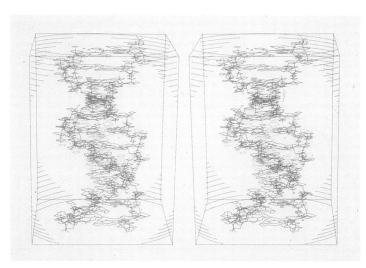

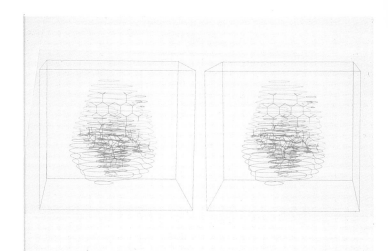

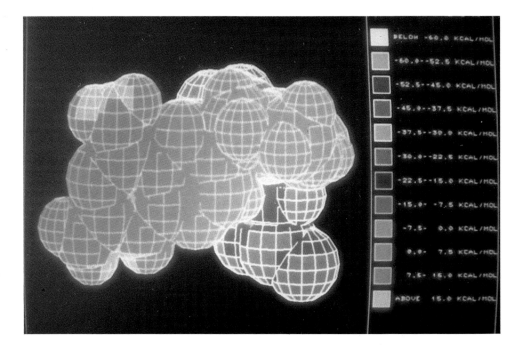

	BELOW -60.0 KCAL/MOL
	-60.0--52.5 KCAL/MOL
	-52.5--45.0 KCAL/MOL
	-45.0--37.5 KCAL/MOL
	-37.5--30.0 KCAL/MOL
	-30.0--22.5 KCAL/MOL
	-22.5--15.0 KCAL/MOL
	-15.0- -7.5 KCAL/MOL
	-7.5- 0.0 KCAL/MOL
	0.0- 7.5 KCAL/MOL
	7.5- 15.0 KCAL/MOL
	ABOVE 15.0 KCAL/MOL

Plate 3 (*opposite, above*) Stereoscopic drawing of the molecular electrostatic potential surrounding actinomycin. Contour values -33 kJ mol^{-1} (green), $+33$ kJ mol^{-1} (red). (Taken from Dean & Wakelin, 1980b.)

Plate 4 (*below and opposite*) The molecular electrostatic potential at the methotrexate van der Waals surface; (*opposite*) the potential generated by methotrexate at that surface (guest on guest), (*below*) the potential of DHFR computed on the van der Waals surface of methotrexate (host on guest). The patterns of potential are complementary. (Data from Nakamura *et al*, 1985.)

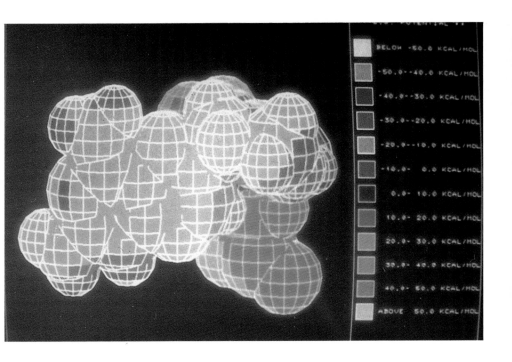

Plate 5: Computer generated hydrogen-bonding regions at a protein surface. The partial globes with longitude and latitude lines indicate the accessible surface. Atoms in the amino acid residues are coloured: red = oxygen, blue = nitrogen, white = carbon, a yellow disk marks out the region swept by the hydrogen atom of a hydroxyl group. Magenta dots signify the region where an approaching acceptor atom could be located; green dots indicate where a donor could be situated: (Taken from Danziger, 1985 b.)

field that points towards the nucleic acid is associated with the *L*-threonine carbonyl oxygen atom, which is available for hydrogen-bond formation with the 2-amino group of guanine in the nucleotide sequence GpC.

With actinomycin and DNA, the pattern match between the electrostatic fields of the binding surfaces of the two molecules can easily be followed by eye. However, with a structurally more complex binding site it is more difficult to discern a pattern match. A graphical technique developed by Komatsu *et al.* (1984) maps the electrostatic potential of the drug molecule onto its own molecular contact surface (guest on guest) and maps the potential generated by the binding site onto the drug molecule's surface (host on guest). The two surfaces to be compared are then structurally identical and relationships between the magnitudes of the

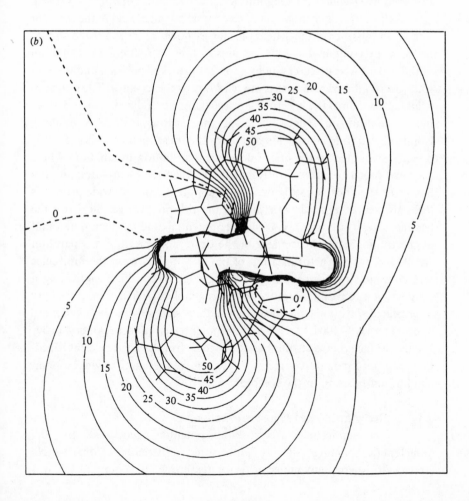

electrostatic potentials can be readily determined. An example of the electrostatic potential of methotrexate (guest on guest) and that generated by the binding site from DHFR (host on guest) is shown in plate 4. A complementary pattern in the electrostatic potentials can easily be observed. Geometrical patterns of positive potential produced by the guest are matched by negative potential mapped onto the guest by the host and vice versa.

7.3.5   Perturbation of the receptor electrostatic potential by a drug molecule

As a drug molecule reacts with its receptor binding site, the molecular electrostatic potentials are modified by the changing geometrical relationships between the two molecules. An illustration of this effect is provided by ethidium docking into (dC-dG) · (dC-dG) (figure 7.13) (Dean & Wakelin, 1979). The molecular electrostatic potential field of the receptor unit, in the absence of the drug molecule, is shown in figure 7.9 and should be used for comparison with the illustrations in figure 7.13. Ethidium initially at 12 Å from the crystallographically docked position and without roll, yaw or pitched rotations is moved towards the receptor. The potential fields surrounding the receptor are partially neutralized, the -350 kJ mol^{-1} contour no longer spanning the narrow groove. Asymmetry is introduced into the fields between the base planes. This can be directly attributed to the unequal charge distribution in ethidium where the ethyl moiety is more positively charged than the phenyl ring and lies nearer to the phosphate (figure 7.8). The potential in the wide groove is little affected compared with that in the narrow groove. Similarly the leading edge of ethidium shows a larger potential gradient than that found at the trailing edge. Movement of the drug molecule to 6 Å separation results in a further neutralization of the potential between the molecules and contours tighten round the trailing edge of ethidium, indicating a progressive electrostatic capture of the drug molecule. At 1 Å from the fully intercalated position, the fields on either side of the minihelix are approximately -100 kJ mol^{-1} and the potential is balanced. Once van der Waals contacts between the bases and the drug have been established, other intermolecular forces may operate to force the ligand to jockey for an equilibrium position in the receptor.

7.4   Receptor-induced ligand orientation

The molecular electrostatic potential generated by the dinucleotide binding site is anisotropic. Therefore, from simple electrostatic theory, any molecule travelling through the potential field will experience a variable interaction energy which would tend to influence the orientation of that molecule with respect to the receptor. Molecular reaction dynamics for collisions between diatomic molecules are quite well understood (Levine & Bernstein, 1974) but the interactions between polyatomic molecules cannot yet be studied adequately by experimental methods. Theoretical procedures are still too rudimentary to provide

Figure 7.13. Perturbation of the electrostatic fields of a receptor by an incoming drug molecule using (dC-dG)·(dC-dG) and ethidium as a model. Contour interval −50 kJ mol^{-1}. (Taken from Dean & Wakelin, 1979.)

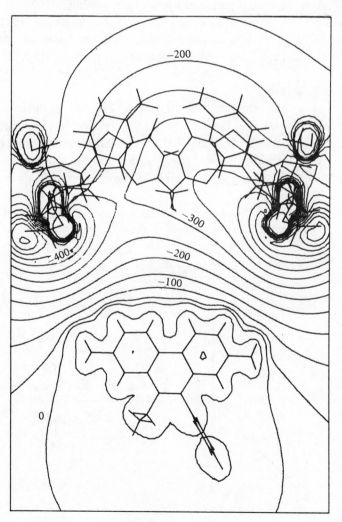

accurate enough interaction potentials for a molecular dynamics, or Monte Carlo, simulation. Thus one is limited to studying the separate components of rotation singly without being able to follow the trajectory clearly. Nevertheless, it is possible to gain valuable insights into orientational effects on a drug molecule placed in an anisotropic potential field. A greatly simplified scheme for the rotation of ethidium near to a restricted model of the binding site has been proposed (Dean & Wakelin, 1979).

A model for the dinucleotide receptor was drastically simplified to include only anionic charges of two methanoate molecules to represent the non-ester linked phosphate oxygen atoms. The base pairs and furanose rings were excluded. Semi-empirical quantum-mechanical PCILO calculations were then performed in a supermolecule framework with a limited number of rotations round the three coordinate axes of the drug molecule (figure 7.14). Rotations round the x-, y- and z-axes are equivalent to the nautical roll, yaw and pitch motions.

7.4.1   Roll rotation

For roll rotation the roll axis lies in the plane of the phenanthridinium ring, normal and midway between the line joining the amine group (figure 7.14). With ethidium separated from its model binding site by 12 Å the roll rotational energy differences are small and no more than 12 kJ mol^{-1} (figure 7.15). Minima in the energy profile appear in the

Figure 7.14. Coordinate framework for rotation of ethidium close to its receptor. The large dots mark the positions of ionic charges.

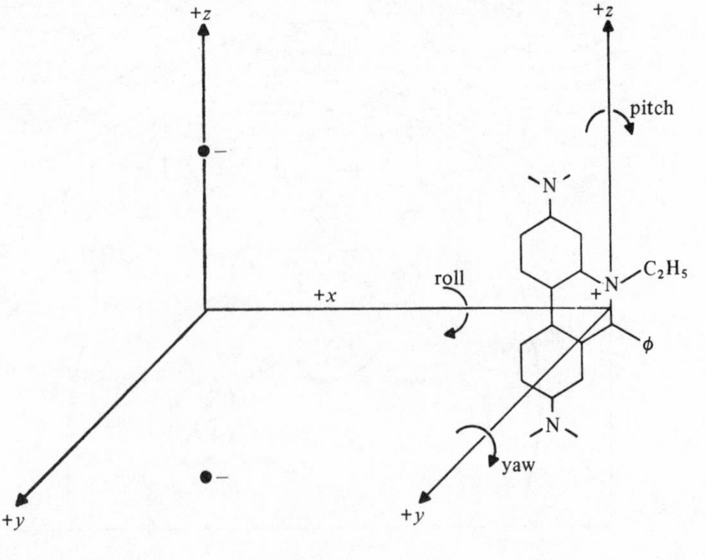

region 0°, 180° and 360° with maxima at 90° and 270°. Thus at 12 Å the phenanthridine ring shows a preference for lying with an orientation in the intercalation plane. As the drug molecule is brought closer to the docked position at 6 Å, the energetic preferences are much more pronounced with two effective minimum energy positions at 0° and 180°; these two positions are favoured by about 56 kJ mol^{-1}. Therefore, as ethidium approaches its receptor the drug molecule experiences considerable orienting effects by the fields of the anionic charges. These forces cause the chromophore to rotate into the plane containing the anionic charges. The electrostatic field generated by (dC-dG) · (dC-dG) shown in plate 1 is dominated by spherical regions associated with the phosphate. With roll rotation, ethidium reaches a minimum energy position when the two phenanthridinium amino groups are embedded in these spherical regions and the plane of the drug molecule is aligned with the line joining the charges. The alignment is within 5° of the intercalation plane found in the crystal structure. The two alignments are equivalent and they do not seem to be affected by the position of the ethyl group.

7.4.2   Yaw rotation

Rotation of ethidium about an axis normal to the plane of the ring atoms is a yaw rotation. The rotational energy profile for yaw, with the roll

Figure 7.15. The energy profile for ethidium executing a roll rotation at 12 Å (filled circles) and 6 Å (open circles) from the fully docked position. (Taken from Dean & Wakelin, 1979.)

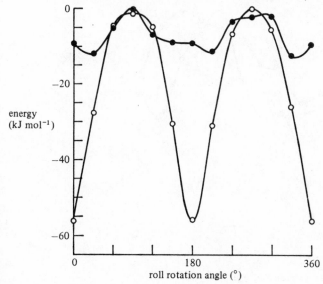

energy
(kJ mol^{-1})

roll rotation angle (°)

rotation fixed at the minimum position, is shown in figure 7.16. Energy differences for yaw rotation are much greater than roll rotation and at 12 Å separation the energy differences are 25 kJ mol^{-1}. The pattern of energy changes does not show the pseudosymmetrical nature of roll rotations. Presumably this asymmetrical character reflects the charge distribution of the ethyl and phenyl groups on the drug molecule. A broad minimum is found in the range 0–150° with the minimum at 120°; at this position the phenyl ring is normal and distal to the line joining the anionic charges. The most unfavourable conformation is found at 300° where the phenyl ring points to one anionic charge and the ethyl group to the other charge. This orientation would make a non-intercalative attack on the receptor. On closer docking, at 6 Å, the most favoured orientation is at 90°, this has the line joining the amino groups parallel to the line joining the anionic charges. These yaw rotations bring the drug molecule into a suitable alignment for docking.

7.4.3   Pitch rotation

Pitching describes the motion of the drug molecule when it is rotated round the z-axis of figure 7.14. The rotation energy is a minimum when the pitch angle is 0°, that is, when the phenyl ring points away from the intercalation site. The energy minimum is −7 kJ mol^{-1} at 12 Å and

Figure 7.16. The energy profile for ethidium executing a yaw rotation at 12 Å (filled circles) and 6 Å (open circles) from the fully docked position. (Taken from Dean & Wakelin, 1979.)

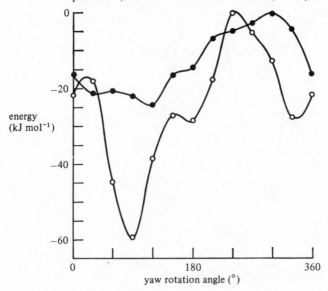

rises to $-54\,\mathrm{kJ\,mol^{-1}}$ at 6 Å.

The three rotations for ethidium of roll, yaw and pitch all suggest that the most favourable orientation of the drug in the field generated by two anionic charges is with the phenyl and ethyl groups pointing away from the receptor on the trailing edge, and that ethidium glides towards its intercalative binding site in the best orientation for a productive collision. These observations would suggest that if the drug molecule is free to rotate within the electrostatic field generated by the receptor, then the receptor could induce an orientation change in the ligand. This notion of receptor-induced ligand orientation may be a significant effect in the docking manoeuvre at a drug receptor. The work needs to be extended from the *in vacuo* simulation presented here to a model which includes freely moving solvent molecules before the significance of the notion can be assessed.

7.5   Electric fields and ligand orientation

The *in vacuo* model for receptor-induced ligand orientation suggests significant changes in interaction energy as the receptor and drug molecules come together. Even with a grossly simplified model, the interaction energy can be partitioned in the quantum-mechanical calculations into the underlying energetic terms. Two terms appear to dominate the roll rotation effect for ethidium; 45% of the interaction energy is correlated with electrostatic and short-range repulsion energies and about 40% of the interaction can be attributed to electron delocalization. An approximately similar partition of the interaction energy is found for yaw rotations. A different, but expected, partition is observed for the translation of ethidium from 12 Å to 6 Å from the docked position; 71% of the energy gain is electrostatic with a small but significant delocalization energy of 19% of the total interaction energy change. Thus the long-range electrostatic interaction is predominant at large separation distances and is a significant factor in changing the orientation of the drug molecule during docking. The electrostatic potential surrounding a receptor is anisotropic and the gradient field associated with the electrostatic potential is a force field and can be represented as a vector field. Thus it should be possible to search for correlations between the vector electric field of the receptor and the orientation of a nearby drug molecule.

7.5.1   The electric field

The electrostatic potential $V(r)$ at position r has already been defined (section 3.5.1); the electric field $\mathscr{E}(\mathbf{r})$ is then given by

$$\mathscr{E}(\mathbf{r}) = -\nabla V(\mathbf{r}) \tag{7.7}$$

This equation can be approximated by

$$\mathscr{E}(\mathbf{r}) = - \sum_{i=1}^{3} d^{-1} \, \delta V_i(\mathbf{r}) \mathbf{e}_i \tag{7.8}$$

where d is a small distance along the axis i (Cartesian notation x, y, z) and the change in potential along axis i, δV_i, is given by

$$\delta V_i(\mathbf{r}) = V(\mathbf{r} + d\mathbf{e}_i) - V(\mathbf{r}) \tag{7.9}$$

Therefore, $\mathscr{E}(\mathbf{r})$ is the electric field strength arising from the electron distribution at position r; it is also a vector quantity in contrast to the scalar electrostatic potential, since the value is dependent on the direction in which the field is computed. This method of calculation has been used by Peinel, Frischleder & Birnstock (1980) to study the electric field round water molecules and to indicate the importance of similar studies for our understanding of molecular orientation. The unit of electric field strength is V m^{-1} whereas the electrostatic potential is usually expressed as kJ mol^{-1}.

At each position, the electric field is associated with a resultant vector. These vector directions can be plotted on a grid with an arrow at each position; the map produced is termed the direction field. The direction in which the arrow points is defined conventionally as the direction in which a proton would move if placed at the grid point. The direction field at any position is normal to the equipotential surface passing through that point. Another representation of the direction field is by plotting the pattern of field lines. The field line is the direction in which a proton would move if it were allowed completely free movement to travel through the direction field. The strength of the field in any particular region is related to the number density of field lines passing through the region. With a complex geometrical structure there may be a region of space where adjacent field lines move in opposite directions; these are termed critical points and may be either sinks, sources or hyperbolic critical points.

7.5.2   *Direction fields surrounding base pairs of nucleic acids*

Direction fields have been charted for the alternating (G-C) polymer of B-DNA (Dean, 1981a) and for the homopolymers poly(dG) · poly(dC) and poly(dA) · poly(dT) (Lavery, Pullman & Pullman, 1982). Figure 7.17 illustrates the direction field in the plane midway between the dinucleotide dGpdC at the centre of one helical turn of poly(dG-dC) · poly(dG-dC); the adjacent base pair gives the local sequence dCpdG. Fields at the edge of the map point towards the helix axis. For the sequence dGpdC, the fields in the major groove fall into three regions. A central band, approximately 9 Å wide, converges towards the position where the guanine oxygen atoms overlap. On either side of this band the field diverges away from the major groove towards the phosphate groups. There are three similar regions in the minor groove. Cations would be attracted in the direction of the arrows and would be expected to attack the dGpdC sequence in both grooves. In contrast, the pattern of the direction field for dCpdG lacks the central band directed towards the middle of the base pairs in each groove and instead the fields diverge towards the phosphate groups. Directional effects are more pronounced when a distance-dependent dielectric function is included in the calculations (figure 7.18). In this figure the dielectric function is approximately linear from 1 D to 76 D over a distance of 0–17 Å from the atomic charge. The zone where the cytosine amino groups come close together shows a local repulsive field

Figure 7.17. The direction field in a plane midway between adjacent base pairs in a poly(dG-dC)·poly(dG-dC) sequence; (a) the sequence GpC, (b) the sequence CpG. The field is computed with a constant dielectric. (Taken from Dean, 1981a.)

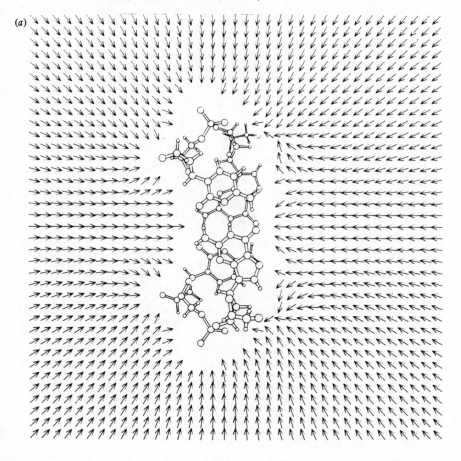

(a)

towards a cation as the direction lines are reversed and point away from the helix axis for about 5 Å from the molecular surface. There are, therefore, quite distinct differences in direction field pattern between the two G-C sequences. A consequence of this observation would be that in the major groove there would be an alternating pattern of attraction and repulsion in the direction fields. This pattern might offer an electric field 'finger print' of the local binding site and enable an attacking molecule to distinguish energetically between different dinucleotide sequences.

7.5.3   Direction fields along the helix axis of nucleic acid

Direction field maps shown in figure 7.19 are constructed along the helix axis and the planes are orthogonal to each other. When interpreting these maps, care must be taken in relating atomic positions to

(*Figure 7.17 contd.*)

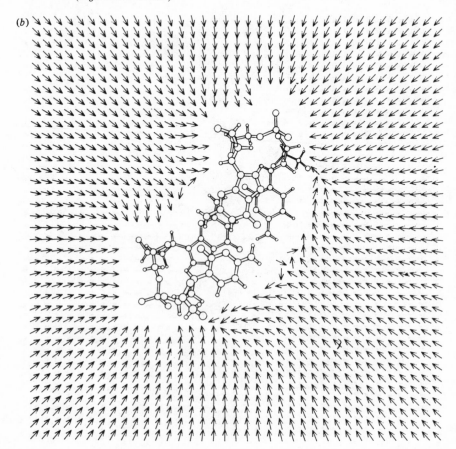

the field lines since only a few atoms lie near to the mapping plane. The fields are dominated by the phosphate groups. However, when the section plane cuts across a groove, distinctions between local sequences can be observed. For example, in the triplet dCpdGpdC (third, fourth, and fifth base pairs from the bottom of the figure) a different pattern in the major groove is shown between two adjacent dinucleotides; the pattern is consistent with that drawn in figure 7.17 with the arrows directed towards the helix axis in dGpdC and away from the axis in dCpdG.

A comparison between the distribution of the electrostatic potential and the electric field along a helical line of B-DNA poly(dG) · poly(dC) shows

Figure 7.18. The direction field in a plane midway between adjacent base pairs in a poly(dG-dC) · poly(dG-dC) sequence; (a) the sequence GpC, (b) the sequence CpG. The field is computed with a distance dependent dielectric varying from 1 D to 76 D over about 17 Å from an atomic charge. (Taken from Dean, 1981a.)

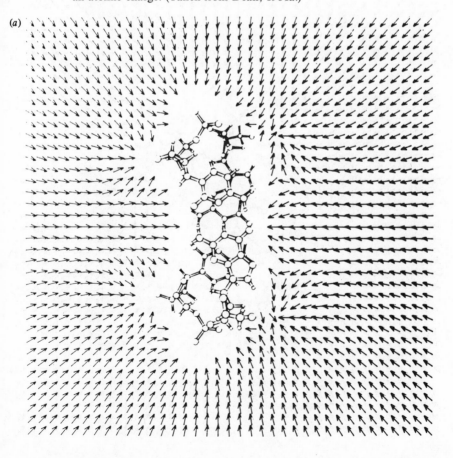

(a)

that the largest values of potential are found in the two grooves; the strongest electric fields are located near to the phosphate groups, whereas small fields of lower magnitude are located close to the carbonyl oxygen atoms of the base pairs (Lavery, Pullman & Pullman, 1982). This difference in distribution can be attributed to the distance dependence of the two properties from a point charge; the field which is a gradient of the potential is more dependent on distance. Thus the potential between two structures can sum significantly to give a profile which is considerably different from the summation of the electric field.

In physiological solutions counterions such as sodium, potassium, magnesium and calcium may be bound to the anionic phosphate groups and could screen the potential and electric field associated with charged DNA. A simple test model to examine the screening effects of counterions

(*Figure 7.18 contd.*)

(b)

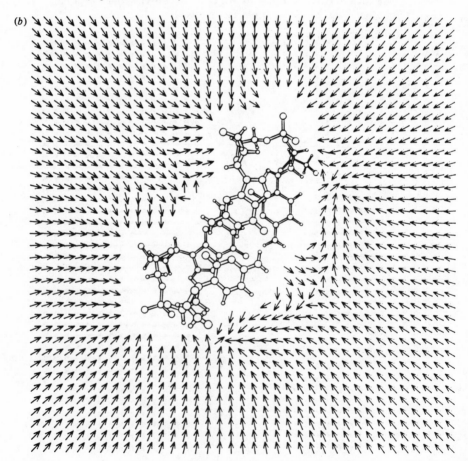

can be constructed where a sodium cation is bound to DNA in the bridging position between adjacent phosphate groups on each strand of the double helix. Screening of the anionic charges by the counterions reduces the magnitude of the potential without changing their pattern of distribution; the effect is as though all that has happened is to reduce the phosphate anionic charge. More noticeable is the effect of counterion screening on the electric fields. The geometrical arrangement of phosphate and sodium ions produces a strand of zwitterions along the sugar phosphate backbone. The direction field is therefore markedly changed in the base planes near to the

Figure 7.19. The direction field of a single turn of B-DNA poly(dG-dC) · poly(dG-dC) mapped onto two perpendicular planes passing through the helix axis. The field is computed with a constant dielectric. (Taken from Dean, 1981a.)

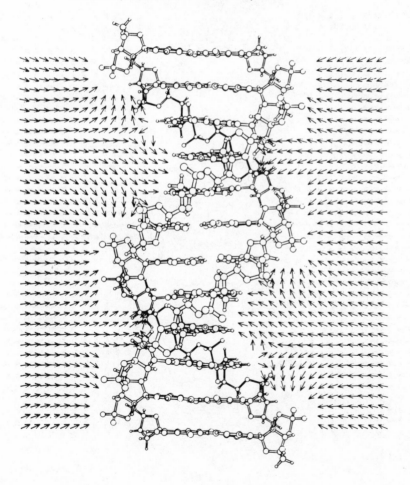

backbone atoms (Lavery, Pullman & Pullman, 1982). Screening decreases the difference in potential in the two grooves of poly(dG)·poly(dC) but increases the distinction for poly(dA)·poly(dT); the effect on the electric fields in the two grooves for each sequence is the opposite.

7.5.4   Ligand orientation in the receptor's electric field

Drug–receptor recognition is often characterized by a stereospecific binding reaction between the two molecular species. The final event in the recognition process, for a productive collision to occur, is an accurate geometrical positioning of atoms in the drug molecule with

(*Figure 7.19 contd.*)

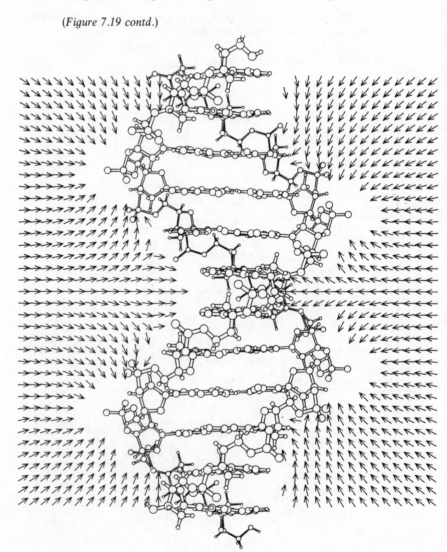

respect to those of the receptor binding site. This spatial juxtaposition may be brought about by initial chance alignments of the two interacting coordinate systems, or orientation-dependent forces may operate to modify the alignment as the reaction proceeds. In the model study of a dinucleotide receptor and ethidium, it has been shown that as the drug molecule approaches the receptor its rotatory freedom is curtailed even though atomic collisions may not occur. Electrostatic components of the total interaction energy account for about 40–50 % of the energy changes. How does this electrostatic component modify the ligand orientation? Simple electrostatic theory can throw light on this problem. A polar molecule has a dipole moment and the behaviour of a dipole rotating in an electric field is well understood.

Ethidium is a positively charged molecule and its dipole, which is small, passes through its centre of mass. The *p*-carboxyphenyl derivative of ethidium has a larger dipole moment since the molecule is a zwitterion and this derivative has been used to investigate the orientation of the dipole with respect to the local direction field of the receptor (Dean, 1981b). A single turn of helical B-DNA with all the base pairs stacked was used as a model receptor; the electrostatic potential and electric fields could be calculated at any point in space. The dielectric of the intervening medium between drug and receptor was either kept constant or allowed to vary, with distance, in the atom-pair calculation of the molecular electrostatic interaction energy to model dielectric anisotropy effects. The minimum energy orientation at seven locations round the helix axis in the plane midway along the helix is shown in figure 7.20; the local dinucleotide drawn in the figure is a dGpdC. The major axis of the ligand is tilted against the helical turn in the major and minor grooves. The carboxyphenyl group points directly away from the helix for all positions, suggesting that the molecule is strongly aligned with the dipole positive end and aimed at the centre of the helix. However, in the sugar–phosphate region the molecule is turned so that the minor axis is parallel to the helix axis and a phenanthridine amino group is towards the phosphate anionic charge. When dielectric anisotropy is included in the calculation of interaction energy, positional differences are clearly observed. The heteroaromatic ring is turned into the inspection plane in the sugar–phosphate region, with a possible interaction between the ligand amino groups and the anionc charge. The orientations shown in figure 7.20 are only for one plane, the midplane along the helix; however, at different locations of the plane, the orientation effects can vary and these are most noticeable at the end of the helix where the additive effects on the electrostatic potential show greatest variation.

At any particular location the orientation restriction can be computed from the rotamer population by converting the orientation energy into the Boltzmann distribution. The probability, P_i, of finding a molecule with a particular energy E_i, is given by:

$$P_i = n_i/N = \{\exp(-E_i/kT)\}/z \tag{7.10}$$

where n_i is the number of molecules with energy, E_i, out of a population of N molecules, and z is a normalizing constant for the phase space. The difference between the minimum and maximum orientation energy can be quite large, as shown for *p*-carboxyphenylethidium in figure 7.21. The

Figure 7.20. Stereoscopic drawing of the orientation of ethidium around the local sequence GpC at the centre of one turn of poly(dG-dC)·poly(dG-dC); (*a*) orientation with a homogeneous dielectric constant, (*b*) orientation in a distance-dependent dielectric. (Taken from Dean, 1981b.)

(*a*)

(*b*)

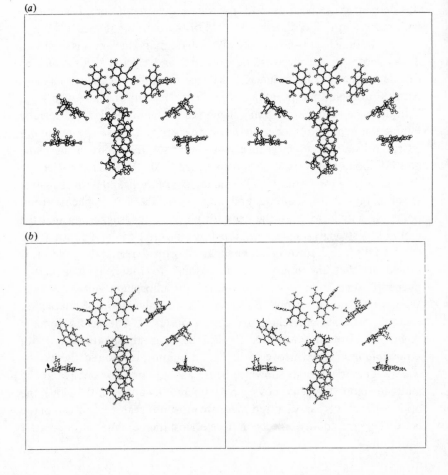

minimum energy positions are all attractive and the smallest energy maxima are found in the grooves of the DNA. The worst orientations show strong repulsion near to the sugar–phosphate backbone. Repulsive forces in that region can produce a large degree of orientation restriction (figure 7.22). Approximately 15–37% of the total possible orientation space is restricted and unlikely to be occupied significantly in a large population of molecules.

Both ethidium and its carboxyphenyl derivatives show orientational preferences, mediated by electrostatic forces at various locations round the receptor. It is possible to resolve the mutual electrostatic interactions separately into drug and receptor components by comparing the direction of two associated vectors at the minimum energy orientation. The direction field points the way that a proton would move, and if this is calculated for the receptor at the centre of mass of the ligand then the angle between the

Figure 7.21. Minimum and maximum orientation energies for *p*-carboxyphenylethidium at various positions round a complete turn of poly(dG-dC) · poly(dG-dC). Letters indicate the positions of the inspection planes; A = 0 Å (ie at the centre of the helix), B = − 3.4 Å, C = − 13.6 Å, D = − 17 Å. Single letters denote curves for the minimum energy, double letters are for the maximum energy. For curve A the angle ϕ for the position of the ligand is set to 0° and the molecule is found in the wide groove (see figure 7.20). Energies are calculated using a distance dependent dielectric. (Taken from Dean, 1981b.)

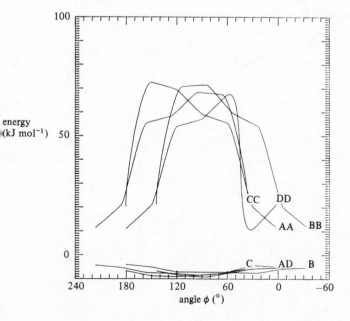

positive end of the ligand dipole and the field can be readily calculated. The angular discrepancy between these two vectors is plotted as a scattergram for p-carboxyphenylethidium in figure 7.23. A maximum discrepancy of 180° would occur if the field line and the positive end of the dipole pointed in opposite directions and a chance alignment would give a mean angular discrepancy of 90°. For this ligand the mean is $19.9 \pm 1.6°$. Therefore, at the minimum energy orientation, the two vectors tend to lie parallel and the correlation between both vectors for ethidium and its p-carboxylated derivative are good, considering the approximations used. This procedure may be useful in forecasting the alignment of drug molecules in the vicinity of the receptor. The only weakness is that the most favourable rotation round the dipole axis cannot be determined by inspection.

The extent of the correlation between the ligand dipole and the local direction field of the receptor is determined by the uniformity of the electric field. If the field exhibits significant anisotropy within the area covered by the molecular surface of the ligand, then orientation correlation with the field will be between each bond dipole moment and the field passing through the centre of each bond. This more complex relationship needs

Figure 7.22. Fractional orientation restriction for p-carboxyphenylethidium at different positions, defined in the legend to figure 7.21, round a turn of poly(dG-dC)·poly(dG-dC). The restriction is calculated using a distance dependent dielectric. (Taken from Dean, 1981b.)

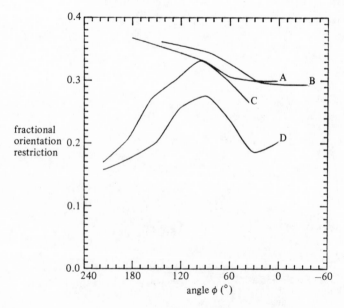

further investigation. The most extreme case of anisotropy in the electric field would arise where the spatial region contained a critical point; in this instance a slight change in the position of the ligand would be expected to exert a large change in orientation. One might speculate that critical points in the field at a binding site could have an important role in precise positioning of a drug molecule, and orientation selection, before formation of the drug–receptor complex.

7.6 Stability of the drug–receptor complex

In the two previous sections we examined the factors which influence the orientation of the drug molecule near to its receptor, although interatomic separations were not close enough to cause steric repulsions. This section will examine the stability of the drug–receptor complex after it has formed. The intervening step between these two stages, which is the reaction pathway, has been omitted. Reaction pathways, even in a simple receptor model like the dinucleotide intercalation example, inevitably involve geometrical re-arrangements in either the binding site or the drug

Figure 7.23. Scattergram of the angular discrepancy between the direction field line and the dipole direction for *p*-carboxyphenylethidium at positions defined in the legend to figure 7.21. The discrepancy is calculated using a distance-dependent dielectric. (Taken from Dean, 1981b.)

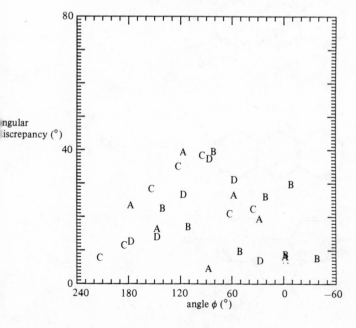

molecule to accommodate the ligand. The presence of multiple torsion angles makes the task of unravelling the pathway too formidable with present-day methods. A multi-dimensional potential hypersurface would be needed to portray the pathway before all the features of the surface could be examined to route the reaction path.

7.6.1   Interaction energies for drug–dinucleotide complexes

In section 7.2 we discussed the pioneering work of Nuss, Marsh & Kollman (1979) on unstacking the dinucleotide. We shall now return to their work to consider the energies of the unstacked receptor interacting with three intercalative drug molecules: ethidium, proflavine and 9-aminoacridine. The molecular structures of the drug molecules are shown in figure 7.24. All have flat triple-ringed structures, the phenanthridine and acridine rings intercalating between the base pairs. Ethidium intercalates from the minor groove, proflavine enters from the major groove and 9-aminoacridine can enter from either side.

The energies of the drug–receptor complexes computed from an empirical potential function are shown in table 7.3 for the dinucleotide sequences dGpdC and dCpdG. There are two main components to the

Figure 7.24. Molecular structures of (a) proflavine, (b) 9-aminoacridine, (c) ethidium.

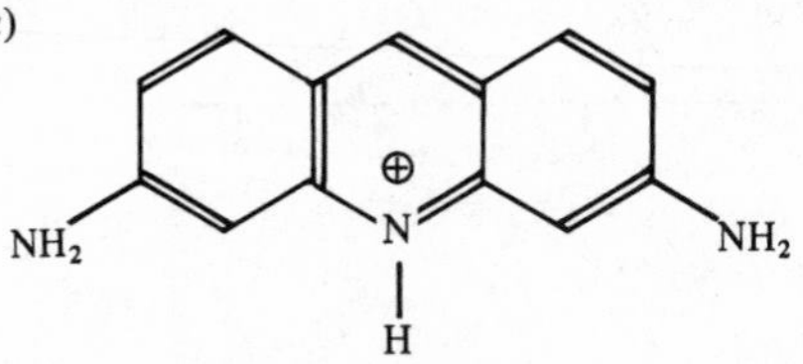

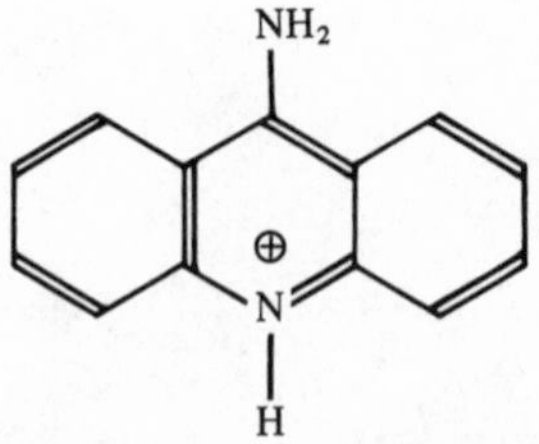

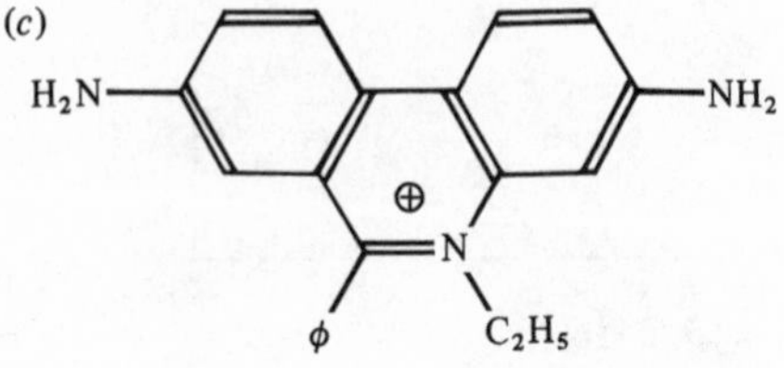

Table 7.3. *Interaction energies between ethidium, 9-aminoacridine and proflavine at the dinucleotide binding site. These are the energies for the formation of a drug–receptor complex* $(kJ\,mol^{-1})$

		Ethidium	Proflavine	9-Aminoacridine	
	Energy component	minor groove	major groove	minor groove	major groove
GpC sequence	Dinucleotide unstacking energy	121.6	118.2	122.3	119.6
	Drug–dinucleotide interaction energy	−680.1	−659.4	−636.6	−626.3
	Net stabilization energy	−558.5	−541.2	−514.3	−506.7
CpG sequence	Dinucleotide unstacking energy	91.7	96.8	103.9	88.7
	Drug–dinucleotide interaction energy	−673.3	−657.4	−673.3	−620.1
	Net stabilization energy	−581.6	−560.6	−569.4	−531.4

These values are taken from Nuss, Marsh & Kollman (1979) using an optimized geometry.

stabilization energy: the energy of unstacking the dinucleotide and the interaction energy of the drug intercalated between the unstacked dinucleotide. The net stabilization energy is the total energy of the formation of the drug–receptor complex. For each of the three drug molecules the energy of the formation of the drug–receptor complex is consistently larger for the dCpdG than for the dGpdC sequence. Variations in the energy of unstacking the dinucleotides for a particular sequence are small, $4.2\,\mathrm{kJ\,mol^{-1}}$ for dGpdC and $15.2\,\mathrm{kJ\,mol^{-1}}$ for dCpdG. These differences in unstacking within a sequence are due to variations in torsion angles along the sugar–phosphate backbone for the different optimized geometries. Therefore, in the formation of these drug–receptor complexes, the most significant interaction is between the drug and the dinucleotide bases and the backbone. The three drug molecules show a difference of about $54.4\,\mathrm{kJ\,mol^{-1}}$ in their interaction with each sequence and this accounts for the stability of the different drug–receptor complexes.

These results from computation of the interaction energies can be compared with experimental measurements of the stability of each complex. However, it must be remembered that the results displayed in table 7.3 are for *in vacuo* calculations; experimental measurements are made in solution, thus we can only compare the order of relative stabilities of the complexes. The computed values indicate that the order of binding affinity is ethidium > 9-aminoacridine > proflavine for the most favoured sequence dCpdG. This order is found for drug interactions with DNA in solution and so the correlation is encouraging. Furthermore, experimental values for drug binding to different sequences show that pyrimidine-3',5'-purine sequences are preferred to the reverse order, a tendency which is borne out by the interaction energy calculations. This difference in sequence selectivity may be due to the electrostatic interactions between the bases in the two sequences; the unstacking of a dCpdG is stabilized electrostatically by $16.7\,\mathrm{kJ\,mol^{-1}}$ whereas unstacking of dGpdC has an electrostatic energy which is destabilized by $4.2\,\mathrm{kJ\,mol^{-1}}$.

7.6.2   Drug-induced conformational changes in the dinucleotide receptor

The optimization of the conformational energy employed by Nuss, Marsh & Kollman (1979) enabled them to examine the best conformation of the receptor for each drug molecule. As the dinucleotide relaxed, the torsion angles changed to give the best fit with the rigid ligands. Torsion angles for dCpdG with the intercalated drug molecules are shown in table 7.4. Only small differences in torsion angles are observed for ethidium and proflavine intercalated into the receptor; the largest changes are about $13°$ at torsion angles ω and ε. Similarities in torsion angles for 9-

aminoacridine intercalated in the major groove are shown by the two other drugs. However, intercalation of 9-aminoacridine from the minor groove shows very marked differences at ϕ and ψ, that is, along the chain of atoms C4′–C5′–O5–P–O3, with differences of 48° and 32° respectively compared with the complex intercalated from the major groove. Smaller differences for ϕ and ψ are found on the other half of the duplex for the 9-aminoacridine complex.

These observations provide a clear demonstration that conformational changes are induced in the binding site by different drug molecules. In general, these changes are small but larger conformational shifts can be accommodated even though the conformational energy changes are still small ($< 16.8 \, \text{kJ mol}^{-1}$). Thus the picture emerges that where the binding site has extensive conformational freedom, the atoms can be moulded round a rigid molecule to achieve the minimum energy arrangement. A molecular dynamics simulation of intercalation linked to an animated display of interatomic distances along the backbone of the nucleic acid could, like that shown in section 4.2.1, indicate the conformational route taken in fitting the receptor round the drug molecule. Work along these lines is needed to give a clearer picture of the final steps in the docking manoeuvre.

7.7 Intramolecular changes during docking

Molecules which approach each other to distances of 12 Å, or approximately their molecular diameter, can, through their respective

Table 7.4. *Torsion angles for the model receptor dCpdG with intercalated drug molecules*

	Drug molecule intercalated			
			9-Aminoacridine	
Torsion angle	Ethidium minor groove	Proflavine major groove	minor groove	major groove
χ	217	223	218	220
ω	289	275	280	290
ϕ	149	152	90	138
ψ	73	73	110	77
θ	65	66	54	60
ε	153	166	174	156
χ'	93	99	101	104

Torsion angle nomenclature is given in figure 7.3.
Data taken from Nuss, Marsh & Kollman (1979).

electric fields, influence their relative orientation. This electrostatic effect can be strong in *in vacuo* conditions and the whole molecule may be shifted. But perhaps an even more important consequence of close intermolecular approach is the effect of their respective molecular fields on each other's electron distributions. So far, in this chapter, we have considered the molecular interactions by keeping the electron distribution constant as a static array of point charges. Whilst this is a useful ploy for simplifying computational work, the real situation with regard to the electron distribution in the two molecules is likely to be much more complex. In chapter 3 we outlined the importance of induction and polarization effects that are introduced by placing a molecule in an electric field. With an anisotropic electric field, only quantum-mechanical studies can hope to elucidate the intricate changes in electron distributions. What is the magnitude of intermolecular field effects on the electron distribution in a drug molecule approaching its binding site, and how are the energies of the molecular orbitals modified?

7.7.1   Modification of the electron distribution in the drug molecule

Ethidium docking into the dinucleotide receptor unit provides a convenient model to investigate the changes in electron distribution within the drug molecule placed in the electric field of the receptor (Dean & Wakelin, 1979). Again, the receptor is approximated only by two anionic charges oriented as in the non-ester linked phosphate groups; the nucleotides are omitted and we confine our analysis on the drug molecule to the two amino group nitrogen atoms since these show the greatest re-distribution of charge. Molecular orbital calculations are performed within the PCILO framework.

In the isolated drug molecule there is little difference in the charges of both nitrogen atoms, approximately $2me$ (table 7.5). As the drug molecule is brought towards the receptor anionic sites in an orientation favourable for intercalation, at 12 Å there is a decrease in charge of $25me$ for the sp^2 nitrogen atom but with a change of less than $1me$ for the sp^3 nitrogen atom. There is, therefore, a differential effect of the electric field on the ethidium molecule even though the field is approximately symmetrical in this region of the nitrogen atoms. Closer approach of the drug molecule to 6 Å separation from the anionic charges enhances the drop in charge on the sp^2 nitrogen atom by $12me$ and $(3-4)me$ on the sp^3 nitrogen atom. Rotation of ethidium from its most favourable roll orientation to the worst roll orientation again perturbs the charge distribution on each nitrogen atom. An unfavourable yaw rotation has a similar effect.

These observations raise the question: why are the two nitrogen atoms affected differently? The greater fluidity of electronic charge centred on the sp^2 hybridized nitrogen atom may result from the fact that the lone-pair electrons reside in π-bonding p orbitals and are resonant with the π electrons of the heteroaromatic system. On the other hand, the lone-pair of the sp^3 nitrogen atom occupies a σ-orbital that lies on one side only of the phenanthridinium plane and is not orthogonal to it. This apparent differential inductive effect of a receptor field on the electron distribution in different parts of a drug molecule needs further investigation; more accurate *ab initio* calculations could examine the generality of the effect. Nevertheless, the point to note is that the electron distribution in this example is not static, and molecular model studies should take into account the shifts in electronic distribution in going from free molecules to bound drug–receptor complexes.

7.7.2 Changes in orbital energies in the drug molecule

When a drug molecule is placed in an electric field to approximate that of the receptor, the electron distribution is perturbed. This perturbation will be reflected in the energies of the molecular orbitals. Two orbitals are of particular interest for determining reactivity, the highest occupied molecular orbital (HOMO) and the lowest unoccupied molecular orbital (LUMO). In ethidium, the HOMO is found to be the lone-pair of the sp^2 hybridized 3-amino nitrogen atom, the LUMO is the π-antibonding orbital in the $C{=}N$ bond of the phenanthridinium ring. In the isolated drug molecule the orbital energies are: HOMO $-18.5\,eV$ and LUMO $-3.1\,eV$. Thus ethidium is a good acceptor and a bad electron donor, as one would expect for a positively charged molecule. Figure 7.25 illustrates the change in energy of the two orbitals as ethidium moves towards two anionic charges representing the dinucleotide receptor. *In*

Table 7.5. *Net electron charges on the amino group nitrogen atoms of ethidium at energy extremes of docking*

	Ethidium amino nitrogen atom	
Energy extreme	sp^2	sp^3
Isolated molecule	-0.1903	-0.1921
Roll favourable	-0.1530	-0.1877
Roll unfavourable	-0.1697	-0.1932
Yaw favourable	-0.1530	-0.1877
Yaw unfavourable	-0.1740	-0.1864
Translation 6 Å	-0.1530	-0.1877
Translation 12 Å	-0.1652	-0.1913

vacuo, these changes in orbital energies are detected at large separation distances from the anionic sites. The energy of LUMO swings to a positive value as ethidium approaches its docking position. The changes in the LUMO and HOMO appear to be mediated by the electric fields of the anionic sites, since changes in the orbital energies occur before any charge transfer is observed. The lone-pair electrons on the guanine nitrogen atoms are directionally located normal to the base plane and immediately above the position where the LUMO of ethidium is found in the crystal structure of the ethidium–dinucleotide complex. This relative positioning might be a contributory factor to the stabilization of the drug–receptor complex by providing electron donation from this guanine nitrogen atom into the LUMO of ethidium.

Changes in orbital energies within the drug molecule, and presumably within the receptor, are expected to be even more important in the analogous problem of the enzyme–substrate complex. In the latter

Figure 7.25. Energy levels for the HOMO and LUMO of ethidium as it approaches the receptor. (Taken from Dean & Wakelin, 1979.)

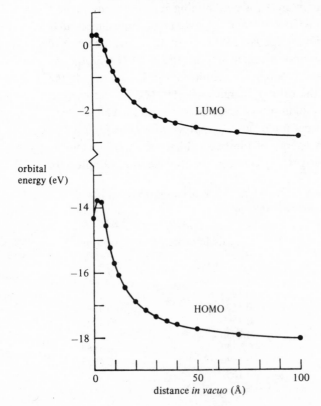

example, a transition state is engendered before particular bonds are formed or cleaved leading to the products of the reaction.

7.7.3 Intramolecular changes in substrate docking at an active site

The binding of a substrate molecule at an enzyme active site causes a loss of translational and rotational entropies of the substrate molecule. Consequently, the correct positioning of the molecule prior to formation of the transition-state complex is, to a large extent, the rate enhancing effect in enzyme catalysis. Precise positioning of a guest molecule at the active site may also affect the reaction pathway towards the transition state by modifying the energies of the molecular orbitals in particular bonding regions. This modification of the substrate by the enzyme during docking and complex formation has been studied extensively by Osman, Weinstein & Topiol (1981) using a model of the carboxypeptidase enzyme and methanamide as a model peptide substrate.

Carboxypeptidase is a zinc containing metalloenzyme that catalyzes the hydrolysis of peptides. Three amino acid residues and the zinc atom define the active site. Arg 145 binds the terminal carboxylate group, Glu 270 and Tyr 248 participate in the catalysis. An active site can be modelled by the complex of $Zn(OH)(NH_3)_2^+$; the methanamide carbonyl oxygen atom is positioned over, and bonds to, the zinc atom and two water molecules are located nearby (figure 7.26).

Figure 7.26. The positional geometry of methanamide in its complex with $Zn(OH)(NH_3)_2^+$ to model the enzyme–substrate complex of carboxypeptidase. (Re-drawn from Weinstein et al., 1983.)

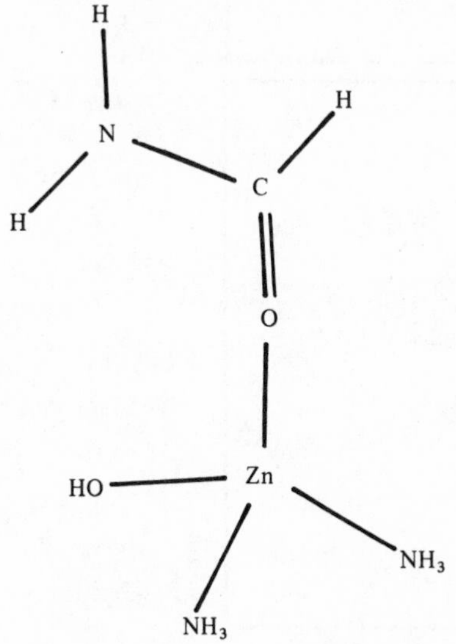

Maps of the electrostatic potential of the methanamide substrate in the absence and in the presence of the active site model show a dramatic modification of the potential around the aldehyde group (figure 7.27). The potential generated by the zinc atom of the active site enhances the magnitude of a local potential maximum lying close to the carbon atom of the carbonyl group and reduces the potential minimum found in the lone-pair region of the oxygen in the isolated substrate. The five-fold increase in

Figure 7.27. The electrostatic potential (kJ mol^{-1}) mapped in a plane 1.5 Å above the methanamide nuclei. (a) In the absence of Zn(OH)(NH$_3$)$_2^+$; (b) in the presence of Zn(OH)(NH$_3$)$_2^+$. The electrostatic potential close to the aldehyde moiety of methanamide is drastically modified by components of the binding site. (Re-drawn from Weinstein et al., 1981.)

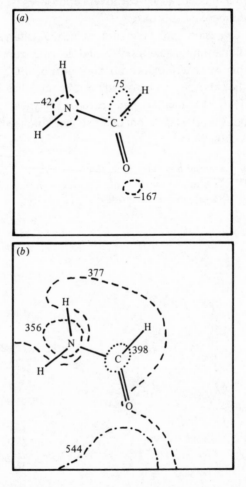

potential near to the carbon atom of methanamide makes the atom strongly positive and increases the susceptibility of this atom to nucleophilic attack, possibly from a neighbouring water molecule or from Glu 270. This striking change in electrostatic potential round the substrate is not simply due to superposition of the two potential fields; there is also a re-distribution of π-charge in the methanamide molecule and the carbonyl bond becomes more polarized with electronic charge migrating from the carbon to the oxygen atom. Since the electron distribution along the carbonyl bond is perturbed, with the carbon atom being made more positive, there is also a change in the electron distribution along the peptide bond. In this state the nitrogen atom of the substrate is rendered more susceptible to hydrogen-bond formation with Tyr 248 or a nearby water molecule. The methanamide molecule is then held, by both ends, ready for hydrolytic cleavage of the peptide bond (Weinstein et al., 1983).

Thus these calculations provide very clear evidence of the importance of the molecular electrostatic potential in recognition between a substrate and its active site. The fields generated by the site dominate the intermolecular interaction with the substrate by forcing the orientation of the peptide into the correct alignment with the receptor. Alignment of the partners then introduces an intramolecular re-distribution of charge. In turn this re-distribution facilitates further hydrogen-bond formation for precise aligning of substrate within the active site. At the same time, these intramolecular changes enhance the carbon atom of the scissile bond to nucleophilic attack. All that now has to be studied is the formation of the transition-state complex and the geometrical pathway to the formation of reaction products. To achieve this objective, accurate *ab initio* calculations representing the complete amino acid residues, the zinc atom, water molecules and substrate will be needed and the computation may have to wait until computing technology is improved for bigger and faster computers.

7.8 Software for docking interactions

One of the major growth areas in the application of molecular graphics to molecular interactions has been in the variety of attempts to simulate the docking manoeuvre. Many programs are in existence and are constantly being updated as new methods of representation are developed. The potential use of this approach is enormous for the drug industry; the computer becomes a tool in the hands of the medicinal chemist to design new ligands to fit known structural binding sites. Modification to the developing molecular structure can be made interactively. The consequences of these modifications, on molecular recognition and fitting into the receptor site, can readily be assessed before the costly process of molecular synthesis. Obviously this graphical facility helps to prune the tree of possible molecular structures and saves time, and money, in the design of new molecules by rational procedures.

If the putative drug molecule and the binding site were rigid structures, then all that would be needed in the docking interaction would be a computation of the total molecular energy with a change in position and orientation of one partner in the interaction with the site kept in a fixed position. The direction of movement of the labile molecule is then determined by the energy gradient. The total energy can be calculated from van der Waals interactions, hydrogen bonds, dipole–dipole, dipole–charge and charge–charge interactions. In principle the solution to the problem *in vacuo* is quite simple. However, if flexibility is introduced into the molecular structure of each partner then a solution to finding the minimum energy configuration is much more tedious. Furthermore, intramolecular changes do take place as docking proceeds and these would need to be accounted for if docking leads to a transition-state complex being formed. The computational problem then becomes less manageable as the number of variables in the solution increases. One way to circumvent these problems is to study the docking manoeuvre with an interactive graphics facility that incorporates rapid empirical energy calculations.

DOCKER is an interactive program written by Busetta, Tickle & Blundel (1983) to calculate molecular interaction energies during docking. Two molecules, or fragments, of up to 400 atoms each can be manipulated interactively on a graphical display unit. The molecules can be represented in skeletal form or by an accessible surface display. Standard geometry packages and conformational energy calculations are included to provide manipulation of the molecular structures as the conformation is changed. A bell warning device, included in the program, indicates to the user when a position is disallowed because of short interatomic contacts. Suspicious contacts are illuminated on the display by 'blinking bonds'. Geometrical operations to shift the molecule can be called from a menu. Changes in solvation energy during docking are computed from the accessible molecular surface. After a suitable juxtaposition of the guest molecule with the host has been performed by interactive steps, there is the facility to allow both molecules to relax to find the local minimum energy conformations in the fully docked state.

The DOCKER program, although it has obvious limitations, is set to become the paradigm routine for modelling docking manoeuvres. The interactive capabilities are flexible and can easily be added to as new subroutines are written.

8

Ligand design techniques

8.1 The problem of design

Ligand design is a half-way house on the road to drug design; we are concerned only with constructing a molecule with pharmacological activity. For a ligand to be of some therapeutic value the ratio of the dose that produces possible toxic effects to that producing therapeutic effects should be high. Drug design is therefore much more complicated than ligand design. Effective design of novel drug molecules requires a multidisciplinary study including drug metabolism, distribution, pharmacokinetics, toxicology, clinical evaluation and risk–benefit analysis. These separate studies are obligatory developments *after* the initial design of the ligand by molecular manipulations. An integrated approach to the design of new drugs, in contrast to novel ligands, is outside the scope of this book.

'Drug design: fact or fantasy?' was the title of a recent symposium (Jolles & Wooldridge, 1984) and it expresses the unease felt by many medicinal chemists working in the practical world of the drug industry. How close are we to using strict design criteria for developing novel ligands? Historically speaking, most novel drugs have been discovered rather than designed according to rational principles dictated by the diseased condition. Thus the alkaloids and antibiotics have provided a natural wealth of readily available material for the chemist to propose minor modifications to change the potency of a derivative. Careful analysis of the effects of structural changes on drug potency gave rise to the assignment of correlations between structure and potency and the emergence of the discipline of QSARs. The relationship between structure and activity is multivariate and includes many physicochemical parameters which may have a bearing on potency. The collection of multi-dimensional data for QSAR studies has meant that greater reliance has to be placed on adequate computing techniques to define the relationships between the variables. Although considerable progress has been made in identifying relationships between particular variables and potency, this whole research area is still in its infancy.

Two general rational methods for drug design can be discerned. Firstly there is the modification of a known lead compound to develop a family of congeners, eg the family of penicillins all related to a basic structural unit, or the sulpha drugs springing from sulphanilamide. Secondly, there is the development of design principles based on a knowledge of structural properties of the receptor. This more recent approach is only possible where high-resolution crystallographic coordinates of the binding site are available.

8.1.1   The biochemical lead

If ligand design is to proceed by modification of the lead compound then the first task is to identify the lead molecule. The search for a lead compound is easier if a particular biochemical reaction is suspected of playing a key role in the pathology of the disease. Once the enzyme can be identified then its properties can be studied and modifications to the natural substrate, or product inhibitor, can be made to alter the catalytic rate of the enzyme. For example, if we need to develop an enzyme inhibitor there is a wide choice of strategies for the medicinal chemist. A single substrate enzyme is the simplest case for the design of an inhibitor: the substrate may be modified to a non-substrate with a similar binding affinity for the enzyme to give an inactive competitive inhibitor. A substrate analogue may be partially metabolized to a transition state but not proceed any further in catalysis; this analogue would be a transition-state inhibitor. An analogue of the product may act as a product inhibitor by binding tightly to the active site. If an allosteric regulator is involved in enzyme activity this may be mimicked to modify the activity. Coenzyme binding may also be a target for competitive inhibition. The mechanism of analogue binding to the active site may also be manipulated; most substrate binding is non-covalent so this is easily matched by analogue manipulation. However, very potent enzyme inhibitors can be made if covalent binding is introduced. Perhaps the simplest form of this type of inhibitor is the substrate analogue with a reactive moiety, such as an alkylating group, attached to it. There is then a good chance that the analogue will be found covalently at the active site. A more sophisticated approach would be to develop a 'suicide' inhibitor where a substrate analogue contains a reactive grouping that is only revealed by enzyme activity and is then able to bind covalently to the active site. This brief list of ways to develop an enzyme inhibitor has some obvious parallels in the design of analogues to antagonize a drug receptor with a known agonist structure.

8.1.2 Combinatorial modifications to a lead structure

Although the broad principles of analogue design, as outlined, are straightforward, the sheer choice of possible analogues is almost unlimited because chemical group addition can proceed in a combinatorial manner. For example, the molecular structure shown in figure 8.1 is drawn with five possible positions for substitution by five chemical moieties included in the attached list. The number N of possible molecular structures is given by combinatorial analysis for n substituents at r positions without repetition by the equation

$$N = \sum_{i=1}^{r} n!/(n-i)! \tag{8.1}$$

and 325 different molecules could be generated from the structure shown in figure 8.1. If substituent repetition is allowed then

$$N = \sum_{i=1}^{r} n^i \tag{8.2}$$

and this would lead to 3905 different congeners from the parent molecule. Similarly, 100 substituents would give 10^{10} different possible structures;

Figure 8.1. Combinatorial molecular substitution. Five possible positions for substitution with groups A–E. The number of possible compounds, N, formed by substitution, with or without repetition, is given by equations (8.1) and (8.2). Groups to be substituted at each position can be any one of A, B, C, D, E.

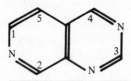

i	N without repetition	N with repetition
1	5	5
2	20	25
3	60	125
4	120	625
5	120	3125
	$\sum_{i=1}^{5} = 325$	$\sum_{i=1}^{5} = 3905$

thus, even for the one parent compound used as an example, the number of possible congeners is practically unlimited. The problem facing the ligand designer is how to optimize the search through the tree of all possibilities to arrive at the most effective pharmacological compound by the shortest route. Obviously blind synthesis and testing of each compound is not a practical proposition. The search tree needs to be drastically pruned to eliminate non-productive branches. Rational methods for ligand design are an attempt to grapple with this problem.

8.1.3   Combinatorial reduction by feature selection

QSAR studies, over many decades, have demonstrated that for a congeneric series certain molecular features are desirable for high binding affinity. These features vary from one series to another. Characterization of molecular structure in terms of features therefore provides one way to prune the tree for searching for the most potent compounds in the series, since molecules without these features are unlikely to be active. Molecular features must firstly be defined and then, secondly, assigned some index value. This procedure is commonly applied to substituent moieties, or fragments of them, to form a database of substituents with associated parameters. The basic logic of the use of features to reduce the combinatorial problem of ligand design is built on the notion that the presence of a feature, f_m, has a certain probability, P_m, that leads to the generation of a quantity of activity A_n so that

$$f_m \xrightarrow{\;P_m\;} A_n \tag{8.3}$$

Features are sets of molecular and geometrical properties which may be: structural fragments, groups of fragments, or sets of electronic properties within geometrical relationships. Therefore, in order to define the probability of a particular feature leading to activity the first task is to assign a description to the compound, the second is to identify the features and, the third, to evaluate the features statistically to obtain a probability value for the feature contributing to activity.

The molecular structure of a compound can be described in its most basic form as: the atom types, their topological linkage properties (bonds) and the distance matrix for the set of atoms suitably modified to remove ambiguity in the chirality problem (three-dimensional geometry). Associated with this structural representation are the more intimate molecular properties revealed by molecular orbital calculations such as orbital energies, electron distributions, bond dipole moments and atomic hybridization moments, the molecular electrostatic potential and electric fields. These are all of great importance in drug–receptor interaction. An extensive survey of information theoretic indices for the characterization of chemical structures is given by Bonchev (1983).

Feature identification is based on the principles of pattern recognition and has two components. An object, or set of objects, can be classified so that they fall into groups: this partition develops a rule for the classification. Pattern recognition then uses this rule to allocate a new object into a particular class. Thus the reliability of the pattern recognition method is strongly dependent on the development of the rules; these rules in turn rely heavily on the size and complexity of the sample initially used for classification. The distribution of substituents on a molecule must be carefully selected to optimize the solution of classifying the objects. The classes of objects which form sets may show intersection where common descriptors are shared by groups of compounds, or they may form part of a subset. Pattern recognition methods applied to QSAR are fraught with problems for the unwary and Wold, Dunn & Hellberg (1984) estimate that about 50 % of published papers in the field are based on fortuitous chance correlations.

After features have been identified, each must be assigned a probability value describing whether the feature contributes to the activity of the molecule. This problem is complex and a rigorous treatment is given by Golender & Rosenblit (1983).

8.1.4 Emerging pathways for rational design of ligands

Rational methods for ligand design can be arbitrarily partitioned into three practical divisions, depending on whether the molecular structure of the binding site is known or not. If the site structure is not known, the designer has to work blindly and use statistical deductions to decide whether he is on the right track in producing potent congeners. In this case the design criteria will be based firstly on the correlation between ligand potency and physicochemical parameters in the molecular congeneric series; secondly, some deduction of structural parameters based on statistical correlations will become apparent. Thirdly, if the structure of the binding site is known, then much of the molecular jigsaw is assembled and all we need to do is construct a ligand to fit the geometry and the chemical properties of the site-points. A rough measure of binding affinity can then be estimated from the calculated interaction energy between the ligand and its binding site. Statistical trends and structural constraints can be used rationally to limit the search tree for possible congeners. Thus in a progressive research program the ligand designer can learn to improve his efficiency in designing better compounds by acquiring rules from his expanding knowledge base. This type of evolving answer to a problem is an ideal candidate for solution by an algorithmic artificial intelligence system (see section 9.2.4). Expert systems can be used to suggest synthetic pathways for making the putative ligand molecule from a database of readily available chemical starting points and a rule base of synthetic reactions (see section 9.2.1). Computer-aided molecular synthesis is rapidly becoming established as a further factor in design strategy because it enables one to assess the feasibility of synthesis in a choice of compounds for an initial training set for feature selection.

This chapter will be devoted predominantly to the problem of designing a ligand to fit a known structural binding site. Attention will be drawn to the development of very sophisticated computing methods as an adjunct to the design problem. Where available, these methods reduce the drudgery in design and are elegant examples of the application of acquired knowledge about molecular interactions between ligands and their receptors.

8.2   Ligand design to fit a known structural site

This type of ligand design is largely based on model building procedures. The first example of ligand design by this method was the deceptively simple, but foundational, work of Goodford and his colleagues at the Wellcome laboratories. Their work will be illustrated here because it employs all the logical principles of ligand design for fitting to a structural binding site.

8.2.1   The diphosphoglycerate binding site on haemoglobin

Haemoglobin is a tetrameric protein consisting of two identical α subunits (α_1, α_2) and two identical β subunits (β_1, β_2). The main biological function of haemoglobin is to transport oxygen through the blood system to a tissue and to remove carbon dioxide from the tissue. The haemoglobin molecule exists in two main structural states. In the deoxy form a cleft is formed between the β_1 and β_2 subunits. This structural cleft binds 2,3-diphosphoglycerate (DPG) and stabilizes the deoxy conformation; in the oxy form the β_1 and β_2 subunits are shifted and the DPG binding crevice is closed.

Once a binding site has been identified, the first step in designing a molecule to fit it is to survey the geometry and molecular features that compose the site. This survey can only be achieved satisfactorily if good x-ray crystallographic coordinates are available for the site. The structure resolution must be good enough to identify the coordinate positions of all the heavy atoms. Beddell *et al.* (1976) used the crystallographic coordinates of haemoglobin and coordinates from a crystallized complex of deoxy haemoglobin with DPG. The complex defined the exact location of the binding site. A structural model for the site can then be built to scale. Molecular graphics has now overcome the need to construct a physical model; a graphical representation of objects can be built more accurately and manipulated more quickly. A schematic representation of the DPG binding site is shown in figure 8.2. The β_1 and β_2 subunits are positioned beside each other by a two-fold rotation. In the oxyhaemoglobin form the two subunits slide apart slightly, whereas in the deoxy conformation the subunits are directly adjacent. At the DPG binding site there are four basic groups contributed by each subunit: the terminal amino group Val 1, His 2, Lys 82, His 143. There are also two more basic residues nearby, Lys 144 and His 146. This whole site region will therefore generate a molecular electrostatic potential which is strongly positive and attractive towards anions. A comparison of crystallographic coordinates between deoxyhaemoglobin and the ligand complex shows that DPG induces small conformational changes at the binding site (Arnone, 1972). The most notable change occurs in the A helix. The terminal phosphate of DPG attracts the NH_3^+ of Val 1 and His 2 bringing the A helix closer to the E helix, near to the EF join. The DPG molecule, being asymmetric, can bind in two orientations with its carboxylate group attached to Lys 82 of β_1 or β_2; the unbridged lysine appears to have an anion bound to it. DPG lies in the gap along the two-fold axis between the subunits. Each non-ester linked

Figure 8.2. The DPG cleft between two β subunits of deoxyhaemoglobin.

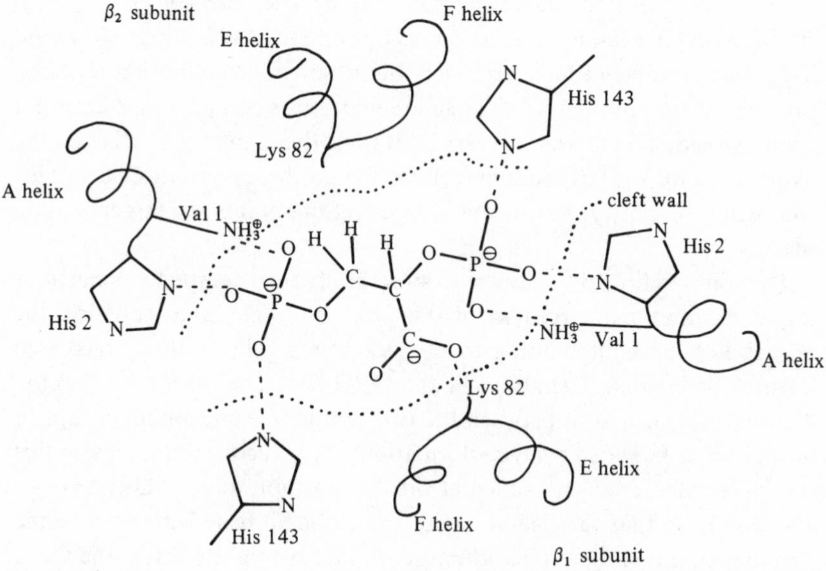

phosphate oxygen atom is salt-bridged to either NH_3^+ Val 1, or His 2 or His 143; the carboxylate is linked to Lys 82 making seven salt bridges in all. An important point to note from this geometrical survey is that the site has some flexibility, and movement between the site-points occurs as the ligand is bound. Therefore, in designing a molecule to fit a geometrical set of site-points, a small amount of positional freedom may be expected and we should not condemn a putative suitable geometry if it is slightly larger than the prospective ligand geometry.

After a careful survey of the binding site, the designer can make all the interatomic distance measurements between the possible site-points in preparation for constructing a molecule with the required dimensions to fit the site. Furthermore, for each possible site-point a note must be made of the distinguishing feature of the point, ie whether it is a hydrogen-bonding acceptor or donor, an atom with a high residual charge or whether a particular region of the receptor is likely to be hydrophobic.

8.2.2   Construction of a ligand to fit the DPG site

The dimensions and the chemical groupings at the site determine the ligand structure that can be made to fit. However, one other important factor has to be borne in mind. If the site lies intracellularly, the ligand must have sufficient lipid solubility to cross the cell membrane. This restriction rules out the possibility of developing a charged molecule as a putative ligand structure. Where the binding site is located extracellularly, then charged groups may be considered for building the ligand.

In carrying out the survey of the binding site, Beddell et al. (1976) restricted their examination to the site-points for DPG alone. It would have been possible to have extended the survey to neighbouring residues, since any set of site-points that include some points occupied by the natural ligand would, if a compound were built to fit them, inhibit binding by the natural ligand. We have seen in section 5.5 that there is strong evidence for competitive inhibitors having differing sets of site-points at a target binding site.

The construction of a ligand with structurally rigid components can limit the conformational freedom within the molecule and enhance the probability for tight binding to the site-points. This strategy has been adopted in building ligands to fit the DPG site. The dibenzylyl skeleton (figure 8.3(b)) would fit between the two terminal valine amino groups in the preferred extended trans conformation. Two reactive groupings, in this case aldehydes, could be added at the 4,4' position to give dibenzyl-4,4'-dialdehyde so that the ligand could form a Schiff base with both valine terminal amino groups. This structure would bind tightly across the DPG site. The only drawback to this idea was the discovery that the ligand was insoluble in water.

Two methods to increase the aqueous solubility were tried. Firstly, a carboxylate group was added to one of the aromatic rings (figure 8.3(c)) in a position to link with Lys 82 in a similar way to DPG. Secondly, the aldehyde groups were replaced by a bisulphite grouping (figure 8.3(d)). Possible methods of binding to the site are shown in figure 8.4. Finally a composite compound, containing both features (figure 8.3(e)), was synthesized to give the possibility of extensive binding to the DPG site-points.

8.2.3   Testing the putative ligands for biological activity

Model building of ligands can easily become a pipe-dream; the acid test is whether the ligand has any biological activity. *In vitro* tests on

Figure 8.3. Compounds synthesized with the objective of being bound to the DPG site on haemoglobin; (a) DPG, (b) dibenzylyl skeleton, (c) compound 1, (d) compound 2, (e) compound 3.

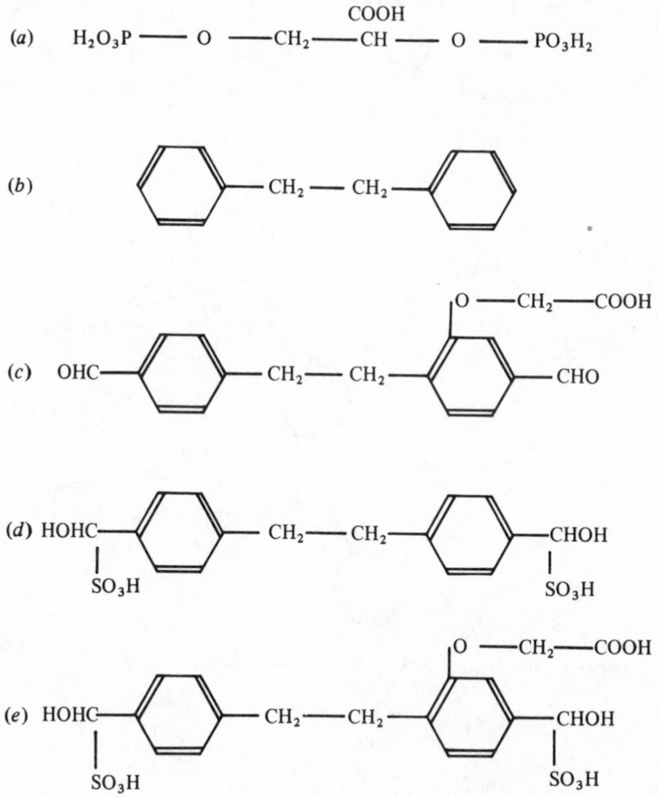

(a)   H₂O₃P —— O —— CH₂—— CH(COOH) —— O —— PO₃H₂

(b)   [benzene ring]— CH₂ — CH₂ —[benzene ring]

(c)   OHC—[benzene ring]— CH₂ — CH₂ —[benzene ring with O—CH₂—COOH]— CHO

(d)   HOHC(SO₃H)—[benzene ring]— CH₂—— CH₂—[benzene ring]—CHOH(SO₃H)

(e)   HOHC(SO₃H)—[benzene ring]— CH₂——CH₂—[benzene ring with O—CH₂—COOH]—CHOH(SO₃H)

Figure 8.4. Schemes for the binding of dibenzylyl derivatives to haemoglobin site-points at the DPG cleft; (a) compound 1, (b) compound 2, (c) compound 3.

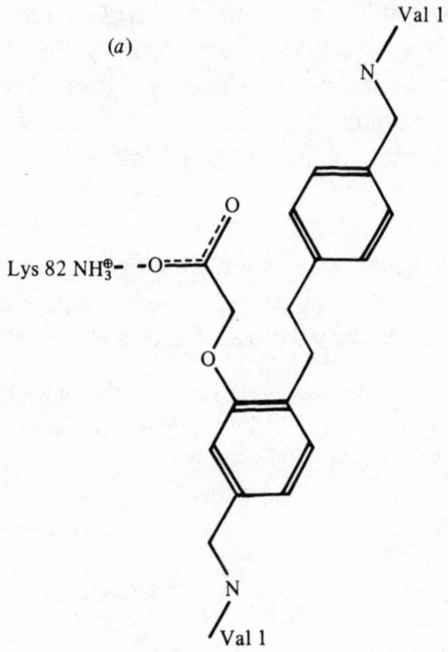

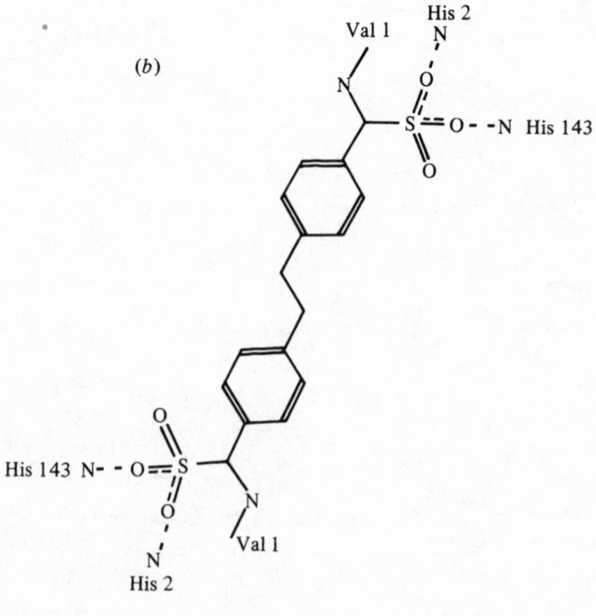

purified receptor material provide the most stringent controlled tests, and thus avoid ambiguity that might arise with cellular or *in vivo* experiments.

The biochemical function of DPG binding to haemoglobin is to decrease the oxygen affinity of haemoglobin. Oxygen affinity can be measured simply by using oxygen dissociation curves, a right shift in the curve indicating a fall in the oxygen affinity. All three dibenzylyl compounds showed right-shifted oxygen dissociation curves and the order of potency was: compound 1 < compound 2 < DPG < compound 3. This order may reflect the number of possible bonds linking the ligand to the DPG site.

Confirmation that the bis-arylhydroxysulphonic acids do bind at the DPG site was provided by NMR studies (Brown & Goodford, 1977). Peaks in the NMR signal associated with the C2 and C4 protons of surface histidine residues could be identified and studied in the absence, and in the presence, of the ligands. Protonation of the histidine ring is perturbed by the close proximity of an anionic group; the positively charged protonated form of the ring is stabilized with respect to the neutral form. pH changes can be used to modify protonation of the ring. DPG and compounds 2 and 3 affect the NMR peaks produced by the C2 and C4 protons. This observation suggests that they act at the same site and that the signal is derived from His 2 and His 143. The long lifetimes of the ligand–receptor complex for compounds 2 and 3 are consistent with a covalent binding mode predicted at the site.

(c)

8.2.4   Ligand binding at genetically modified sites

The site-points for the attachment of DPG to normal human haemoglobin are Val 1, His 2, Lys 82 and His 143. Genetically different forms of haemoglobin exist with modified residues at three of these site-points (table 8.1). Therefore, it may be expected that the binding affinities of the haemoglobins for the ligands would show significant variations from normal human haemoglobin. These affinities were measured by Beddell *et al.* (1979) and a measured value for the free energy of binding, ΔG, was obtained from the relationship

$$\Delta G = -RT \ln K_D \tag{8.4}$$

where K_D is the affinity constant of the ligand for the deoxy haemoglobin and R and T have their usual thermodynamic meaning. Binding to the site-points is by salt bridges and reversible covalent links. Ligand-point–site-point bonding is shown in table 8.2 for DPG, compound 1 and compound 3. If the affinity of the ligand for its binding site can be expressed as a linear summation of the number of ionic and covalent bonds by the equation

$$\Delta G = a_1 N_I + a_2 N_C + a_3 \tag{8.5}$$

where a_1, a_2, a_3 are regression coefficients and N_I and N_C are respectively the numbers of ionic and covalent bonds, then it should be possible to calculate the free energy of binding, $\Delta G_{calculated}$. The coefficients have calculated values of $a_1 = -3.14 \text{ kJ mol}^{-1}$, $a_2 = -6.78 \text{ kJ mol}^{-1}$ and $a_3 = -8.29 \text{ kJ mol}^{-1}$. The relationship between $\Delta G_{measured}$ and $\Delta G_{calculated}$ is linear, passing through the origin and with a slope of unity. It is easy to criticize the crudity of this approach since it neglects van der Waals interactions, hydrophobic interactions and conformational changes; nevertheless, these results do suggest that this general method may be useful for predicting relative affinities of compounds at binding sites

Table 8.1. *Naturally occurring genetic modifications to the DPG binding site of haemoglobin*

Haemoglobin type	Residue modification	Normal human haemoglobin residue
Raleigh	β acetylalanine 1	β val 1
Diabetic (HbA$_{IC}$)	β val 1 + glucose	β val 1
Foetal minor	β acetylvaline 1	β val 1
	β ser 143	β his 143
Foetal major	β ser 143	β his 143
Horse	β gln 2	β his 2

Data taken from Beddell *et al.* (1979).

Table 8.2. *Ionic and covalent binding to genetically modified site-points for DPG on haemoglobin*

	Site-point	Normal human	Raleigh	Diabetic	Foetal minor	Foetal major	Horse
DPG	Val 1	I				I	I
	Val 1	I				I	I
	His 2	I	I	I	I		
	His 2	I	I	I	I		
	Lys 82	I	I	I	I		
	His 143	I	I	I		I	I
	His 143	I	I	I		I	I
Compound 1	Val 1	C			C	C	C
	Val 1	C			C	C	C
	Lys 82	I	I	I	I	I	I
Compound 3	Val 1	C			C	C	C
	Val 1	C			C	C	C
	His 2	I	I	I	I	I	
	His 2	I	I	I	I	I	
	Lys 82	I	I	I	I		
	His 143	I	I	I		I	I
	His 143	I	I	I		I	I

I = ionic bonding, C = covalent bonding.
Data taken from Beddell *et al.* (1979).

with known structures. Any deficiencies in this simple model could be rectified, perhaps, by improved energy calculations.

8.3   Ligand design in the absence of information about site structure

In the case where the molecular geometry of the binding site is known in detail, as described in the previous section, ligand design is analogous to the assembly of a three-dimensional jigsaw puzzle. Geometrical pattern and chemical pattern, furnished by moieties at the site-points, provide clues for reasoning about constructing a molecule to give a good fit to the site. The shape of the accessible surface of the site provides the constraints for geometrical fitting of the ligand. Chemical groupings complementary to the site-points are suggestive of suitable putative ligand-points. Once an array of ligand-points in three-dimensional space has been established, all that needs to be done is to hang the chemical ligand-points onto an appropriate molecular spacer skeleton and a putative ligand has been constructed. The principle clues for reasoning are provided by geometrical and chemical patterns that are easily recognized by the medicinal chemist.

Where the site structure is not known, these clues for reasoning are absent; ligand design by rational principles is therefore much more difficult. In this case, the chemist has to deduce possible structural and chemical properties of the binding site by statistical analysis of molecular and geometrical parameters. These parameters are obtained from a large number of ligands that produce the response being examined. Chemometrics is the discipline that examines these statistical relationships. An intermediate goal in ligand design, for this type of problem, is to obtain an approximate map of ligand-points complementary to the unknown receptor map; this would indicate the positions and chemical properties of possible site-points. We shall only consider in this section the case where the response is a single, clearly defined variable. There are responses, such as experimental models for the evaluation of drug action against schizophrenia, that are multivariate; in that case special techniques are needed to evaluate the goodness of the test response as a model in themselves before any QSAR studies can be undertaken (Mager, 1985).

8.3.1   Chemometrics

The activity of a drug molecule at its binding site is determined by many parameters. Which, and how many, parameters adequately account for the activity? Chemometrics attempts to sort and classify, by pattern recognition, the parameters that contribute to activity. Many of the principles of multivariate analysis are employed in this process. To begin with, let us assume that a group of molecules having a particular activity, say an antagonist action, are members of a congeneric series; that is, they are composed of some basic structural unit with only substituent variations at particular positions. Antagonist activity is a single dependent variable which can be measured unambiguously. Suppose that a number, n, of compounds have been synthesized and their activity measured; each compound is known as an object. Each object will have associated with it a number, p, of independent variables. These independent variables may be geometrical or structural properties of the object, eg pK_a, dipole moment, $\log_{10} P$ ($P = $ oil/water partition coefficient), σ (Hammett constant), etc. A table can be constructed with each object as a row and the values of the independent variables arranged in the columns. This table forms the data matrix X; a corresponding table can be built for the dependent variables – the data matrix Y. In the simplest type of study, the data matrix Y has only one column; in more complex analyses there may be more than one dependent variable. For example, for a series containing drugs of mixed function, we might have measured affinity, antagonist activity and agonist activity, perhaps on different test systems, as dependent variables. From the data matrix X we want to extract those features which adequately describe the matrix Y. Extensive reviews of the techniques of pattern recognition in drug design can be found in Stuper, Brugger & Jurs (1979), Wold, Dunn & Hellberg (1984), Wold *et al.* (1984), Clementi (1984), Kowalski (1984), Lewi (1984) and Franke (1984).

Selection of an initial set of compounds, called the training set, is crucial for the design of the experiment. If there are five substituents to be added at five different positions, there are 3905 possible compounds to be synthesized (equation (8.2)). This is too large a number to be tested; we may only have a budget to synthesize 30 compounds. How shall we select, from all possible compounds, 30 that would be representative in a multivariate problem? A bad experimental scheme, and thus one to avoid, would vary only one substituent at the five possible positions in the mistaken belief that an immediate clue would be provided about a column in X that would determine the entry in Y. In multivariate analysis multiple sites should be varied in a single compound, therefore a random allocation of substituents at each position should be attempted. This can be simply achieved using an integer random number generator for substituents labelled 1–5 for each of the five positions until we have the desired number, say 30, compounds.

8.3.2 *Multivariate analysis*

Suppose that we can initially classify the compounds (objects) into a single class, then this class might be a congeneric parent structure with only substituent variations distinguishing each object. Let there be n objects. For each object we define a set of p independent variables; these variables can be expressed as numerical properties of each object. Within each variable the same scale is used to describe the objects. The objects with their accompanying variables describe the data matrix X which can be written as

Variables

$$
\begin{array}{cccccc}
 & X_1 & X_2 & \ldots & X_j & \ldots & X_p \\
1 & x_{11} & x_{12} & \ldots & x_{1j} & \ldots & x_{1p} \\
2 & x_{21} & x_{22} & \ldots & x_{2j} & \ldots & x_{2p} \\
\text{Objects} \vdots & \vdots & \vdots & & \vdots & & \vdots \\
i & x_{i1} & x_{i2} & \ldots & x_{ij} & \ldots & x_{ip} \\
\vdots & \vdots & \vdots & & \vdots & & \vdots \\
n & x_{n1} & x_{n2} & \ldots & x_{nj} & \ldots & x_{np}
\end{array}
$$

where each object is described by a row of variables, and each variable is a column of objects. We therefore have an $n \times p$ data matrix X, and x_{ij} is the jth variable of object i. Each row or column can then be handled as a vector. The mean value of variable i is given by $\bar{x}_i$

$$\bar{x}_i = n^{-1} \sum_{r=1}^{n} x_{ri} \tag{8.6}$$

and its variance s_{ii} is

$$s_{ii} = n^{-1} \sum_{r=1}^{n} (x_{ri} - \bar{x}_i)^2 \tag{8.7}$$

If we wish to estimate the covariance s_{ij} between two variables i and j then

$$s_{ij} = n^{-1} \sum_{r=1}^{n} (x_{ri} - \bar{x}_i)(x_{rj} - \bar{x}_j) \tag{8.8}$$

The covariance matrix S for p variables is

$$S = (s_{ij}) \tag{8.9}$$

formed from all elements generated by equation (8.8). It must be remembered that the covariance s_{ij} is dependent on the scale and origin of the variables, but this can be standardized by autoscaling to a unit variance with a mean of zero. Autoscaling of element x_{ij} to x'_{ij} is achieved by

$$x'_{ij} = (x_{ij} - \bar{x}_j)/S_j \tag{8.10}$$

The sample correlation coefficient, r_{ij}, between two variables i and j is given by

$$r_{ij} = s_{ij}/(s_i s_j) \tag{8.11}$$

and is invariant of scaling: the sample correlation matrix R is

$$R = (r_{ij}) \tag{8.12}$$

Dependent variables are contained in a data matrix Y and analogous operations can be performed on contents of the matrix Y. The objective of multivariate analysis is to relate the data contained in X to that contained in Y. This analysis proceeds on the basis that the variables are linearly related and can be expressed by a suitable linear combination of variables. Not all the independent variables in X may be needed to describe the relationship with a particular dependent variable in Y; X may contain redundant information. Two problems therefore arise and are of particular interest in chemometrics. Firstly, we need to examine the relationships between X and Y as a multiple linear regression problem. Secondly, we need to extract the principal components that account for the variance in the data matrix and discard some of the variables in X that do not contribute significantly to the relationship between X and Y.

8.3.3 Multivariate regression analysis

Let the data matrix Y of responses for each object contain r dependent variables, and the data matrix X contain p independent variables; each matrix has n objects. Each object i is denoted by

$$y_{i1} \cdots y_{ir}; \quad x_{i1} \cdots x_{ip}$$

For each dependent variable y_k a linear model has to be fitted of the form

$$y_k = b_{1k}x_1 + b_{2k}x_2 + \cdots + b_{pk}x_p \tag{8.13}$$

where the parameters $b_1, \ldots, b_p$ are regression coefficients. This equation has to be fitted to the n objects for each variable y_k and is usually carried out by the method of least squares. For each data point i

$$y_{ik} = b_{1k}x_{i1} + b_{2k}x_{i2} + \cdots + b_{pk}x_{ip} + e_i \tag{8.14}$$

where $i = 1, 2, \ldots, n$ and the system of n linear equations has to be solved with the condition that

$$\sum_{i=1}^{n} e_i^2$$

is minimized. If the regression constant is required it may be extracted by inserting a dummy variable of 1.0 into the data matrix so that a coefficient is generated for this term; this coefficient then becomes the regression constant.

8.3.4 Principal component analysis

The objective of principal component analysis is to reduce the number of variables in the data matrix, by selecting linear combinations of component variables, whilst retaining as much information as possible about variables that may be correlated. Therefore, the procedure attempts to expose linear correlations where they exist between the variables. If there are p variables, we wish to determine a number k of variables, where $k \leqslant p$, which are the principal components that account for a given percentage of the variance. The number of principal components, k, is an abstract mathematical identity and does not indicate which variables in p dimensions are the components. The actual variables have to be selected by a process of elimination.

Principal component analysis can proceed from the covariance matrix (autoscaled) S, or the correlation matrix R. The first principal component is a linear combination of the p variables

$$Y_1 = a_{11}X_1 + \cdots + a_{p1}X_p \tag{8.15}$$
$$= \mathbf{a}'_1 \mathbf{x}$$

whose sample variance

$$S^2_{Y_1} = \sum_{i=1}^{p} \sum_{j=1}^{p} a_{i1}a_{j1}s_{ij} = \mathbf{a}'_1 \cdot \mathbf{S}\mathbf{a}_1 \tag{8.16}$$

is greatest for all coefficient vectors that are normalized so that

$$\mathbf{a}'_1 \cdot \mathbf{a}_1 = 1$$

Similarly, the second principal component is given by

$$Y_2 = a_{12}X_1 + \cdots + a_{p2}X_p \tag{8.17}$$

whose coefficients are subject to the constraints that $S^2_{Y_2}$ is a maximum and

$$\mathbf{a}'_2 \cdot \mathbf{a}_2 = 1 \quad \text{and} \quad \mathbf{a}'_1 \cdot \mathbf{a}_2 = 0$$

Each principal component explains progressively a smaller proportion of the sample variance as the number p of components is reached. This explanation is taken from Morrison (1978); alternative methods of expression are given by Mardia, Kent and Bibby (1979). Standard computer packages are available for computing the principal components from the data matrix (SAS, 1982; GENSTAT, 1983; Alvey, Galway & Lane, 1982).

Principal component analysis can be interpreted geometrically. Each object has p coordinates in p-space. A plot of the objects autoscaled will produce a swarm of points in p-space with the mean at the centre of the coordinate system. For three-space, a swarm of points may be illustrated as in figure 8.5. The principal components Y_1, Y_2, Y_3 are drawn passing through the origin. The component Y_1 is the first principal component since it explains the largest proportion of the total variance. The second principal component, Y_2, lies at right angles to Y_1 in the direction of the second largest proportion of variance. Similarly, Y_3 is perpendicular to the plane defined by the lines Y_1 and Y_2. For the approximate ellipsoidal distribution illustrated in the figure, Y_1 corresponds to the major axis of the ellipsoid and is described by the regression line fitted to the swarm of points. The principal component Y_2 is the regression when Y_1 has been stripped from the points and is a minor axis of the ellipsoid. Similarly the third principal component Y_3 is the other minor axis. If two successive components account for equal proportions of the total variance, the distribution of points in that plane will be enclosed in a circle. Furthermore, if all principal components are equal, the points will be isotropically distributed.

Each principal component contains the coefficients expressed as a linear combination of the p variables (equation (8.15)). Not all variables contribute significantly to the variance in the matrix S; a number, $(p-k)$ variables, must be rejected so that only k variables are retained to describe the variance adequately. A $p \times p$ matrix is formed from the coefficients of each principal component as column vectors, with the variables as rows. The principal component p contributes the least variance to S; the coefficient with the maximum absolute value in column p is identified and the corresponding variable discarded. Column $p-1$ is considered next and the variable with the highest coefficient, which has not previously been discarded, is selected to be eliminated. The procedure is repeated until k variables remain. Thus the original data matrix can be reduced from p

Figure 8.5. An ellipsoidal swarm of data points in three dimensions; the variables are X_1, X_2, X_3 and the principal components are Y_1, Y_2, Y_3.

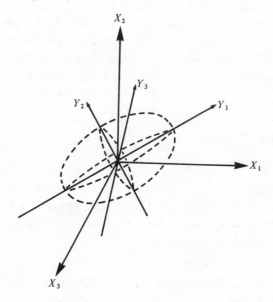

variables to k identified variables that contribute significantly to the variance.

Principal component analysis is of great value in ligand design where the molecules (objects) in a training set contain many descriptors (variables) and the designer seeks to find an underlying trend in the data matrix that can then be correlated with biological activity.

8.3.5   Descriptors of molecular structure

The data matrix used in chemometrics consists of numerical elements assigned to each variable for every object. If the multivariate changes introduced by each substitution are single atom replacements, then little overall change in molecular shape, at least in terms of the bond skeleton, is introduced. However, the substitution of small atomic assemblies (moieties or groups) will undoubtedly change the shape of congeners. How are shape changes to be represented in the data matrix? The objects of the data matrix may only show a loose structural relationship and may not form a strictly congeneric series; in this case can we elaborate a description of the geometry that can be encoded into a data matrix? Although these two questions are related, they reflect different levels of complexity in the problem of molecular representation in multivariate regression analysis.

The example illustrated in figure 8.1 shows the parent molecule of a congeneric series for which there are five positions for substitution and five possible substituent groups for each position. Let there be p substitution sites $(1, \ldots, j, \ldots, p)$ and m substituents $(1, \ldots, k, \ldots, m)$ then any compound i can be represented by an $m \times p$ matrix where the element jk is assigned a value a_{ijk} of 1 or 0 depending on whether substituent k is found at position p or not. The relationship between geometry and biological activity then has two structural components, μ for the parent structure and a contribution Z_{jk} from substitution. Empirical studies of structure–activity relationships suggest an equation of the form

$$\log_{10} A_i = \mu + \sum_{j=1}^{p} \sum_{k=1}^{m} a_{ijk} Z_{jk} \qquad (8.18)$$

where A_i is the measured biological activity of compound i. This equation can be solved by linear regression for n compounds, provided that the distribution of groups in the compounds is randomly allocated in a well-designed experiment. The minimum value, n, of compounds needed to solve the system of linear equations is

$$n = 6 + mp \qquad (8.19)$$

and in the example of figure 8.1, n would be 31. The procedure outlined here is that of Free & Wilson (1964). In practice, the utility of the method relies heavily on the choice of substituent types used in a modification of the lead compound. Furthermore, the method is restricted to a congeneric series.

For structurally diverse compounds, a different approach to structure–activity relationships is needed. A rational test synthesis procedure cannot be envisaged at the start of the analysis; raw data has to be extracted from the literature and other sources. Molecules that fall into a particular activity class need to have their molecular geometry expressed in a form suitable for analysis by computer. The objective is to map regions of three-dimensional space that are correlated with activity; the ligand-points generated by mapping may then be used to construct new parent molecules. Each molecule can be described by a connectivity table for its structure and a distance matrix for the geometry. The distance matrix is a two-dimensional representation of interatomic distances and does not distinguish between optical isomers; the matrix has been described in detail in section 4.2. Connectivity tables can be partitioned into submatrices to describe molecular fragments. This property is used extensively in searching databases of molecular structures for the occurrence of fragment types (see section 2.5). The connectivity tables and the distance matrices contain all the information needed to describe and compare the geometry of molecules in the training set for which structure–activity relationships are required. A full atom-by-atom comparison is rarely practical because the search procedure is combinatorial and rapidly exceeds available computing time. A search limited to carefully selected substructural fragments in the set of molecules can be a feasible solution for finding fragments with a common geometrical positioning. This limitation of precise structural fragment comparison can be partly overcome if fragments have similar known chemical properties. For example, hydrogen-bonding groups might be considered as a substructural class divided into three groups, pure donors, pure acceptors, or mixed function groups able to be donors and acceptors (Danziger & Dean, 1985). Structural correspondence between molecules can then be assessed on the numbers of atoms, or groups, matched and their superimposed geometries. Figure 8.6 shows a match for six hydrogen-bonding moieties in saxitoxin compared with tetrodotoxin. The atoms are matched to a tolerance of 0.4 Å as an rms of the difference distance matrix for the six labelled atoms. If we have n molecules to compare in a pairwise manner, the number C of combinations necessary for a complete analysis is

$$C = n(n-1)/2 \qquad (8.20)$$

These combinations can be expressed in an $n \times n$ matrix M in which only the upper, off-diagonal, elements need be considered. Each element of M has associated with it a distance matrix of matched atoms. Similarities in the geometry between elements of matrix M could readily be identified and perhaps classified into geometrical groups of ligand-points.

8.3.6   Molecular descriptors in linear free enthalpy relationships

The biological activity of a molecule is described only in part by its three-dimensional geometry. The geometry represents the specific fitting of ligand-points to site-points as a matching between sets of points in space. Whilst drug–receptor interaction has this component as an essential ingredient, there is also the contribution from the binding energy between the ligand and its site that is electronic and hydrophobic in nature. Furthermore, if the ligand is to be a useful drug molecule, one has to take into account those factors which determine the transport of the molecule from its place of entry to its site of action in the body. Consider the concentration of a drug immediately after intravenous injection to be $[C]$, then if the transport into a particular tissue is controlled by a function u, the concentration $[D]$ of drug in the tissue will be given by

$$[D] = u[C] \tag{8.21}$$

If the overall equilibrium constant between a drug binding to its receptor and the steps leading to the response is K, then the amount of stimulus, S, produced by a concentration $[D]$ is

$$S = K[D] \tag{8.22}$$

Figure 8.6. A stereoscopic drawing showing tetrodotoxin (solid lines) superimposed over saxitoxin (dashed lines) for a six-atom match. The matched atoms are shaded.

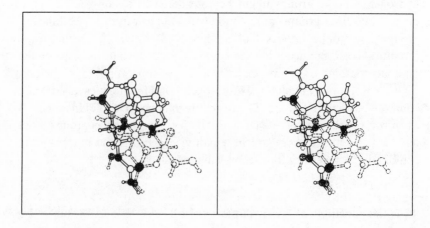

therefore

$$S = Ku[C] \tag{8.23}$$

Suppose S is measured at some constant value, such as 50% of maximum response, and we measure the activity, A, as the inverse of concentration, we have

$$\text{constant} = Ku[C]; \quad A = 1/[C]$$

then

$$A = Ku \times \text{constant} \tag{8.24}$$

Taking logarithms yields

$$\log_{10} A = \log_{10} K + \log_{10} u + \text{constant} \tag{8.25}$$

$\log_{10} K$ can be decomposed into hydrophobic x_1, electronic x_2, and steric x_3 components so that

$$\log_{10} K = \sum_{i=1}^{3} f_i x_i \tag{8.26}$$

The transport function u is predominantly a hydrophobic effect and can be incorporated into x_1, therefore

$$\log_{10} A = \sum_{i=1}^{3} f_i x_i + \text{constant} \tag{8.27}$$

It is found empirically that higher powers of x_i are needed to describe $\log_{10} A$ accurately so that the equation (8.27) can be re-expressed as:

$$\log_{10} A = a + \sum_{i=1}^{3} b_i x_i + \sum_{i=1}^{3} c_i x_i^2 \tag{8.28}$$

Therefore, for any congeneric series of ligands the coefficients in the linear free enthalpy relationship can be obtained by regression analysis (Franke, 1984).

The variables x_i in the linear regression need to be identified. Hydrophobic interactions can, in principle, be computed for ligand interaction with a known structural site in terms of cavity surface area (see chapter 3). When the site structure is unknown, formal calculations are not possible and the ligand designer has to resort to an approximation of hydrophobic interactions by using empirically derived lipid–water partition coefficients. Octanol is used to model lipid and the value $\log_{10} P$ is obtained from equilibrium partition values for the concentration of drug in the octanol and water phases. Within a congeneric series, $\log_{10} P$ is observed to change according to the approximate equation

$$\log_{10} P_X - \log_{10} P_H = k\pi_X \tag{8.29}$$

where P_H is the partition coefficient of the parent unsubstituted molecule, P_X is the partition coefficient of the substituted molecule, k is a constant dependent on the solvent partition system, and π_X is a constant related to substituent X; π_X denotes the free energy change introduced by the substituent between the two solvent systems. For octanol/water solvent systems k is set to 1. The substituent constant π_X for fragments in a congeneric series is additive

$$\log_{10} P = \log_{10} P_H + \sum \pi_X \tag{8.30}$$

Thus it is possible to estimate $\log_{10} P$ from $\log_{10} P_H$ and a table of π values for substituent fragments. Moreover, the parent molecule may also be treated as a conglomerate of fragments so that P_H can be evaluated similarly.

An alternative method for calculating $\log_{10} P$ has been proposed by Rekker (1977). This procedure has the advantage that the constant can be obtained directly from structural units considered as assemblies of atoms not related to a congeneric series. Anomalies such as chain branching and polar groups can be accounted for in a hydrophilic term. Thus

$$\log_{10} P = \sum_{i=1}^{n} a_i f_i + 0.289 \sum kn \tag{8.31}$$

where a_i is a numerical factor indicating the presence of a fragment in the molecular structure, f_i is the hydrophobic fragment constant, n is the number of fragments and kn is a key number for correcting anomalies. More recently Leo & Weininger (1984) have produced a computer program to calculate $\log_{10} P$ automatically from a molecular structure code. This program is incorporated in a versatile medicinal chemistry program package (MEDCHEM).

The effects of the electronic properties of a molecule on its biological activity are more complicated to assess, in the absence of a known structure for a binding site, because the electronic effects are structure dependent. In contrast, $\log_{10} P$ is a linear summation function and is not geometry dependent. Therefore a very large number of different electronic substituent constants have been used to model the electronic effect in structure–activity studies; Franke (1984) lists 22 different constants. Selection of appropriate constants to include in a chemometric analysis should be assessed on chemical grounds dictated by the molecular structures of the series of ligand molecules. The only electronic substituent parameter to be discussed here will be the frequently used Hammett σ constant.

Consider the ionization of benzoic acid in water (figure 8.7) with an equilibrium constant K. It is found experimentally that substitution at the m or p positions alters the equilibrium constant according to the general relationship

$$\log_{10}(K_X/K_H) = \rho\sigma \tag{8.32}$$

where K_X and K_H are the equilibrium constants for the substituted and parent molecule respectively, ρ is the reaction constant for the series being studied and σ is the electronic substituent constant. The constant σ indicates whether the substituent is an electron withdrawing group (high positive value) or an electron donating group (negative value). The ionization of benzoic acid is the standard reaction and ρ for that reaction is set to 1. In essence, each substituent X is compared with a hydrogen atom at the substitution position for its value σ. With substitution at the m or p position in a multisubstituted aromatic ring it is found that the following approximate relationship holds:

$$\log_{10} K_X = \rho \sum \sigma + \text{constant} \tag{8.33}$$

so that σ is an additive parameter. Since σ can vary with position of substitution, the parameter ρ must be given a value related to position as well as to the reaction constant.

The Hammett σ constant has proved to be very useful in drug design and many substituent constants have been measured for different reaction types. Nevertheless, despite its usefulness, the parameter is still a blunt instrument and does not predict electronically active regions in a ligand. For an adequate application of the developing theory of drug–receptor interaction, more precision is needed in locating electronically active points on the ligand molecule. For example, the ligand docking at a binding site was shown in chapter 7 to be carefully controlled through long-range electrostatic effects; it was also shown that particular molecular orbitals, HOMO and LUMO, can be profoundly affected by the relative orientations of the docking partners. Quantum-chemical parameters for ligands can readily be computed. Reactivity is determined by HOMO and LUMO for the frontier orbitals (Fukui, 1975). Electron densities on all atoms can be calculated and partitioned into σ and π components; dipole moments and the direction of the resultant dipole can be deduced from the charge distributions. The practical disadvantage of using quantum-

Figure 8.7. The ionization of benzoic acid is used as the standard for determining electronic substituent constants.

$$H_2O +$$ $\longrightarrow$ $+ H_3O^+$

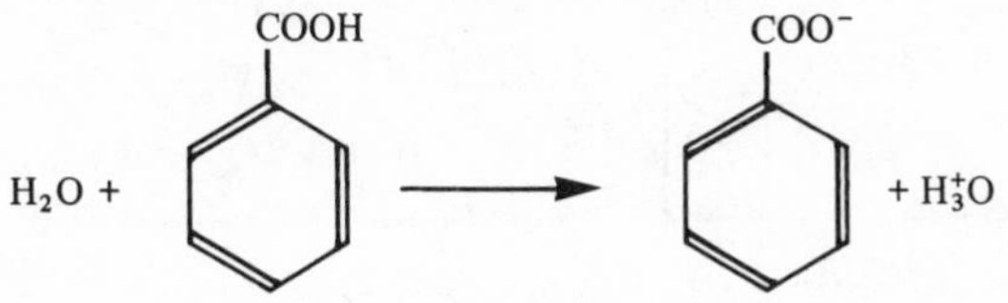

chemical variables in a QSAR study is that all variables need to be calculated for each molecule in the series, and most variables that can be identified are related to one another in a non-trivial way.

Steric properties of a ligand and its substituents are difficult to characterize by a single parameter since each atom is related to its neighbours by three coordinate parameters. Moreover, it can easily be envisaged that steric effects are strongly position-dependent relative to the set of ligand-points; only substituents that collide with the accessible surface of the binding site will have steric parameters showing significant variation with biological activity. The incorporation of conformational flexibility into steric parameters poses a formidable problem. Two parameter types will be outlined here. Firstly, the STERIMOL parameters of Verloop, Hoogenstraaten & Tipker (1976) will be examined since they attempt to characterize the dimensions of a substituent. Secondly, the minimal steric topological difference method (MTD) of Simon *et al.* (1984) will be described because their method attempts to derive parameters from deduced mappings of ligand-points and site-points on the accessible surface of a binding site.

Five STERIMOL parameters characterize the shape of a substituent. A length L depicts the maximum distance from the attachment point of the substituent to the end of the chain including the van der Waals radius of the terminal atom. Perpendicular to the line L are four orthogonal axes B_1–B_4, these describe the cross-section of a box enclosing the substituent; parameters B_1–B_4 are assigned in ascending order of measured widths (figure 8.8). The advantage of STERIMOL parameters is that they can be readily computed and can be given different values for conformational changes. In practice, L, B_1 and B_4, which describe the maximum and minimum dimensions, appear to be the most important of the five

Figure 8.8. The characterization of substituent shape by STERIMOL parameters L, B_1–B_4. Van der Waals radii are used for the calculation.

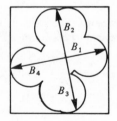

parameters. All five parameters are included in a multiple regression analysis for linear free enthalpy relationships within a congeneric series.

The MTD method is not restricted to a congeneric series of ligand molecules. Each molecule to be investigated is described by a topological graph where the vertices denote atom positions and connectivities represent bonds to produce a network description of each molecule. These topological networks are then superimposed to generate a hypermolecule with minimal topological difference. This hypermolecule is built round a ligand with the highest biological activity – the standard molecule. The hypermolecule lies within the walls of the receptor cavity. Each vertex of a test molecule can be assigned a parameter $\varepsilon_j = -1, 1$ or 0 depending on whether the vertex lies within the receptor cavity, in the wall of the receptor or exterior to the binding site respectively. Ambiguous assignments in parameter values will obviously take place with flexible regions described by low energy conformations. Parameters ε_j relate to the affinity of the drug molecule for the site. Thus suppose we have a molecule i with n vertices, the minimum topological difference MTD_i is given by

$$MTD_i = S + \sum_{j=1}^{n} \varepsilon_j x_{ij} \qquad (8.34)$$

where S is the number of vertices of the standard molecule, ε_j are vertex parameters and x_{ij} denotes whether molecule i has vertex j and is given a value 1 or 0 accordingly. The parameter S is defined as

$$S = \sum_{j=1}^{n} \varepsilon_j, \quad \text{for all } \varepsilon_j = -1 \qquad (8.35)$$

The MTD values can be easily calculated and FORTRAN programs are given by Simon *et al.* (1984).

8.3.7 *Optimization of molecular design from chemometric data*

Suppose that a small number of compounds have been synthesized; each shows some biological activity and chemometric studies have identified the principal components that lead to a particular class of activity. The question that now faces the ligand designer is, how can he optimally design the most potent members of the series? In essence this is a standard problem in combinatorial optimization and if the structure–activity relationship is expressed by a linear regression then the solution can be achieved by appropriate linear programming.

Consider the example outlined at the beginning of this chapter (section 8.1.2) where, in a molecular series, five positions on the parent molecule can be substituted and we limit ourselves to five substituent types. In this case there are 3,905 possible compounds that can be synthesized. One compound will give maximum biological activity; which will it be and how can we find it through the minimum number of syntheses and tests? An efficient solution obviously has clear commercial implications.

The combinatorial problem can be considered geometrically; the set of all solutions may be represented by the vertices of a convex polytope defined by the linear constraints. From any starting position, that is, a vertex on the polytope, we need to find the shortest route along the edges to the goal vertex with the highest associated activity. In our problem the goal vertex is not known explicitly, it has to be found by trial. A full discussion of this type of problem is given by Papadimitriou & Steiglitz (1982). We shall limit our discussion to the simplex algorithm. If the regression equation describing the structure–activity relationship has two independent variables x_1 and x_2, the polytope mapping can be drawn as a set of points as in figure 8.9(a). Each point is a vertex and has a value for the activity associated with it. Similarly for three independent variables x_1, x_2, and x_3, we can draw a three-dimensional figure where the faces form the convex hull of the points (figure 8.9(b)). Higher n-dimensional polytopes can exist for n independent variables but they cannot be represented by a drawing.

The two-dimensional simplex is the simplest to understand but even then there are pitfalls for the unwary. Variables x_1 and x_2 determine the activity, Y, of the molecule. If the initial training set contained i molecules (say 10) in a well-designed experiment, we would be able to plot the positions of these points in x_1–x_2 space and assign a value Y_i to each compound. Theoretical position values in x_1–x_2 space for other molecules in the test set can be assigned from the chemometric parameters and plotted; what is missing is their Y_i value since these molecules have neither been synthesized nor

Figure 8.9. (a) Polytope mapping in two dimensions; (b) polytope mapping in three dimensions showing vertices and faces of the polyhedron.

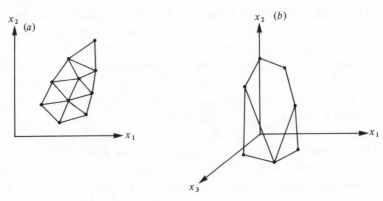

tested. An optimization procedure to reach the goal solution of the compound with the highest activity value, Y_{max}, evolves by a sequential step procedure governed by five rules.

(1) Take three data points from the training set; they form the vertices of the triangle A, B, C (see figure 8.10). Care is necessary in the initial selection of the three points because if they are too far apart in x_1-x_2 space the optimum point may lie within the triangle and rule 2 would lead nowhere. An approximately equilateral triangle is a sensible start.

(2) Take the point of the triplet with the lowest associated Y value, say (A), and reflect this point across the line BC to a position D. From the test set select a molecule with coordinates x_1x_2 closest to D, synthesize this molecule and determine its activity Y_D. This point forms the new simplex BCD. Again the procedure is repeated to find point E and a third simplex is formed. The procedure continues as more potent compounds evolve. The search stops when no more potent compounds can be constructed with parameters x_1 and x_2. In practice the simplex procedure is a gradient search method; if we are looking for increased activity we are measuring graphically $\partial Y/\partial x_1$ and $\partial Y/\partial x_2$.

Figure 8.10. Scattergram of points for the training set and some points calculated for unsynthesized compounds. Points A, B, C form the initial chosen simplex. In this two-dimensional simplex $Y = f(x_1, x_2)$. If Y_A has the lowest value then point D is the nearest point to A reflected through BC. The second simplex is triangle BCD. If Y_B is the lowest value of the second simplex, the point E is selected and triangle ECD forms the third simplex; and so on.

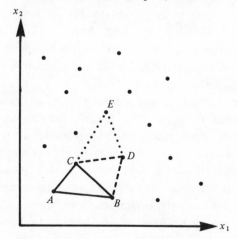

With all gradient methods, pitfalls can occur where the gradient moves onto a ridge or reaches a local but not a global minimum. Rules (3), (4) and (5) are designed to cope with these problems and are extensively explained by Deming (1984); he also provides the equations for optimization by an n-dimensional simplex.

Optimization methods for ligand design are becoming increasingly important in combinatorial problems involving large numbers of possible substituents. Moreover, sequential simplex operations can be carried out when the principal components for activity have been identified for non-linear regression relationships.

8.4   Computer-aided ligand design

The general procedure for designing a ligand to fit a known structural binding site was outlined in section 8.2. The methods explained the basic logic needed to carry out novel design. Manual construction of models of the binding site is a time-consuming business and needs dedicated care, patience and skill to make the structure representative of the true architecture of the site. Once the site has been built as a physical model, visual surveys of the site are needed before the designer can set about making a model molecule with a complementary structure to fit the site. Advances in the application of computer graphics to ligand design, since the early 1980s, have revolutionized the design process. What may have taken man-weeks of human labour can now be achieved in a few seconds of computer time and with high accuracy.

Computer-aided ligand design is still in its infancy but clear trends in the research field are emerging as foundational blocks (Vinter, 1985b). Developments in networking mean that the designer has instant access to databases and also to programs held in other institutions. Collaboration and sharing of research facilities between groups on different continents is becoming possible through satellite networks. This development is particularly important for commercial users where duplication of effort in software design and usage can be eliminated by tightly planned research and development strategies within an international corporation.

The facilities available to a small team of medicinal chemists designing novel ligands are illustrated in figure 8.11. Specific features now in widespread usage for precisely defined problems in ligand design are shown in figure 8.12 to give an overall picture of how rational procedures for ligand design can be put together almost as a menu. Computing tasks can then be performed without the need for local extensive programming. Ready-made packages for various parts of the menu can be bought from specialist software houses; leasing arrangements can be made for updating and maintaining the software. All the equipment that is needed for a drug design group is the computing hardware (a good-size host computer capable of networking and a raster graphics terminal) and the software packages.

8.4.1 Representation of receptor structure

Many, if not most, macromolecular structures are deposited in the Brookhaven Protein Data Bank as lists of coordinates for the atoms (see

Figure 8.11. General facilities available to a computer-aided ligand design group.

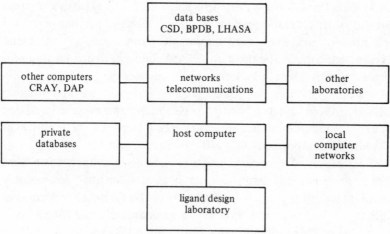

Figure 8.12. Commonly available menus for computer-aided ligand design.

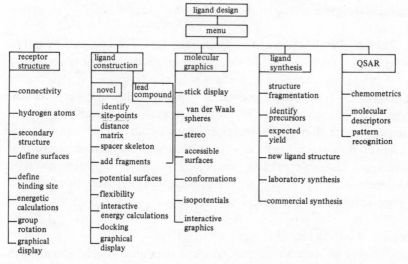

section 2.5.3). This structural data can be bought directly at modest cost. Each magnetic tape of molecular data contains a number of utility routines which are particularly useful for handling problem structures and extracting geometrical parameters. For example, connectivity routines (CONECT) search through the data for interatomic distances which can then be assigned to intramolecular bonds within the macromolecule (Vinter, 1985a). The connectivity program defines the atom–bond topology for all heavy atoms. With this information it is possible to draw the molecule on a graphics display unit since atom type, position and bonding are specified. Hydrogen atom positions are not usually included in Protein Data Bank coordinates; these can be added from private libraries of atom positions for amino acid residues and standard positions expected along protein backbones. Freely rotatable groups such as —OH and —NH$_2$ moieties need to be explicitly defined. Similarly hydrogen atoms for ionizable groups need to be entered with defined pK_a values. Torsion angles from four-atom sequences can be extracted by the program DIHDRL (Abola, 1980). Dihedral angles along the protein backbone define the secondary structure properties. These dihedrals can be plotted as ϕ–ψ plots by the routine FISIPL (Bernstein, 1979).

The extraction of secondary structure from coordinate lists is a non-trivial exercise but this data is frequently needed for analysing possible structural features that are related to α helices or β sheets. An algorithm DSSP can assign secondary structure from coordinate lists (Kabasch & Sander, 1983). This program has a further useful facility in that solvent exposure of the molecular surface can be calculated, a feature which is of great value to the ligand designer since he is predominantly interested in molecular regions on the receptor surface that are accessible to the putative ligand.

The ECEPP program of Browman *et al.* (1975) can be used to extract CNDO (complete neglect of differential overlap) charges for all atoms in a protein. These residual atomic charges form a basic library that can be linked to any set of protein atomic coordinates. Other routines within the program are useful for conformational energy analysis for ϕ–ψ torsion angles and for studying rotational freedom within specified residues.

These routines provide a core of programs for analysing receptor structure. They can then be linked, by calling from the menu, to a molecular graphical display for visual inspection of the structure from any viewpoint and within a graphical window.

8.4.2   Construction of a hypothetical ligand
We will consider the case where the ligand is being designed to fit a

known structural site. There are two possible courses of action here, see figure 8.12. The designer may be at the first stage of the study with no real ideas of a structure to start from, or he may have a lead compound with known biological activity and he seeks to improve upon the activity of the lead structure. In the latter case he may jump down the decision tree of the menu to an appropriate section to continue the development. If there is no lead structure, preliminary information about the binding site is required. The site needs to be defined so in the first instance, hopefully, there will be crystallographic information from substrate–bound cocrystals to define the site unambiguously if the protein is an enzyme; failing this, there might be spectroscopic information that helps to locate particular residues involved in the biochemical function of the enzyme. Further educated guesses at functional binding sites may be provided by studying proteins with known related functions. If the protein contains a structurally coordinated metal ion, this often provides a clue to function and the identification of an active region.

In the search for a novel lead structure, once the binding site region has been identified, the site must be surveyed. The more precise this survey is, in terms of structurally associated features, the more chance there will be of reducing the combinatorial problem and of designing a molecule with high specificity and high affinity. Hydrogen-bonding regions, both as acceptors and donors, form specific and geometrically directed site-points. A DM of these site-points is most useful for offering an accurate mapping to possible ligand-point regions (section 4.2). The accessible surface at the site can be mapped out and displayed on the graphics terminal by using the Connolly (1983b) algorithm (sections 4.3 and 4.8.2). If the radius of the probe sphere is 1.4–1.7 Å this surface will approximate to that in which van der Waals contacts could be made with the ligand. Molecular electrostatic potentials from the binding site mapped onto the accessible surface can be colour coded to define visually the regions of potential minima and maxima (see section 3.5). A ligand molecule with a complementary distribution of potential would be expected to bind tightly to the site and in a specific orientation (section 7.5). Similarly, strong hydrophobic regions on the accessible surface can be noted from the residue associated with the surface region, thus hydrophobic ligand moieties might be expected to enhance binding to corresponding regions in the binding site.

Once the site has been surveyed by these standard computing methods it is then possible to begin to construct seriously a novel ligand molecule. Alas, it is not possible to take the site-point DM and search through the Cambridge Structure Database for an approximate geometrical matching; the combinatorial problem is immense. Intuitive knowledge and trial and error methods have to be used by the ligand designer initially to draw a chemical structure to fit the geometry of the site-points. This is the human bottleneck in the rational design of novel ligands. One rapid aid to design is to use a special interactive graphics tablet to draw the molecular structure. This hand drawing is then linked to computer programs which check the drawing for chemical sense and at the same time the routines insert standard bond lengths and bond angles. An example of chemical graphics operations on a hand-drawn formula is given by Allen & Kennard (1983) (figure 8.13). DM calculations can then be performed on this putative ligand structure to compare ligand-point distances with the distance matrix of site-points.

In practice one of the difficulties encountered in the design of novel ligands is the selection of site-points. Studies of drug–receptor interaction frequently show that only a small subset of all site-points are attached to the ligand in the complex (see section 5.5). Therefore, a judicious choice of a subset of site-points is needed to simplify the ligand construction problem. If one of the possible site-points is charged this will function as an anchor point for an oppositely charged ligand-point and enhance the probability of correct docking through electrostatic interactions. Different subsets of

Figure 8.13. Use of an interactive graphics tablet to input molecular structures: (a) hand-drawn molecular structure: step 1, digitization and connectivity checker; (b) incorrect segments indicated: step 2, corrections applied; (c) corrected structure: step 3, standard bond lengths and angles supplied; (d) molecular structure drawn correctly by the computer. (Re-drawn from Allen & Kennard, 1983.)

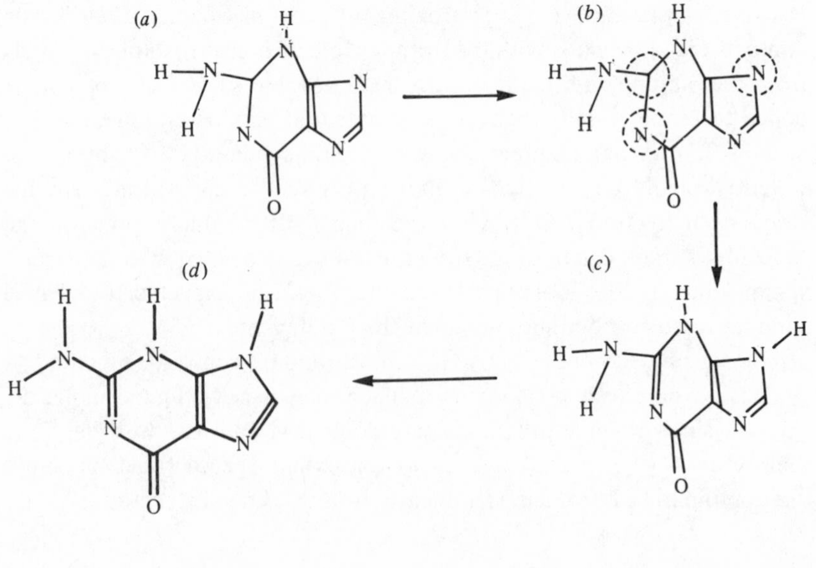

site-points can be used to construct a variety of lead structures. The establishment of a putative ligand structure is similar to having a known lead compound; all that now has to be done is to modify the structure with the hope of making rational improvements to the fit between the ligand and the binding site.

The fit between the putative lead compound and the receptor needs to be tested using energy calculations and interactive docking. Good graphics devices are essential for this. The ligand can be steered into the site and the orientation adjusted; at the same time simple interaction energy calculations can be performed on the ligand and receptor. Intramolecular flexibility can be examined by fixing different rotations round selected torsion angles. Electrostatic potential calculations for points on the accessible surface of the ligand can be compared for complementarity with those on the receptor (Nakamura *et al.*, 1985; Namasivayam & Dean, 1986). This is a useful facility and can draw attention to electrostatic factors that may be involved in high affinity. Furthermore, visual trends revealed by electrostatic potential mapping may provide clues for further modification of the ligand structure.

8.4.3   Molecular graphical displays

Colour molecular graphics plays an essential role in computer-aided ligand design. Two software houses are the major distributors of programs for these design problems: Chemical Design Ltd, Oxford, UK, and Molecular Design Ltd, Hayward, California. In a comprehensive graphical system, such as CHEMGRAF, most graphical options can again be selected by a user-friendly menu.

Molecular drawings can be carried out with various levels of sophistication; combining options is possible and selected manipulation of graphical objects can be achieved by windowing and zooming. The most frugal representation is the stick portrayal of bonds; some assessment of depth can be created by depth cueing and gentle rocking or rotation. Colouring of selected bonds can draw attention to particular regions. A more useful option is to colour the atoms in a ball-and-stick representation or even to provide a space-filling model with shading and highlights so that depth perception is possible. Very realistic displays can be produced by this technique. Windowing options allow the user to restrict his examination to particular molecular regions. Other graphical options might include labelling the atoms, altering the perspective and creating stereoscopic pairs for three-dimensional viewing.

Molecular drawing options form the heart of molecular graphics systems. Nevertheless, what is more useful is the ability, given to the designer by interactive computer graphics, to modify and manipulate picture segments at will. Building primitives in CHEMGRAF include insert atom, add hydrogens, read fragment, join fragment and make bonds. Molecular structures can be modified by other primitives such as delete atom, re-name atom, change bond geometry, delete fragment and break bond. Operations on the primitives can be carried out by moving a cursor around the display screen and using a menu of verbs to perform the task sequentially. A manipulation menu enables the designer to rotate segments round the x-, y- or z-axes, to translate fragments, to perform rotations round specified bonds, select the viewpoint and fit structures together.

Suppose that the ligand being designed has a small number of torsion angles, the user may want to know what is a permissible region for each torsion angle. Conformation analysis and van der Waals energy calculations can be selected to work out the conformational options. Torsion angles can be computed by energy minimization and geometries optimized by molecular mechanics computations. Regions of space can then be allocated to the molecule for torsion angle changes associated with particular groups. These spatial regions are often drawn as nets of arcs in space and are very useful for matching the ligand to the receptor site-points thus giving a graphical display of whether a particular molecule could fit the receptor.

A great variety of techniques are available for drawing molecular surfaces and these have been discussed in section 4.3. Of particular interest are the electrostatic and electron density surfaces (Richards & Mangold, 1983; Dean, 1984); contour mapping onto a three-dimensional surface can show similarities between molecules and complementarity between the ligand and its site.

So far in our discussion of designing ligands to fit binding sites we have thought of the interacting molecules as having static but not necessarily rigid geometrical structures. Molecular dynamics simulations show that considerable movement, particularly in binding sites, is possible (section 4.7). This molecular motion can be studied graphically only if high-performance graphics devices are available. The graphical program CHEMMOVIE is a useful interface from the molecular graphics package CHEMGRAF to these devices. Thus the designer can watch, in a real-time simulation, how the ligand and drug molecule move together in the complex. Observations of this type can reveal the subtle changes of drug-induced conformational changes in the receptor. Techniques to illustrate molecular movement are novel and with time they can be expected to furnish the designer with a wealth of new information about drug–receptor interactions.

8.4.4 Computer-aided ligand construction

Designing a molecule to fit a binding site with high specificity and high affinity is crucial but not the only part of the problem. Is the molecule a feasible chemical structure? Can it be readily synthesized to give an acceptable yield? If these two criteria cannot be satisfied, other molecular designs will be needed.

When ligand design is part of an ongoing project with a known lead compound, then there is usually a large body of information about the synthesis of the lead structure. This information should be stored in a private chemical database. Molecular Design Ltd produce two excellent pieces of software for this purpose. A molecular access system (MACCS) handles graphical input, storage, retrieval and searching of molecular information (Anderson, 1984). In practice, it functions as a management handling system for private chemical databases. Standard chemical search operations are possible, using graphical input, for substructures, stereoisomers and numerical data. A strong feature of the system is a good interactive graphics facility with a menu of verbs to carry out the operations. MACCS can be coupled to a reaction access system (REACCS) which is normally a private database of synthetic reactions. This makes it possible to search for all reactions with a specific product or reactant. Substructural features can be included in the search by reducing them to particular reaction centres and all reactions containing these centres can be listed. Graphical representation of substructures enables all reactions containing substructures to be retrieved. Therefore retrosynthetic procedures, by the method of disconnections, can be scanned as an aid to the synthesis of a putative ligand. They also provide stringent checks for the feasibility of synthesis of a target molecule.

8.4.5 Computer-aided QSAR

Although QSAR programs are the last part of the menu to be described in figure 8.12, they may in fact be used before the start of a molecular graphical ligand design procedure. If the study is not novel, the QSAR of a molecular series will have indicated useful substituents to test by computer-aided drug design. The ADAPT program is an automated pattern recognition system that caters for the needs of most medicinal chemists. Special features and the structure of the program are shown in figure 8.14. Graphical input can be used to create three-dimensional molecular models; structural descriptors are then generated. Other descriptors such as σ, π and additional substituent constants can be added from external databases linked to the program. Pattern recognition techniques available include multiple linear regression, Bayesian discriminants, linear learning machines and classification analysis.

8.4.6   *Inhibitors of angiotensin converting enzyme (ACE)*

A very clear demonstration of the power of computer-aided drug design has been provided by Hassall *et al.* (1984) and Hassall (1985). ACE metabolizes the decapeptide angiotensin 1 to an octapeptide angiotensin 2, which becomes a strong vasoconstrictor. It is believed that inhibitors of the enzyme may be useful in the clinical control of hypertension.

The crystallographic structure of ACE is not yet known; the active site is believed to contain zinc. Derivatives of the dipeptide Ala-Pro are good substrates for ACE. These observations led to the speculation that the dipeptide might fit into the active site close to the zinc atom. Captopril, an analogue of Ala-Pro, was the first inhibitor to be synthesized; the molecule incorporated a thiol group to link it to the zinc atom. Enalopril has a similar structure (figure 8.15). Conformational studies of captopril in solution revealed that the carbonyl oxygen atom of the amide lies predominantly on the same side of the proline ring as the carboxyl group. The dihedral angle of the thiol appears to lie at $\pm 60°$ or $180°$ with respect to the methyl group. This approximate stereochemistry of captopril provided a basis for predicting the positions of three site-points with respect to the ligand-points SH, C=O and COOH. The site-points are of course not points that can be defined exactly but loci at the active site. If hydrogen bonds link the two oxygen atoms to the receptor then the donor atoms

Figure 8.14. Facilities available in the ADAPT program for automated pattern recognition between molecular structures.

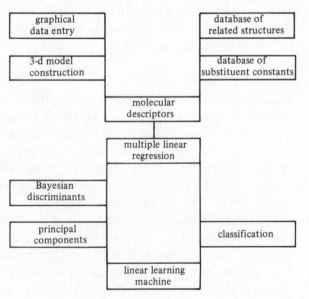

should be located within a specified circular cap radiating from the oxygen acceptor atoms. The position of the zinc atom could be defined similarly. Predicted site loci can then be drawn by computer graphics and the binding site so defined can be used to design other inhibitors.

The site-point loci are believed to be NH_3^+ or $NH \cdot CH(NH_3^+) \cdot NH$ for the positively charged centre reacting with the COOH of captopril, and an NH group on the receptor hydrogen-bonded to the carbonyl oxygen atom with the C=O . . . HN angle of 150–180°; in captopril the thiol group could be rotated round three adjacent torsion angles. A more precise location of the thiol was attempted by building a series of rigid analogues along the three torsion angles and testing for ACE inhibition. The conformations were checked by nmr spectroscopy.

Figure 8.15. Molecular structures of Ala–Pro derivatives designed to inhibit ACE. (Re-drawn from Hassall *et al*., 1984.)

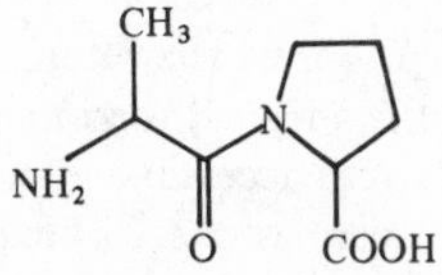

Ala–Pro dipeptide

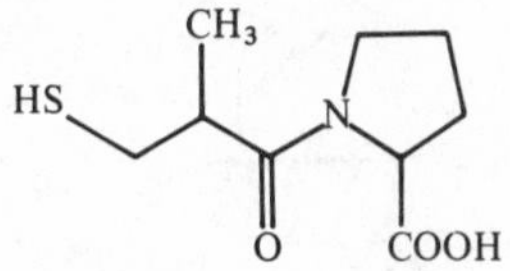

captopril

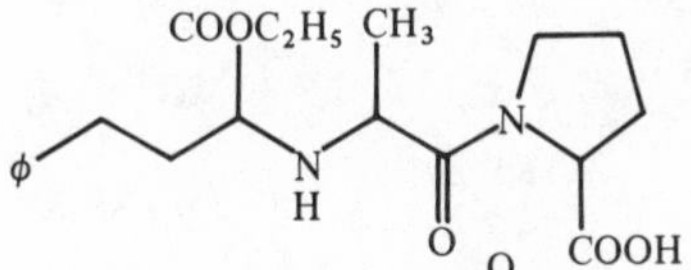

enalopril

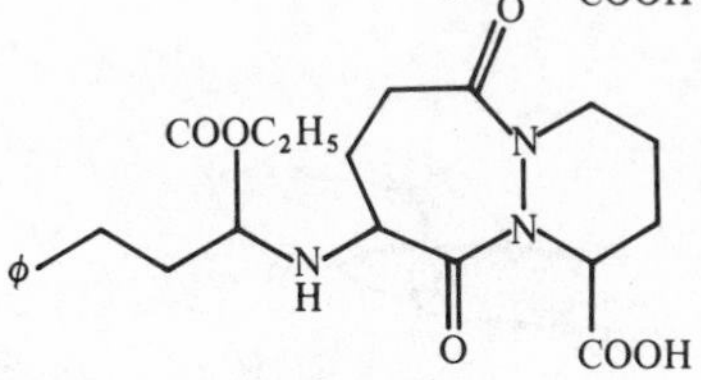

Ro31-2201

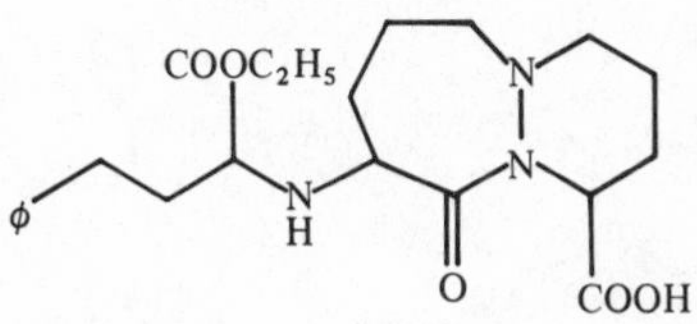

Ro31-2848

An assumption is then made that the most potent rigid analogue has its ligand-points fixed in the active conformation of captopril. This postulate is only valid if a large number of possible conformations have been scanned by a range of rigid analogues. Graphical comparison between captopril and the most potent rigid analogue is then possible by superpositioning the molecular structure of captopril and its analogues and then drawing the region swept out by rotation of the thiol of each compound (figure 8.16). Regions of curved surfaces that intersect may then be regarded as probable conformational positions for the thiol group. Similar procedures can then be applied to the site-point loci corresponding to the two other ligand-points. A final map of all the loci can be produced and for the ACE inhibitor is illustrated in figure 8.17.

Site-point loci mapped in three dimensions can be used to form a model of the site; computer superposition of novel ligands may then be carried out to predict possible good candidates for synthesis and pharmacological

Figure 8.16. The surface swept out by the sulphur atom of captopril: (a) with free rotation about the adjacent torsion bonds $O{=}CH(Me)$ and $HMeC{-}CH_2$; (b) the surface accessible after discounting conformational regions with energies $> 210\,kJ\,mol^{-1}$. (Re-drawn from Hassall et al., 1984.)

(a)

(b)

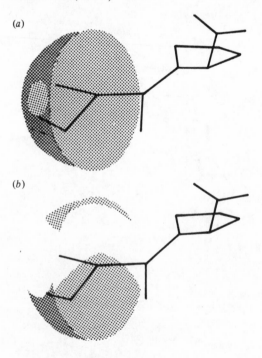

testing. Two other compounds, Ro31–2201 and Ro31–2848, are active inhibitors of ACE, the latter compound has an IC_{50} of 8×10^{-10} mol dm^{-3} (see figure 8.15).

These studies of ACE inhibitors are an elegant example of computer-aided drug design and are an important pointer to future developments in the rational design of novel drug molecules.

Figure 8.17. The binding site region for captopril. Three regions are postulated: a salt-bridge, a hydrogen-bonding region and a thiol link to zinc. (Re-drawn from Hassall *et al.*, 1984.)

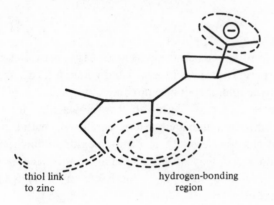

thiol link   hydrogen-bonding
to zinc   region

9

Future studies in drug–receptor interaction

The prediction of future trends in science is never easy since developments rely heavily on experimental luck in associated research fields. With this caveat in mind, this chapter is split into three sections: protein engineering, knowledge engineering and hardware developments for future computational chemistry. These are broad research areas with significant potential applications to the study of drug–receptor interaction. New technologies herald a new way of thinking. This chapter acts only as a pointer to future work and highlights some fascinating studies that are already being developed.

9.1   Protein engineering

In sections 5.1 and 6.4.2 brief mention was made of the important work of genetic cloning and sequencing for receptor materials. Once the gene is known and isolated then receptor material can be grown inside bacteria to yield large amounts of pure receptors. These genes can be inserted into oocytes and the biological activity of the receptors, if they are membrane active, can be studied electrophysiologically. This is a beautifully precise method for studying receptors. Once the genetic code of a receptor is known there is also the possibility that the code can be modified to generate a mutant protein. The effect of an artificially induced mutation can then be explored on both the structure and the function of the protein. Mutations can be induced in the gene at specific positions, thus enabling sequential modifications to be made to protein structure at will; this technique is called site-directed mutagenesis.

9.1.1   Site-directed mutagenesis

Proteins are made from sequences of amino acids. There are 20 amino acids known to be involved in protein function. Each amino acid is coded for by a deoxyribonucleotide triplet; 64 triplets are possible but three triplets code for the termination of the protein chain; thus 61 triplets can code for the amino acids. In protein synthesis the DNA code has firstly to be transcribed to messenger-RNA (mRNA). The mRNA is transported to the ribosome and amino acids are attached to transfer-RNA (tRNA). The ribosome then condenses the amino acids into the polypeptide in the order specified by the code of the mRNA. Replacement of one triplet for another in the DNA sequence will introduce a different amino acid at the position in the protein sequence corresponding to the point of the mutation. Natural mutations give rise to random changes in sequence. Site-directed mutagenesis introduces specific mutations at defined positions within the sequence and therefore offers the possibility of modifying protein structure at will.

A review of oligonucleotide-directed mutagenesis and detailed protocols is given fully by Zoller & Smith (1983). Three criteria are involved in the design of the experimental method: (a) the mutation should be introduced at any desired position in a cloned DNA fragment, (b) mutagenesis must be efficient otherwise screening for mutants would be tedious, (c) identification of the mutant should be simple and not dependent on natural selection or the need to sequence the mutant DNA.

Oligonucleotide-directed mutagenesis makes use of a small oligonucleotide strand of DNA synthesized in the laboratory and containing the mutation in the genetic code. This portion of oligonucleotide is then incorporated into the gene by DNA-polymerase. Synthetic oligonucleotides are short; 8–12 bases long in ϕ X174 (phage) and 14–21 nucleotides for cloned fragments in M13 (phage). These short oligomers serve as primers for DNA synthesis. The mutational mismatch is introduced into the centre of the oligomer to facilitate a greater binding differential between matched duplex and mismatched duplex. Furthermore, a central mutation reduces the probability of the mismatch being corrected by exonuclease. Longer oligonucleotides may be needed to avoid priming at competing sites; this possibility can be determined by computer analysis of the gene sequence of the vector (M13) (Zoller & Smith, 1983).

An example of site-directed mutagenesis is taken from Winter *et al.* (1982) in their work of re-designing the structure of tyrosyl tRNA synthetase. The amino acid sequence Leu–Tyr–Cys–Gly–Phe is coded for by the oligonucleotide sequence 5'CTCTAC$\overset{*}{T}$GCGGGTTTG3'; the desired mutation is to replace Cys (TGC) by Ser (AGC). A complementary strand is therefore synthesized containing a mutation 5'CAAACCCGC$\overset{*}{T}$ GTAGAG3' at that point; the base $\overset{*}{T}$ is the mismatched point and is positioned towards the centre of the sequence. The mismatched triplet TCG is complementary to AGC, the desired code for Ser. Single-stranded template DNA is then primed with the mismatched sequence (figure 9.1); the complementary oligonucleotide strand pairs with the template sequence. The rest of the complementary strand is synthesized by extension of the primer by DNA-polymerase along the single-stranded circular template and ligated to produce the closed circular DNA hetero duplex. Cloned circular double-stranded DNA is then transfected into *E. coli* to yield a mixture of wild-type and mutant phage. The two types of phage are then separated by screening. Enzyme containing the Leu–Tyr–Ser–Gly–Phe sequence is produced by the phage mutant. In order to be absolutely sure that the mutation has been generated at the desired position the DNA sequence in the mutant needs to be checked against that of the wild-type by standard nucleotide sequencing procedures.

Figure 9.1. The procedure for oligonucleotide site-directed mutagenesis of tyrosyl tRNA synthetase; the scheme is taken from Winter *et al.* (1982). (*a*) Single-stranded DNA template. Step 1, addition of primer. (*b*) Single-stranded DNA template with attached primer containing mismatched T*. Step 2, extension by DNA polymerase with ligation. (*c*) Closed circular DNA. Step 3, insertion into *E. Coli* (JM101 strain). (*d*) Wild-type single-stranded DNA. (*e*) Mutant single-stranded DNA.

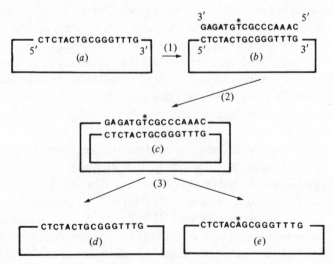

9.1.2 Protein engineering and macromolecular recognition

The intimate molecular details of macromolecular self-assembly from component parts into oligomeric structures have yet to be elucidated. However, this new and powerful technique of site-directed mutagenesis is providing a method for systematically altering protein structures so that functional changes can be determined. An important advance has been provided by recent studies on a repressor protein bound to its operator DNA (Anderson, Ptashne & Harrison, 1985; Wharton & Ptashne, 1985). Here the operator is a 14 base-pair synthetic oligonucleotide and the repressor protein is made by coliphage 434. X-ray crystallographic studies of the repressor–operator complex revealed that the repressor is composed predominantly of α-helical subunits with a folding conformation similar to the four α helices of λ repressor. The operator has a B-DNA conformation. In the complex the repressor is positioned so that the $\alpha3$ helix is found in the major groove of the DNA helix and follows the turn of the helix close to the base pairs. Sequence specific intermolecular contacts are apparent. The $\alpha2$

Figure 9.2. The binding of repressor 434 to the B-DNA operator showing the relative positioning of the $\alpha2$ and $\alpha3$ helices.

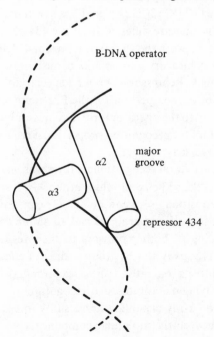

helix lies above the α3 helix and spans across the major groove, possibly making contacts with phosphate groups on opposite strands of the DNA (figure 9.2). This positioning of helices along DNA is found in other protein–DNA complexes and so may be a general feature of protein–nucleic acid assemblies. The α3 helix is not positioned exactly in the centre of the major groove but is displaced by 2 Å from the midline. This off-centre alignment appears to be caused by other repressor–DNA contacts. It is also noteworthy that the α2, α3 and α4 helices have their N-terminal aimed at the phosphate anionic charges along the DNA backbone. This orientation maximizes the ion–dipole interactions between the charges and the protein helices (see section 4.6).

Structural experiments performed by building molecular models of the 434 repressor and operator suggest that sequence specific binding might be altered by mutagenesis along the α3 helix. A simple way of producing a significant change in recognition properties of the repressor is to exchange the 434 α3 helix with an α helix from a different repressor. However, to make this change more subtle, only those residues on the solvent exposed surface, ie that surface which would bind to the operator, were changed for those of the P22 repressor of *Salmonella* bacteriophage. Residues on the inner surface were left unchanged. Thus minimal changes were induced in the α3 helix to circumvent a change in the folding properties of the inner surface of the helix. The original wild-type repressor is termed 434; the re-designed repressor is designated 434R[α3(P22R)] and has five amino acid substitutions along the α3 helix that are different from the 434 α3 helix.

The differences in recognition between the wild-type repressor 434 and its mutant 434R were determined by measuring the affinity of the repressor for the corresponding operator DNA. Repressor 434 bound specifically to its own wild-type operator but not to the operator for the P22 repressor. In contrast, 434R bound specifically to the operator for P22 repressor but not to the operator for 434. Therefore the recognition properties of the two α3 helices appear to have been reversed.

A more detailed analysis of structural recognition features was made by subdividing five substitutions of the α3 helix into three groups, A, B and C. Group A contained residues at position −1, 1 and 2 of the α3 helix; group B had a single substitution of position 5 and group C had an amino acid at position 9. High affinity binding of both repressors to their respective operators was determined by groups A and B; group C did not affect the affinity of either repressor for its operator. These experiments have provided new methods for studying macromolecular recognition. What is now needed is a high-resolution x-ray crystallographic study to examine the intermolecular bonding between the molecular components.

9.1.3   Modifications to the acetylcholine receptor by site-directed
         mutagenesis
         The acetylcholine receptor AchR has five subunits (α_1, α_2, β, γ, δ)
and the genetic code for the protein has been discussed previously (section
5.1). The three-dimensional structure of AchR is as yet unknown. However,
location of the binding site can be deduced by site-directed mutagenesis.
Kao et al. (1984) were the first group to attempt to locate the binding site by
mutation of Cys 192 to Ser 192 on the α subunit protein by changing the
codon TGC to AGC at the appropriate position on the cDNA for the α
subunit. Cys 192 was thought to be involved directly in acetylcholine
binding to AchR. More recently a much larger study of site-directed
mutagenesis of AchR has been carried out by Mishina et al. (1985); 26
separate pieces of the gene sequence were involved in the mutation
experiments involving a total of 196 codons of the cDNA. Mutants were
then tested for acetylcholine activity by expression of the AchR in Xenopus
oocytes. This was achieved by injecting the oocytes with mRNAs for α, β, γ,
and δ subunits in the proportions 2:1:1:1.

Expression of the AchR mutants was then studied by comparing the
electrical behaviour of the oocyte membrane with control injections of
polyA RNA. The electrophysiological test system used by Mishina et al.
was a voltage clamped oocyte in Ca^{2+} free Ringer solution containing
atropine. Acetylcholine was delivered to the cell surface by iontophoretic
application; the inward current at various membrane potentials could
easily be measured. Thus the functional properties of the modified α
subunit could readily be assessed. One further test of functionality was
applied, that was the ability of the α subunit to bind acetylcholine. This
property was measured by the inhibition of the binding of the antagonist
[125]Iα-bungarotoxin by the nicotinic agonist carbamylcholine to AchR in
oocytes.

Of the 26 mutation segments of the α subunit gene mutated, all but six
were segment deletions, with minor replacements; the remainder were
single amino acid residue exchanges. The effects of gene deletions on the
binding of bungarotoxin and responsiveness to acetylcholine are shown in
table 9.1. Gene deletions in all these segments removed the acetylcholine sensitivity compared with the wild-type AchR. Toxin binding
was substantially reduced by gene deletion in segment M1. Mutations in
the 70 codon segment between M3 and MA did not affect the sensitivity to
acetylcholine nor did they significantly affect α-bungarotoxin binding. This
segment between M3 and MA is the putative cytoplasmic region and does
not appear to be important in the function of the α subunit. Mutations in
the hydrophobic segments M1, M2, M3, M4, although reducing the

amount of bungarotoxin bound compared with the wild-type, did not affect the inhibition of bungarotoxin binding by carbamylcholine; thus it may be concluded that the affinity of the binding site for α-bungarotoxin is unaffected by deletions in these segments. In contrast to the results obtained from codon deletions, the exchange of cysteine 192 and 193 for serine residues altered the affinity of the acetylcholine binding site; the oocytes with AchR from these mutants did not respond to acetylcholine but did exhibit bungarotoxin binding activity.

These studies on site-directed mutagenesis of AchR are a good indication of the importance of this new technique for investigations of the structural properties of drug receptors. We can expect many similar studies on other receptors; these developments will open up another field of structure–activity studies. Not only can we modify the structure of the ligand, we now have the capability of modifying the receptor as well.

9.1.4 Redesigning binding sites for structure–function studies

Until recently, variations in the binding site could only be studied using naturally occurring mutations in polypeptide proteins. An example of this procedure was examined in section 8.2.4 where the effect of mutations on the binding of DPG to haemoglobin was studied. However, one of the drawbacks of using naturally occurring mutant proteins is that the active sites are often conserved by the evolutionary process and generally it is found that mutations occur in non-functional regions. Furthermore, with natural mutations the scientist has no control over changing the structure of a particular residue. All these restrictions imposed by natural mutations are overcome by site-directed mutagenesis. Any number of amino acid residues can be changed at specific positions; residues can be deleted, added or exchanged. The ligand-binding site can be

Table 9.1. *The effect of codon deletions in the gene for the putative transmembrane segments of the α subunit of AchR*

Transmembrane segment	^{125}Iα-Bungarotoxin binding as a percentage of wild-type	Response to Ach as a percentage of oocytes tested
M1	0.5	0
M2	17	0
M3	8	0
MA	43	0
M4	23	0

Data taken from Mishina *et al.* (1985).

re-designed at will. Therefore altering the properties of a receptor is now a realistic objective.

The facility for re-designing proteins enables one to test many theories about biomolecular interactions. For example, what structural properties control substrate specificity? What are the strengths of hydrogen bonds from proteins to ligands in solution? What is the mechanism of enzyme catalysis? A short introduction to recent work on re-designing binding sites is given by Perutz (1985) together with examples of the application of the work to structural problems of major biological interest. It is only a matter of time before computational chemistry will provide us with a detailed understanding of the changing patterns in inter- and intramolecular forces for re-designed and modified biomolecular interactions.

Fersht and his colleagues (1985) have carried out an extensive and detailed study of re-designing the binding site of tyrosine in tyrosyl tRNA synthetase. By a judicious selection of mutation points, they were able to make single point replacements of amino acid residues at all positions in the substrate binding site. Many of these wild type amino residues were believed to fix the position of the substrate by hydrogen bonding. Re-designing the binding site enables this hypothesis to be tested by experimental determination of the kinetic constants of the reaction for substrate binding and catalysis.

The aminoacylation of tRNA by tyrosine to yield $tRNA^{Tyr}$ is a two-step reaction. The enzyme first binds tyrosine and ATP,

$$E + Tyr + ATP = E \cdot Tyr\text{—}AMP + PPi \qquad (9.1)$$

tyrosine is then transferred to the appropriate tRNA.

$$E \cdot Tyr\text{—}AMP + tRNA = Tyr\text{—}tRNA + AMP \qquad (9.2)$$

The three-dimensional structure of the wild type enzyme and the tyrosyl adenylate complex are known (Blow & Brick, 1985); 11 amino acid residues in the wild type form hydrogen bonds with the amino acylate. The complex is schematically drawn in figure 9.3. Different types of hydrogen bonds are possible: those between uncharged acceptors and donors, those between charged acceptors and donors and bonds where only one partner is charged. Furthermore, selected hydrogen bonds can be deleted by site-directed mutagenesis without adding steric constraints. For example, tyrosine can be exchanged for phenylalanine so that the phenolic hydroxyl group is eliminated. It is assumed that these small changes in selected side-chains do not affect the conformation of the other residues. An unambiguous test would be to determine the crystal structure for each mutant. A quick check could be performed with a molecular mechanics calculation using the wild type structure as a starting point for each mutant. Any differences between calculated structures could then be observed by molecular graphics. These graphical displays allow the scientist to see by how much the spatial characteristics of the site are altered by mutation. This is an excellent aid for assessing the effects of changes in structure on the function of the binding site because it provides a visual understanding by which mechanistic changes can be interpreted.

In particular Fersht *et al.* (1985) have examined the contribution of hydrogen bonding in solution to the enzyme kinetics of substrate binding and catalysis. A general model is assumed where potential hydrogen-bonding groups on the enzyme or substrate are occupied by water molecules in solution. When substrate is bound, these hydrated hydrogen bonds are exchanged for hydrogen bonds between the enzyme and substrate according to the equation

$$EH—OH_2 + H_2O \ldots B—S = E—H \ldots B—S + HOH \ldots OH_2 \quad (9.3)$$

where E—H is the enzyme donor group and B—S is the substrate acceptor. Therefore the number of hydrogen bonds participating in the reaction is unchanged by complex formation. However, the strengths of the individual hydrogen bonds may differ. If a hydrogen-bonding group on the enzyme is removed by deletion then the space vacated by the group may be filled by a sequestered, but not hydrogen-bonded, water molecule according to the

Figure 9.3. A large network of hydrogen bonds links tyrosyl adenylate to tyrosyl tRNA synthetase. (Re-drawn from Fersht *et al.*, 1985.)

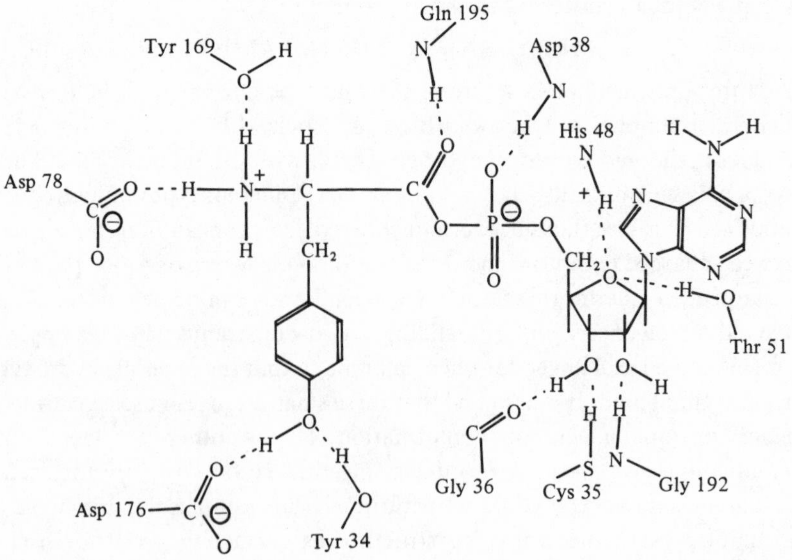

equation

$$EOH_2 + HOH \ldots B—S = [E—B—S] + HOH \ldots OH_2 \qquad (9.4)$$

Tyr 34 can be mutated to Phe 34 so that the hydroxyl group is removed; compared with the wild type this mutant has a K_M increased by a factor of 2 and the enzyme specificity, with phenylalanine as a substrate instead of tyrosine, is decreased 15-fold. The change in binding energy for tyrosine between the two enzyme types is only $2.1\,kJ\,mol^{-1}$. From a number of studies with different mutants, a generalization about binding energy could be made: if a side-chain to be deleted interacts with an uncharged group on the substrate then deletion of the hydrogen-bonding group only causes a small loss of binding energy in the range $2.1–6.3\,kJ\,mol^{-1}$.

Tyr 169 interacts with the quaternary amino group of tyrosine, and Gln 195 binds to the carboxylate oxygen in the phosphate ester link (figure 9.3). In both cases the amino acid residue binds to a charged group in the substrate. The losses in binding energy caused by mutation of Tyr 169–Phe 169 and Gln 195 are 15.6 and $18.8\,kJ\,mol^{-1}$ respectively. This change in binding energy is quite large and can account for a 1000-fold alteration in substrate specificity.

Other ideas about enzyme catalysis can also be examined by protein engineering. For example, Wells & Fersht (1985) have attempted to assess the notion of strain in catalysis by using tyrosyl–tRNA synthetase mutations. Amino acid residues distant from the substrate binding site help to bind ATP preferentially in the transition state of the reaction.

This presentation of hydrogen bonding by Fersht and colleagues is a preliminary work and further measurements need to be performed on all mutants; at the same time there is a need to take into account changes in van der Waals interactions and dispersion energy contributions as well as hydrogen-bonding energies. Nevertheless the work is an important pointer to future use of site-directed mutagenesis in structure–function studies of biological macromolecules. These studies have drawn attention to the importance of charged dipoles contributing strongly to hydrogen-bonded interactions and their role in molecular specificity.

9.2 Knowledge engineering

In general, the great majority of computer programs take in data, usually numerical, and process the data through a series of functions which are precisely defined to yield the result for output. The functions might be mathematical or involve a search through a database. The whole process is solved by standard programming techniques and is based on the premise that there exist precise relationships between the input and the output.

Suppose now that the relationship between input data and desired output is imprecisely understood and in the past the problem has only been handled (not solved in the strictest sense) by a human expert. In these circumstances the reasoning process, and the knowledge acquired by the expert, has to be coded into a form to be manipulated by a computer. Most reasoning processes work by question and answer. The answers can be either of the yes–no form or allocated some intermediate value on an arbitrary scale. The expert has within his mind a rule-base to chart his way through the problem to the goal. This rule-base interacts with acquired knowledge and is constantly being updated by experience. Knowledge engineering is a branch of artificial intelligence that seeks to emulate the specific work of an expert.

In most cases of knowledge engineering the human expert is not the same person as the knowledge engineer. The first problem in building an expert system is therefore to produce a fruitful communication between people from very different academic disciplines. An expert's jargon has to be translated fully into sets of rules and a knowledge base so that each rule, as well as a cascade of rules, can be tested separately. This process is termed knowledge acquisition and has a number of components. The knowledge engineer has to extract definitions, relations, algorithms, facts and heuristic strategies from the expert (figure 9.4). Having acquired that knowledge, the engineer has to arrange it. If it is factual then those facts have to be entered into a flexible database; if the knowledge is logical then it must be represented by ordered rules. These are all preliminary steps and form the groundwork for building the expert system. Most expert systems can be represented by two components: the inference engine and the knowledge base.

Special languages like LISP and PASCAL have been developed to handle non-algorithmic data. The database of facts and rules can therefore be coded efficiently into the knowledge base of the expert system. The inference engine is written to manipulate symbolic logic in order to process the rules and facts. Symbolic logic is simply a convenient notation for performing logical operations, rather like algebra in mathematics.

The reasoning process emulated by the inference engine may consist of conditional statements chained forwards or backwards. For example, in the LHASA program (see section 9.2.3) which is used to deduce chemical synthesis pathways, each reaction is represented by a chain of conditions. The conditional statements, when translated into English, might be of the form

If atom A is carbon and
if atom B is carbon and

if the bond between A and B is of order 2 and

if the number of hydrogen atoms attached to A is less than the number attached to B.

Then HBr will add to the double bond

to yield —ABr—BH— (Markownikoff's rule)

This sequence would be forward reasoning about the addition of a halogen to an alkene. The rule has a number of premises and one action. In backwards chaining the strategy is reversed and the search is directed to finding the correct premise. A state-of-the-art approach to building expert systems can be found in books by Hayes-Roth, Waterman & Lenat (1983) and Buchanan & Shortliffe (1984).

Knowledge engineering has recently begun to be applied to drug design. In particular, commercial use has been made of synthesis planning programs for novel drug molecules. An ideal candidate for knowledge

Figure 9.4. Flow diagram for the construction of an expert system.

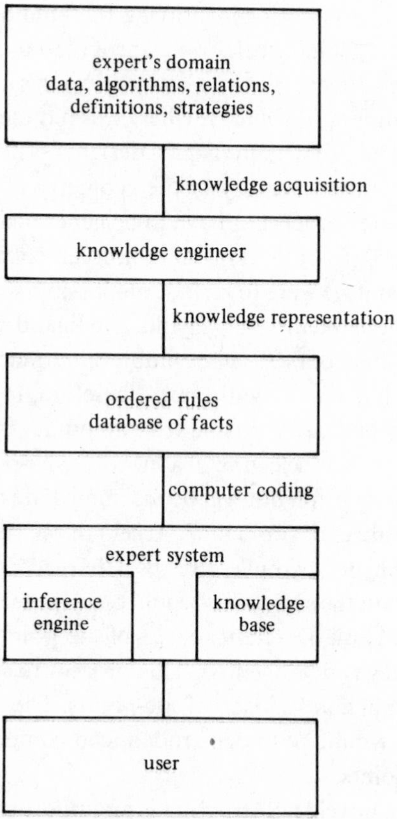

engineering would be QSAR studies in molecular design. Many man-years of work are necessary before an expert system can be built to perform as well as the human expert. Nevertheless, some expert systems are becoming available that open up specialist expertise to a wide range of users.

9.2.1   Drug design using computer-aided synthesis

In chapter 8 we considered some of the principles of drug design. If we have a novel lead compound on which a number of ligand-points have been identified to produce changes in affinity of the drug for its binding site, then modification of these ligand-points generates a combinatorial number of putative compounds that need to be studied. Some types of modification may not be expected to enhance drug binding, or potency, and these possible structural changes can be weeded out by chemometrics. Drug design by modification of a lead structure is a conceptually simple procedure.

Consider now the case where we have no lead compound, but the structure of the cellular target macromolecule is known to atomic resolution and we have strong reasons for identifying the binding–active site. In other words, we have the geometrical arrangement of a complete set of site-points on the accessible surface. We can also identify regions of high electrostatic potential by computation and, further, we may be able to locate hydrophobic regions. The problem for a drug designer is now much more challenging; structures complementary to the geometry of the site-points as well as macromolecular properties have to be generated to yield precisely positioned ligand-points built onto a molecular spacer. One difficulty in this problem is that if we were to include all possible site-points in a method for generating a complementary ligand, the ligand would in many cases be too large and possess undesirable pharmacokinetic properties; it might also be difficult to synthesize. Therefore, in *de novo* drug design, a small number of closely positioned site-points is usually taken as a target for drug design. The selection of a subset of site-points has to be made with care. The first determinant will be the immediate chemical environment covered by the subset of site-points. Access to the site-points from solution should be unimpeded by bulky groups. Charged hydrogen-bonding acceptors and donors in the site-points would contribute to strong interaction with the putative ligand. Different subsets of site-points should be surveyed and their advantages documented. This procedure should in theory lead to a small number of good subsets of site-points. The next step in designing a drug molecule would be to construct a lead compound for each selected subset of site-points.

The creation of a promising novel lead structure is an important step in drug design; however, commercial considerations may determine whether the putative lead compound is worth synthesizing. If molecules are to be congenerically derived from the parent, the basic skeletal structure of the parent must be capable of synthesis by known methods and, furthermore, it must be possible to scale up these methods for industrial synthesis. There is no point in refining a molecular structure to have 'perfect' pharmacological properties if the cost and effort of synthesis is too high. Ideally, the putative lead compound should be synthesized from readily available precursors by a few well-known, and easily achieved, synthetic steps.

9.2.2　Synthesis planning

　　From molecular graphical studies performed round a subset of site-points, the medicinal chemist should be able to construct a molecular structure that would be expected to bind to the site and, hopefully, have a high affinity and structural specificity. This parent, or lead molecule, we shall term the synthetic target. How do we assess the feasibility of synthesizing the target molecule? The standard procedure is to plan the synthesis retrosynthetically. We start from the target and break it down into smaller precursors until a sufficiently simple starting point for synthesis has been achieved. An immediate difficulty in this procedure arises; there may be many possible ways to perform subsequent fragmentation steps. A logical tree structure of possible retrosynthetic routes can be generated. The skilled synthetic chemist would ideally choose the simplest and cheapest route, but in practice this is often difficult since the logic tree is generated combinatorially. For each node on the tree a synthetic procedure is necessary. A database of reaction procedures and fragmentation steps is essential. Computer-aided synthesis planning can reduce the drudgery of searching through the logic tree for an optimum pathway.

9.2.3　Logic and heuristics applied to synthetic analysis (LHASA)

　　The reaction tree, generated retrosynthetically, expands combinatorially. At the heart of the LHASA system is a mechanism to reduce the combinatorial problem (Long, Rubenstein & Joncas, 1983; Johnson, 1985). This device functions in a two-fold way: firstly, by user interaction the most appropriate node at each level may be selected to yield a single pathway; secondly, the user can pre-select defined constraints to operate at each level so that a limited number of pathways are automatically generated through the search tree. The choice of which procedure to use has to be made carefully, taking into account the chemist's knowledge of the specific problem.

A short example of the uses of LHASA should give the reader an overall picture of the strategy behind the program before we examine some of the components in more detail. The LHASA routines are structured in a processing menu along with computer graphics and a purpose-built chemical language CHMTRN. The routines are linked to a database of reactions that can be used to transform the target molecule into precursors. Subgoal transforms are of three types: interchange, addition, or removal of groups to form a subgoal structure. A goal transform converts a subgoal structure to a goal structure which is the precursor. For example, suppose the chemist wishes to synthesize the six-membered ring compound 4-methyl-hydroxymethylcyclohexane as his target molecule (figure 9.5) (Long, Rubenstein & Joncas, 1983). A useful goal transform is the Diels–Alder reaction to synthesize a six-membered ring from small readily available components. The goal precursors would be methylacrylate and 2-methyl-1,3-butadiene but this reaction alone is not applicable; other subgoal transforms are needed to define the other precursor steps. A double-bond is introduced by a functional group addition and a methyl ester is interchanged for the hydroxyl group. The computer program searches for these intermediate subgoal steps.

Figure 9.5. The LHASA procedure; 4-methylhydroxymethylcyclohex-ane is the target. Goal precursors are obtained retrosynthetically; the Diels–Alder reaction is a specified goal transform. (Re-drawn from Long, Rubenstein & Joncas, 1983.)

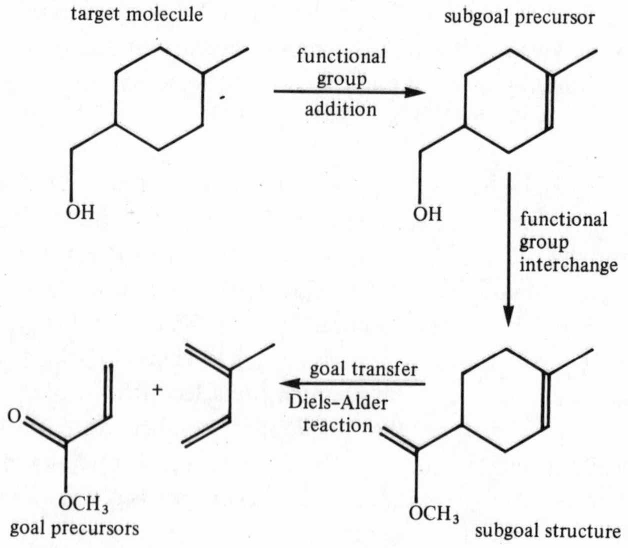

Chemical input to the program is graphical. The target molecule is drawn with a graphics tablet and a menu of atom labels; stereochemical centres are identified and bond directions are drawn in standard notation to give a good representative picture of the molecular structure. Then follows an intelligent but automated analysis of the molecular structures; this routine works in an analogous way to a chemist's perception. Fragment substructures are automatically assessed from the drawing and held in memory along with other relationships such as six-membered rings and important synthetic features.

LHASA is very flexible and its decision-making procedure is grouped into three strategies: group-oriented strategies are disconnective, re-connective, unmasking or opportunistic; bond-oriented strategies describe bridge fusing and appendages; long-range strategies include ring formation, halolactonization, etc. Other options are also available for stereochemical control, molecular mechanics, and various house-keeping routines. The reaction database contains many hundreds of examples that the search-tree routine can call upon. Each reaction is known as a transform.

Each transform is in fact a subroutine of decision statements. If the decision does not satisfy certain criteria, an exit is made from that branch and control switches to another node transform to be examined in sequence down a further branch of the search tree (figure 9.6). However, if the criteria are satisfied then a reaction mechanism is proposed and control is transferred back to the calling routine for further processing. Management of control for selecting the transforms depends on which strategy was adopted at the start of the retrosynthetic pathway. Satisfactory criteria for the probability of a reaction achieving the desired result can be estimated numerically by ascribing a value for each qualification, and subtracting this value of the qualifier from the initial rating if the qualification is not met.

Figure 9.6. Decision making using hierarchical trees. Each node represents a decision to be made, in this case the answer is of the type: yes or no. Eight final nodes are possible at level (3); each is reached by a different pathway. Branches may terminate before the maximum level.

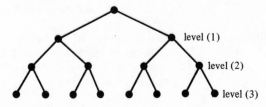

This is the heuristic step in the analysis; a logic flow within a transform is outlined in figure 9.7. If more than one transform can, in theory, perform the reaction, the value of the rating at the end of the transform can be used to decide between alternative reactions. Thus different retrosynthetic routes to the starting materials are possible and each can be ranked according to the summation of the ratings. When the decision procedure is complete, and this depends on the strategy option chosen, the pathway through the decision tree can be displayed and the transformations along the route of subgoals may be inspected graphically.

LHASA is a very useful program in synthetic medicinal chemistry; it is constantly being upgraded to extend the facilities provided and to include new transforms as they appear in the literature. The program system LHASA could mildly be termed an artificial intelligence program – or more specifically an 'expert system' since it can only solve the problem by going through the options that experts have built into it. LHASA cannot suggest an entirely new chemical reaction and so is not intelligent in that sense, nor can the program 'learn' from past experience.

9.2.4   Artificial intelligence in QSAR

In QSAR studies the medicinal chemist is concerned with discerning the causal relationships between structure and biological activity. Once these links have been discovered, he should be able to improve the design of a

Figure 9.7. The ordered structure of a transform in LHASA.

Start	Transform *n*	
	Reaction description	Already used to select transform *n* from initial chemical 'perception'
	Rating 50	
	Qualifiers	If...
		then...
		subtract 10 from rating
		If...
		then...
		subtract 5 from rating
		If rating less than cut-off value then EXIT
	Conditions	Specifically related to fragment
	Mechanism	Perform the reaction of transform *n*, manipulate atoms and bonds to convert the target to the precursor.
End		

new molecule. One of the key steps in this logical procedure is to define the potential descriptors; the bottleneck in analysis is to discriminate which properties to include in the study and which to omit. In many cases the choice of descriptors that have to be considered is based on qualitative human judgement. Normally, it is only after this pre-selection procedure has been carried out that the quantitative methods of assessment, outlined in chapter 8, are used to validate chosen descriptors by further tests. It is this initial choice of descriptors that is such a key factor in QSAR and, because it requires expert knowledge, it should be possible to model the logic by an expert system. Klopman (1984) has provided an outline strategy for building an artificial intelligence program for structure–activity studies. The rationale behind the program falls into two parts: (a) the development of a method to transcribe chemical structure into a computer code and (b) the construction of a program to select descriptors from data. These two aspects will now be described briefly.

9.2.5 Representation of molecular structure by computer codes

Wittgenstein's dictum 'limits of my language are limits of my mind' is very pertinent to the problem of manipulating representations of molecular structure in computer memory by symbolic logic. When we look at a model of a molecular structure, the atoms are usually colour coded and we appreciate the interconnections between them by locating bonds; three-dimensional depth perception helps us to discern geometric spatial relationships. Our minds are very good at spotting similarities and differences between objects, partly because we have evolved to discriminate between friend or foe. However, the computer has to be taught every step in the logical process of discrimination before it can answer a query about structure. Electronic computers can only operate by manipulating symbols, therefore we have to represent three-dimensional molecular geometry by a string of symbols. If a language for structure notation can be written without ambiguity, it should then be possible to search a sequence of symbols coding for one molecule and make a comparison with the symbols coding for a different molecule.

The development of a language for structure representation is a continuously maturing objective. Golender & Rosenblit (1983) outline four factors in chemical structure that are constantly determining the evolution of an efficient language: (a) topological relationships between chemical groups within a molecule, (b) the relationship between structural formulae and electron density in space, (c) the problem of extracting different types of information from a structural representation and (d) the development of software for structure manipulation.

Many languages for structure representation follow some form of vector notation. Functional groups might be represented by a numerical or letter code so that if we had 1-fluoro-3-carboxy-5-nitrobenzene and the codes were $1 = C_6H_6$, $2 = F$, $3 = COOH$, $4 = NO_2$, the notation might read 1,(1), 2,(3),3,(5),4 where the number in brackets is the substituent position in the aromatic ring. A common scheme that operates by this type of vector string is the Wiswesser Line Notation; groups have unique letter codes and number symbols. Substructural fragments combined in a vector notation are useful for searching molecular databases and for descriptor representation of chemical groups within a QSAR framework. Their disadvantage is that they do not convey stereochemical and geometric information of three-dimensional structure.

Three-dimensional structure notation can be handled in matrix form. The molecule is represented by a connectivity matrix for bonds and a distance matrix for interatomic distances. Although this matrix form for structural notation has larger requirements of memory space, it is more adaptable to changes in the structure of molecules and is independent of the limitations of fragment notation. Manipulation of distance matrices can enable conformational information to be compared between molecules (Crippen, 1981). In QSAR studies we may have a set of active compounds and want to identify descriptors within the set; at the same time we may wish to extract, from a large database of structures, similar features containing the suspected descriptors. Searching through distance matrices would be very expensive in computer time. A commonly used method to reduce this problem is to carry out a rapid screening of the database first; likely candidates can be identified by determining their intermolecular similarities in connectivity. DM checking then narrows the search to retrieve the required molecules (Crandell & Smith, 1983; Carhart, Smith & Venkataraghavan, 1985).

9.2.6   Automatic selection of structural descriptors for QSAR

The subjective element of human bias must be removed, if at all possible, from the selection of structural descriptors. Klopman's (1984) method for computer automated structure evaluation (CASE) takes the training set of drug molecules and decomposes them into connected linear fragment units; each unit is 3–12 heavy atoms long. To facilitate this fragmentation, the molecules are expressed in a linear code called the KLN code (Klopman & McGonigal, 1981). All possible routes through the molecule have to be examined. An example of fragment generation for acetylcholine is shown in figure 9.8. Supplementary codes indicate the type of bonds and whether side-chains are attached. The KLN code is derived from a connectivity matrix.

In the CASE method each fragment is given a label, active or inactive, according to the activity assigned to the compound from which the fragment was taken. Fragments are then classified statistically into possible descriptors or non-descriptor fragments by a method which is similar to the linear discriminant function for two groups. The classification procedure assumes a binomial distribution function for all fragments. If the distribution of the fragment type falls within the random distribution function, the fragment is classified as a non-descriptor. However, in the case where a fragment type lies outside the expected random distribution and if the fragment type is associated with activity, then that fragment is classified as a possible descriptor.

Various tests on the CASE procedure have been carried out. The program works efficiently to retrieve the fragment descriptors for carcinogenicity with polycyclic aromatic hydrocarbons and for N-nitrosamines. Insecticidal activity of ketoxamine carbamates can be attributed to particular fragments. Overall, the selection of descriptors by CASE techniques works well and removes the arbitrary element in the

Figure 9.8. KLN code for heavy atom fragments in acetylcholine.

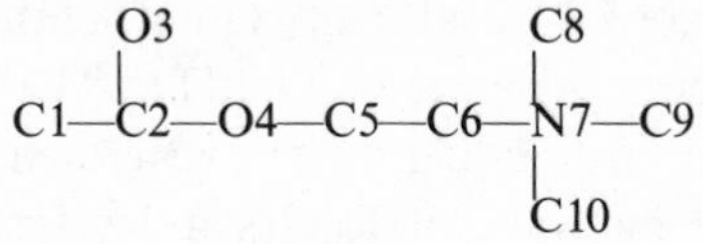

```
     O3              C8
     |               |
C1—C2—O4—C5—C6—N7—C9
                     |
                     C10
```

3 atoms	4 atoms	5 atoms
1 2 3	1 2 4 5	1 2 4 5 6
1 2 4	3 2 4 5	3 2 4 5 6
3 2 4	2 4 5 6	2 4 5 6 7
2 4 5	4 5 6 7	4 5 6 7 8
4 5 6	5 6 7 8	4 5 6 7 9
5 6 7	5 6 7 9	4 5 6 7 10
6 7 8	5 6 7 10	
6 7 9		
6 7 10		

6 atoms	7 atoms
1 2 4 5 6 7	1 2 4 5 6 7 8
3 2 4 5 6 7	1 2 4 5 6 7 9
2 4 5 6 7 8	1 2 3 4 5 6 7 10
2 4 5 6 7 9	3 2 4 5 6 7 8
2 4 5 6 7 10	3 2 4 5 6 7 9
	3 2 4 5 6 7 10

allocation of structural descriptors by hand. At the moment the method does not include other forms of molecular descriptors such as lipophilicity or extra parameters in QSAR studies.

9.2.7   Automatic allocation of hydrogen-bonding site-points to macromolecular surfaces

In the process of building a drug molecule to fit a known macromolecular structural binding site, the putative site must be surveyed (section 8.2). Hydrogen-bonding positions for both donor and acceptor atoms can be defined by the HSITE program of Danziger (1985a, b). The program takes the molecular coordinates from the Brookhaven Protein Data Bank as input. Acceptor and donor atoms for hydrogen bonding are identified at the molecular surface. Regions of possible hydrogen bonding on acceptor atoms are allocated close to the points where the lone-pair electrons would be found on the acceptor atom. Specific loci for hydrogen bonding are then defined empirically from the work of Taylor, Kennard & Versichel (1983, 1984a, b). Similarly, for a macromolecular hydrogen-bonding donor group, the position of an acceptor atom for a putative bond with a ligand can be defined. The HSITE program has the facility to handle rotatable groups such as the hydrogen of the hydroxyl group; constraints on rotation are accounted for. Putative hydrogen-bonding loci at site-points can then be displayed by molecular graphics at the accessible surface of the protein macromolecule. Plate 5 illustrates an example of the surface display of hydrogen-bonding regions generated by HSITE. The program has been tested to see if the hydrogen-bonding regions generated coincide with the location of hydrogen-bonded water molecules in the crystal at the surface of myoglobin and plastocyanin. The results of the tests support the usefulness of HSITE in predicting surface hydrogen-bonding regions.

The examples given in this section form an introduction to knowledge engineering in studies of drug–receptor interaction. The goal is to design novel drug molecules by automatic reasoning processes. This research field is in its infancy but nevertheless provides a pointer towards the development of artificial intelligence expert systems for drug design based on a detailed knowledge of drug–receptor interactions.

9.3   Computations for drug–receptor interaction by supercomputers

Throughout this book the whole emphasis has been on the use of computers to examine problems in the molecular theory of drug–receptor interaction. The computer is the experimental tool by which theory may be investigated and ideas tested. In many respects this approach to drug–receptor interaction has opened up novel ways of thinking about each of the discrete molecular steps and has led to new and practical developments in drug design. However, this very experimental tool is in the process of undergoing a radical new design. Most computers of the 1970s and early 1980s are serial processing machines. This means that they function as a single pipeline; information is taken in at the beginning of a program and manipulated sequentially until the data is in its final form for output. The operation of a hand-held calculator mimics this process; the computer is capable of carrying out only one operation at a time. Serial computations can become a bottleneck in large iterative calculations because of this limitation of being able to execute only one instruction at a time. Suppose we have a one-dimensional array of length 100, eg $A(100)$, and we wish to add a constant C to each element in A.

$$DO \ 1 \ I = 1,100$$

$$1 \ A(I) = A(I) + C$$

A serial computer has to perform 100 sequential additions to increment the values in A by C; let us say this takes 100 time units. Now if we had 100 processors working in parallel so that each was assigned a different element A and the value of C, then the computation for all additions takes only one time unit. Therefore by introducing multiple processors working in parallel the computing time for this repetitive calculation can be reduced proportionally. Instead of having a single pipeline, the computing effort is shared by the multiple pipelines.

9.3.1 *Parallel processing*

The application of parallel processing to the development of supercomputers has been reviewed recently by Mokhoff (1984), Hindin (1985) and Lackey, Veres & Ziegler (1985). The concept of parallel processing is still developing; however, five trends in computing design can be observed.

(*a*) Pipelining is now frequently built into microcomputers and large mainframes. The execution of the program is broken into segments; if these are independent they can be controlled by separate processors.

(*b*) In multiple execution units there may be separate processors for specific dedicated tasks, for example there may be a processor for floating point arithmetic and one for integer instructions. Machine performance is therefore made more efficient if these operate in parallel. High-level languages are unchanged and program compilation handles the instruction traffic between the processors.

(c) Vector processors are used in supercomputers and need special compilers.

(d) Multiprocessing is usually different from the previous architectures; many computers are linked to a global memory. Their advantage is that multiple users can work simultaneously.

(e) In many scientific applications, it is frequently found that loops in FORTRAN programs often contain calls to subroutines and conditional jump or terminator statements. Pipelining or vectorizing cannot handle these statements effectively and bottlenecks arise in the program. Concurrency control can shuttle parts of programs around at the compiler stage to use several processors simultaneously.

9.3.2   Distributed array processors (DAP)

Serial pipeline architectures have essentially isolated parallel processors. DAPs are composed of arrays of processing elements. In the ICL (International Computers Limited) DAP, the processing elements (PEs) are linked in a square array 64×64 elements. Each element is joined to its four neighbours (figure 9.9); those elements at the boundary are also connected to corresponding elements at the opposite boundary. Computers of this type belong to a class where a single instruction is given

Figure 9.9. The arrangement of processing elements in a 64×64 array of a DAP. The elements are interconnected to their nearest neighbours.

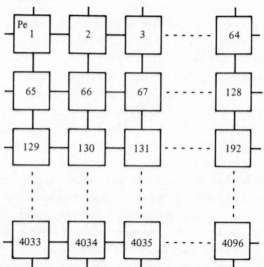

to all the PEs, each PE then obeying the instruction according to the data stored in the element. On the ICL DAP each processing element has 4K RAM (random access memory) and the total storage is 2M bytes; input and output is controlled by a host machine. The PEs can operate in two ways, either as vectors of 64 elements in parallel or on matrices 64×64 elements in parallel. Elements can be switched off if not necessary. Programming is carried out in a special high-level language DAP FORTRAN to make use of the special DAP architecture.

The ICL DAP is ideally suited to studying problems that have cyclic or periodic boundary conditions. In fact all types of lattice calculations up to four dimensions can be readily modelled on the DAP (Pawley & Thomas, 1982). A review of these new types of supercomputer, together with changes in computer algorithms, is given by Paddon (1984).

9.3.3 Molecular dynamics calculations on DAPs

Computations for problems in drug–receptor interaction have not yet been carried out on DAPs. There are probably two reasons for this. Firstly, the processing advantages of a DAP are not widely appreciated; secondly, the FORTRAN code needs to be re-written to make full use of the subtleties and flexibility of the DAP-FORTRAN language. However, the architecture of the DAP makes it particularly suitable for calculations that can be linked to a large array of grid points. This design for the communication between the PEs employing a square lattice arrangement makes lattice calculations by parallel processing quite easy to achieve. Therefore, to make full use of the DAP, problems need to be accommodated within a lattice framework.

One area in which DAPs are being readily utilized is in molecular dynamics calculations on large numbers of molecules. Serial processing machines can only model effectively small numbers of molecules, for example 32 or 108 molecules for the plastic phase of solid methane. In contrast, the DAP can model 4096 molecules simultaneously and compute a molecular dynamics cycle on all molecules in less than 4% of the time required by a CRAY-1 computer (Dove & Pawley, 1983, 1984). Indeed, the performance of the DAP for this type of repetitive calculation is so good that it is quicker to generate huge quantities of data rather than to manipulate the data files.

The liquid state can be simulated on a DAP by setting up a lattice structure within the sample space. Diffusing molecules can move through these cells. Care must be taken to circumvent the possibility of two molecules occupying the same cell. This is easily overcome if the cell size for the lattice is reduced below the volume of a molecule; there would then be numerous empty cells. Each cell is then allocated to one PE. At the end of each cycle the PEs are updated depending on whether the molecule has moved or not. Pawley & Dove (1983) present a method for liquid simulation. This method could provide an opening for studying molecular interactions in solution. These procedures are certainly worth testing for models of drug–receptor interaction in the presence of large quantities of solvent. This will be possible when the DAP architecture has been expanded to an array of 128 × 128 PEs and where each PE has a storage of 16K RAM. Very detailed simulations of drug–receptor interaction through solvent will then become feasible.

POSTSCRIPT

This book has touched on some molecular considerations that are believed to be the foundations of drug–receptor interaction. Many of the ideas about the interaction have been drawn from a variety of examples of different drug molecules and binding sites. However, it can easily be seen that no single model has been studied to anything like completeness. This is an unsatisfactory state of affairs and could, perhaps, be rectified by more coordinated research efforts between groups using a multidisciplinary approach. Probably the main deficiency in molecular simulations of drug–receptor interactions is the lack of an adequate method for taking into account the solvent environment. One consequence of this is that hydrophobic interactions, which are expected to be very important, are poorly understood. It is essential for scientists working in molecular pharmacological theory to focus their attention on solvent related problems.

The final chapter has been looking forward to developments that might be expected in drug–receptor interaction. Those advances that were itemized are, of course, not the only ones that will occur; they are simply those that the author eagerly anticipates. Protein engineering, knowledge engineering, and new computer architectures should provide a wealth of fascinating new insights into detailed mechanisms of drug–receptor interaction. The events that take place after formation of the complex are still hazy; much biochemical and biophysical spadework is needed to relate chemical stimulus to physiological response. It is harder to forsee, over the next decade, significant progress in a molecular understanding of stimulus–response coupling.

How much of this book will live up to the claims implicit in its title? In the preface it was made clear that the principal objective was to relate recent work in computational chemistry to drug–receptor interaction. This approach can only be accepted with the caveat that the immaturity of contemporary research findings has not yet stood the test of time. In many respects my own personal impression of the research outlined in this book is similar to the pleasure generated by tasting young wines from the Moselle; some of the research appears racy and exciting, but other pieces of work are elegant and austere, and augur well for laying down as an investment for future years.

REFERENCES

Abola, E. E. (1980). *DIHDRL: A Program to Generate Complete Torsion Angles.*
Brookhaven: Protein Data Bank.

Allen, F. H. & Kennard, O. (1983). The Cambridge database of molecular structure.
Perspectives in Computing, **3**, 28–43.

Allen, F. H., Bellard, S., Brice, M. D., Cartwright, B. A., Doubleday, A., Higgs, H.,
Hummelink, T., Hummelink-Peters, B. G., Kennard, O., Motherwell, W. D. S.,
Rogers, J. R. & Watson, D. G. (1979). The Cambridge Crystallographic Data Centre:
computer-based search, retrieval, analysis and display of information. *Acta
Crystallographica*, **B35**, 2331–9.

Allinger, N. L. (1977). Conformational analysis. 130. MM2. A hydrocarbon force field
utilizing V_1 and V_2 torsional terms. *Journal of the American Chemical Society*, **99**,
8127–34.

Altona, C. & Sundaralingham, M. (1972). Conformational analysis of the sugar ring in
nucleosides and nucleotides. A new description using the concept of pseudorotation.
Journal of the American Chemical Society, **94**, 8205–12.

Alvey, N., Galway, N. & Lane, P. (1982). *An Introduction to Genstat.* London: Academic
Press.

Amidon, G. L., Yalkowsky, S. H., Arik, S. T. & Valvani, S. C. (1975). Solubility of non-
electrolytes in polar solvents. 5, Estimation of the solubility of aliphatic
monofunctional compounds in water using a molecular surface area approach. *Journal
of Physical Chemistry*, **79**, 2239–46.

Amos, R. D. (1979). An accurate *ab initio* study of the multipole moments and
polarizabilities of methane. *Molecular Physics*, **38**, 33–45.

Anderson, J. E., Ptashne, M. & Harrison, C. C. (1985). A phage repressor–operator
complex at 7 Å resolution. *Nature*, **316**, 596–601.

Anderson, S. (1984). Graphical representation of molecules and substructure search
queries in MACCS. *Journal of Molecular Graphics*, **2**, 83–90.

Ariens, E. J. (1954). Affinity and intrinsic activity in the theory of competitive inhibition.
Part 1: Problems and theory. *Archive pour Internationale Pharmacodynamie*, **99**, 32–49.

Ariens, E. J. & de Groot, W. M. (1954). Affinity and intrinsic activity in the theory of
competitive inhibition. Part 3: Homologous decamethonium derivatives and
succinylcholine esters. *Archive pour Internationale Pharmacodynamie*, **99**, 193–205.

Ariens, E. J. & Simonis, A. M. (1954). Affinity and intrinsic activity in the theory of competitive inhibition. Part 2: Experiments with para amino benzoic acid derivatives. *Archive pour Internationale Pharmacodynamie*, **99**, 175–87.

Ariens, E. J., van Rossum, J. M. & Koopman, P. C. (1960). Receptor reserve and threshold phenomena. 1, Theory and experiments with autonomic drugs tested on isolated organs. *Archive pour Internationale Pharmacodynamie*, **127**, 459–78.

Arnone, A. (1972). X-ray diffraction study of binding of 2,3-diphosphoglycerate to human deoxyhaemoglobin. *Nature*, **237**, 146–9.

Artymiuk, P. J., Blake, C. C. F., Grace, D. E. P., Oatley, S. J., Phillips, D. C. & Sternberg, M. J. E. (1979). Crystallographic studies on the dynamic properties of lysozyme. *Nature*, **280**, 563–8.

Arunlakshana, O. & Schild, H. O. (1959). Some quantitative uses of drug antagonists. *British Journal of Pharmacology*, **14**, 48–58.

Bader, R. F. W., Anderson, S. G. & Duke, A. J. (1979). Quantum topology of molecular charge distributions. 1. *Journal of the American Chemical Society*, **101**, 1389–95.

Bader, R. F. W., Nguyen-Dang, T. T. & Tal, Y. (1979). Quantum topology of molecular charge distributions. 2, Molecular structure and its change. *Journal of Chemical Physics*, **70**, 4316–29.

Baker, D. J., Beddell, C. R., Champness, J. N., Goodford, P. J., Norrington, F. E. A., Smith, D. R. & Stammers, D. K. (1981). The binding of trimethoprim to bacterial dihydrofolate reductase. *FEBS Letters*, **126**, 49–52.

Barnard, E. A., Miledi, R. & Sumikawa, K. (1982). Translation of exogenous messenger RNA coding for nicotinic acetylcholine receptors produces functional receptors in *Xenopus* oocytes. *Proceedings of the Royal Society, London* B, **215**, 241–6.

Barrett, A. N. (1983). The determination of biomolecular charge distributions using interactive computer graphics. *Journal of Molecular Graphics*, **1**, 71–4.

Barry, C. D. & North, A. C. T. (1972). The use of a computer controlled display system in the study of molecular conformations. *Cold Spring Harbor Symposium*, **36**, 577–84.

Beddell, C. R. (1984). Dihydrofolate reductase, its structure, function and binding properties. In: *X-ray Crystallography and Drug Action*, ed. A. S. Horn & C. J. De Ranter, pp. 169–93. Oxford: Clarendon Press.

Beddell, C. R., Goodford, P. J., Norrington, F. E., Wilkinson, S. & Wootton, R. (1976). Compounds designed to fit a site of known structure in human haemoglobin. *British Journal of Pharmacology*, **57**, 201–9.

Beddell, C. R., Goodford, P. J., Stammers, D. K. & Wootton, R. (1979). Species differences in the binding of compounds designed to fit a site of known structure in adult human haemoglobin. *British Journal of Pharmacology*, **65**, 535–43.

Beeman, D. (1976). Some multistep methods for use in molecular dynamics calculations. *Journal of Computational Physics*, **20**, 130–9.

Ben-Naim, A. (1980). *Hydrophobic Interactions*. New York: Plenum Press.

Bernstein, F. C. (1979). *FISIPL: A Program to Produce Phi/Psi Plots on Printer*. Brookhaven: Protein Data Bank.

Bernstein, F. C., Koetzle, T. F., Williams, G. J. B., Meyer, E. F., Brice, M. D., Rogers, J. R., Kennard, O., Shimanouchi, T. & Tasumi, M. (1977). The Protein Data Bank: a computer-based archival file for macromolecular structures. *Journal of Molecular Biology*, **112**, 535–42.

Birdsall, B., Bevan, A. W., Pascaul, C., Roberts, G. C. K., Feeney, J., Gronenborn, A. & Clore, G. M. (1984). Multinuclear nmr characterization of two coexisting conformational states of the *Lactobacillus casei* dihydrofolate reductase–trimethoprim–NADP[1] complex. *Biochemistry*, **23**, 4733–42.

Birdsall, B., Burgen, A. S. V. & Roberts, G. C. K. (1980a). Binding of coenzyme analogues to *Lactobacillus casei* dihydrofolate reductase and ternary complexes. *Biochemistry*, **19**, 3723–31.

Birdsall, B., Burgen, A. S. V. & Roberts, G. C. K. (1980b). Effects of coenzyme analogues on the binding of *p*-aminobenzoyl-*L*-glutamate and 2,4-diaminopyrimidine to *Lactobacillus casei* dihydrofolate reductase. *Biochemistry*, **19**, 3732–7.

Black, J. W. & Leff, P. (1983). Operational modes of pharmacological agonism. *Proceedings of the Royal Society, London B*, **220**, 141–62.

Blaney, J. M., Jorgensen, E. C., Connolly, M. L., Ferrin, T. E., Langridge, R., Oatley, S. J., Burridge, J. M. & Blake, C. C. F. (1982). Computer graphics in drug design: molecular modelling of thyroid hormone prealbumin interactions. *Journal of Medicinal Chemistry*, **25**, 785–90.

Blow, D. M. & Brick, P. (1985). Aminoacyl-tRNA synthetases. In: *Biological Macromolecules and Assemblies*, Vol. 2, ed. F. A. Jurnak & A. McPherson, pp. 422–69. New York: Wiley.

Blundell, T. L. & Johnson, L. N. (1976). *Protein Crystallography*. New York: Academic Press.

Bolin, J. T., Filman, D. J., Matthews, D. A., Hamlin, R. C. & Kraut, J. (1982). Crystal structures of *Escherichia coli* and *Lactobacillus casei* dihydrofolate reductase refined at 1.7 Å resolution. 1, General features and binding of methotrexate. *Journal of Biological Chemistry*, **257**, 13650–62.

Bolis, G. & Clementi, E. (1977). Analytical potentials for the interaction between biomolecules. 3, Reliability and transfer of pair-potentials. *Journal of the American Chemical Society*, **99**, 5550–7.

Bonaccorsi, R., Scrocco, E., Tomasi, J. & Pullman, A. (1975). Ab initio molecular electrostatic potentials. Guanine compared to adenine. *Theoretica Chimica Acta, Berlin*, **36**, 339–44.

Bonchev, D. (1983). *Information Theoretic Indices for Characterization of Chemical Structures*. Chichester: Research Studies Press, John Wiley & Sons.

Breitwieser, G. E. & Szabo, G. (1985). Uncoupling of cardiac muscarinic and β-adrenergic receptors from ion channels by a guanine nucleotide analogue. *Nature*, **317**, 538–40.

Brickenkamp, C. S. & Panke, D. (1973). Polyhedral clathrate hydrates XVII. Structure of the low melting hydrate of n-propylamine: a novel clathration framework. *Journal of Chemical Physics*, **58**, 5284–95.

Brickman, J. (1983). Shaded surfaces in molecular raster graphics. *Journal of Molecular Graphics*, **1**, 62–7.

Browman, M. J., Carruthers, L. M., Kashuba, K. L., Momany, F. A., Pottle, M. S., Rosen, S. P. & Rumsey, S. M. (1975). ECEPP: empirical conformational energy programs for peptides. *QCPE*, **11**, 286.

Brown, F. F. & Goodford, P. J. (1977). The interaction of some bisarylhydroxysulphonic acids with a site of known structure in human haemoglobin. *British Journal of Pharmacology*, **60**, 337–41.

Buchanan, B. G. & Shortliffe, E. H. (1984). *Rule-Based Expert Systems: the Mycin Experiments of the Stanford Heuristic Programming Project*. Reading, Massachusetts: Addison–Wesley.

Buckingham, A. D. (1978). Basic theory of intermolecular forces: applications to small molecules. In *Intermolecular Interactions: from Diatomics to Biopolymers*, ed. B. Pullman, pp. 1–68. London: John Wiley.

Burgen, A. S. V. (1966). The drug–receptor complex. *Journal of Pharmacy and Pharmacology*, **18**, 137–49.

Burkert, U. & Allinger, N. L. (1982). *Molecular Mechanics*. American Chemical Society Monograph no. 177.

Busetta, B., Tickle, I. J. & Blundell, T. L. (1983). DOCKER, an interactive program for simulating protein receptor and substrate interactions. *Journal of Applied Crystallography*, **16**, 432–7.

Cambridge Crystallographic Data-Centre User-Manual. The Chemical Laboratories, Lensfield Road, Cambridge, U.K.

Carhart, R. E., Smith, D. H. & Venkataraghavan, R. (1985). Atom pairs as molecular features in structure-activity studies: definition and applications. *Journal of Chemical Information and Computing Science*, **25**, 64–73.

Carozzo, L., Corongiu, G., Petrongolo, C. & Clementi, E. (1978). Analytical potentials from *ab initio* computations for the interaction between biomolecules. 4, Water with glycine and serine zwitterions. *Journal of Chemical Physics*, **68**, 787–93.

Chothia, C. (1974). Hydrophobic bonding and accessible surface area in proteins. *Nature*, **248**, 338–9.

Chothia, C. (1984). Principles that determine the structure of proteins. *Annual Reviews of Biochemistry*, **53**, 537–72.

Chou, P. Y. & Fasman, G. C. (1977). β-Turns in proteins. *Journal of Molecular Biology*, **115**, 135–75.

Chou, P. Y. & Fasman, G. D. (1978). Empirical predictions of protein conformation. *Annual Reviews of Biochemistry*, **47**, 251–76.

Clark, A. J. (1926a). The reaction between acetylcholine and muscle cells. *Journal of Physiology*, **61**, 530–46.

Clark, A. J. (1926b). The antagonism of acetylcholine by atropine. *Journal of Physiology*, **61**, 547–56.

Clark, A. J. (1933). *The Mode of Action of Drugs on Cells*. London: Edward Arnold.

Clark, A. J. & Raventos, J. (1937). The antagonism of acetylcholine and of quaternary ammonium salts. *Quarterly Journal of Experimental Physiology*, **26**, 375–92.

Claudio, T., Ballivet, M., Patrick, J. & Heinemann, S. (1983). Nucleotide and deduced amino acid sequence of *Torpedo californica* acetylcholine receptor γ subunit. *Proceedings of the National Academy of Sciences*, **80**, 1111–5.

Claverie, P., Daudey, J. P., Diner, S., Geissner-Prettre, C. P., Gilbert, M., Langlet, J., Malrieu, J. P., Pincelli, U. & Pullman, B. (1972). PCILO: Perturbative configuration interaction using local orbital method in the CNDO hypothesis. *QCPE*, **11**, 220.

Clementi, E. (1980). Computational aspects for large chemical systems. *Lecture Notes in Chemistry*, Vol. 19. Berlin: Springer-Verlag.

Clementi, E. & Corongiu, G. (1979). Iso-energy contour maps for the interaction of water with DNA double helix in the B conformation. *Chemical Physics Letters*, **60**, 175–8.

Clementi, E. & Corongiu, G. (1980). A theoretical study on the water structure for nucleic acid bases and base pairs in solution at $T = 300$ K. *Journal of Chemical Physics*, **72**, 3979–92.

Clementi, E., Cavalonne, F. & Scordamaglia, R. (1977). Analytical potentials from *ab initio* computations for the interaction between biomolecules. 1, Water with amino acids. *Journal of the American Chemical Society*, **99**, 5531–45.

Clementi, S. (1984). Statistics and drug design. In *Drug Design: Fact or Fantasy?* ed. J. Jolles & K. R. H. Wooldridge, p. 73. London: Academic Press.

Cohen, F. E. & Sternberg, M. J. E. (1980). On the use of chemically derived distance constraints in the prediction of protein structure with myoglobin as an example. *Journal of Molecular Biology*, 137, 9–22.

Cohen, F. E., Richmond, T. J. & Richards, F. M. (1979). Protein folding: evaluation of some simple rules for the assembly of helices into tertiary structures with myoglobin as an example. *Journal of Molecular Biology*, 132, 275–88.

Colquhoun, D. & Hawkes, A. G. (1977). Relaxation and fluctuations of membrane currents that flow through drug-operated channels. *Proceedings of the Royal Society, London* B, 199, 231–62.

Colquhoun, D., Dionne, V. E., Steinbach, J. H. & Stevens, C. F. (1975). Conductance of channels opened by acetylcholine-like drugs in muscle end-plate. *Nature*, 253, 204–6.

Connolly, M. L. (1983a). Solvent accessible surfaces of proteins and nucleic acids. *Science*, 221, 709–13.

Connolly, M. L. (1983b). Analytical molecular surface calculation. *Journal of Applied Crystallography*, 16, 548–58.

Connolly, M. L. (1985). Molecular surface triangulation. *Journal of Applied Crystallography*, 18, 499–505.

Cook, D. B. (1978). *Structures and Approximations for Electrons in Molecules*. Chichester: Ellis Horwood Ltd.

Corongiu, G. & Clementi, E. (1981a). Simulations of solvent structure for macromolecules. 1, Solvation of B-DNA double helix at $T = 300$ K. *Biopolymers*, 20, 551–71.

Corongiu, G. & Clementi, E. (1981b). Simulations of solvent structure for macromolecules. 2, Structure of water solvating Na^+–B–DNA at 300 K and a model for conformational transitions induced by solvent variations. *Biopolymers*, 20, 2427–83.

Crandell, C. W. & Smith, D. H. (1983). Computer-assisted examination of compounds for common three-dimensional substructures. *Journal of Chemical Information and Computing Science*, 23, 186–97.

Crippen, G. M. (1981). *Distance Geometry and Conformational Calculations*. Chichester: Research Studies Press.

Dale, H. H. (1956). A brief introduction to the published work of Paul Ehrlich. In *The Collected Papers of Paul Ehrlich*, Vol. 1, ed. E. Himmelweit, p. 10. London: Pergamon Press.

Danziger, D. J. (1985a). HSITE: a program to calculate hydrogen-bonding regions round a protein. *Journal of Molecular Graphics*, 3, 102–3.

Danziger, D. J. (1985b). The location of hydrogen-bonding sites around macromolecular surfaces. M.Phil. Thesis, University of Cambridge, U.K.

Danziger, D. J. & Dean, P. M. (1985). The search for functional correspondences in molecular structure between two dissimilar molecules. *Journal of Theoretical Biology*, 116, 215–24.

Dashevsky, V. G. & Sarkisov, G. N. (1974). The solvation and hydrophobic interaction of non-polar molecules in water in the approximation of interatomic potentials: the Monte Carlo method. *Molecular Physics*, 27, 1271–90.

Dean, P. M. (1981a). Drug–receptor recognition: electrostatic field lines at the receptor and dielectric effects. *British Journal of Pharmacology*, 74, 39–46.

Dean, P. M. (1981b). Drug–receptor recognition: molecular orientation and dielectric effects. *British Journal of Pharmacology*, 74, 47–60.

Dean, P. M. (1983). Molecular recognition in drug research. *Perspectives in Computing*, **3**, 14–27.

Dean, P. M. (1984). Graphical methods for the analysis and display of the molecular electrostatic potential surrounding a drug or its binding site. In *QSAR in Design of Bioactive Compounds*, chaired M. Kuchar, pp. 253–64. Barcelona: J. R. Prous.

Dean, P. M. & Wakelin, L. P. G. (1979). The docking manoeuvre at a drug receptor: a quantum mechanical study of intercalative attack of ethidium and its carboxylated derivative on a DNA fragment. *Philosophical Transactions of the Royal Society, London* B, **287**, 571–604.

Dean, P. M. & Wakelin, L. P. G. (1980a). Electrostatic components of drug–receptor recognition. 1: Structural and sequence analogues of DNA polynucleotides. *Proceedings of the Royal Society, London* B, **209**, 453–71.

Dean, P. M. & Wakelin, L. P. G. (1980b). Electrostatic components of drug–receptor recognition. 2: The DNA-binding antibiotic actinomycin. *Proceedings of the Royal Society, London* B, **209**, 473–87.

Deisenhofer, J. & Steigmann, W. (1975). Crystallographic refinement of the structure of bovine pancreatic trypsin inhibitor at 1.5 Å resolution. *Acta Crystallographica*, **B31**, 238–50.

Del Bene, J. & Pople, J. A. (1970). Theory of molecular interactions. 1, Molecular orbital studies of water polymers using a minimal Slater-type basis. *Journal of Chemical Physics*, **52**, 4858–66.

Deming, S. N. (1984). Experimental design: response surfaces. In *Chemometrics. Mathematics in Statistics and Chemistry*, ed. B. R. Kowalski, pp. 251–66. Dordrecht: D. Reidel.

Dickerson, R. E. (1983). Base sequence and helix structure variation in B and A DNA. *Journal of Molecular Biology*, **166**, 419–41.

Dickerson, R. E. & Drew, H. R. (1981). Structure of a B-DNA dodecamer II. Influence of base sequence on helix structure. *Journal of Molecular Biology*, **149**, 761–86.

Dickerson, R. E., Drew, H. R., Conner, B. N., Kopka, M. L. & Pjura, P. E. (1983). Helix geometry and hydration in A-DNA, B-DNA and Z-DNA. *Cold Spring Harbor Symposium. Quantitative Biology*, **47**, (**P1**), 13–24.

Dickerson, R. E., Drew, H. R., Conner, B. N., Wing, R. M., Fratini, A. V. & Kopka, M. L. (1982). The anatomy of A- B- and Z-DNA. *Science*, **216**, 475–85.

Dove, M. T. & Pawley, G. S. (1983). A molecular dynamics simulation study of the plastic crystalline phase of sulphur hexafluoride. *Journal of Physics, C., Solid State Physics*, **16**, 5969–83.

Dove, M. T. & Pawley, G. S. (1984). A molecular dynamics simulation of the orientationally disordered phase of sulphur hexafluoride. *Journal of Physics, C., Solid State Physics*, **17**, 6581–99.

Drew, H. R. & Dickerson, R. E. (1981). Structure of a B-DNA dodecamer III. Geometry and hydration. *Journal of Molecular Biology*, **151**, 535–56.

Edmonds, D. T., Rogers, N. K. & Sternberg, M. J. E. (1984). Regular representation of irregular charge distributions, applications to the electrostatic potentials of globular proteins. *Molecular Physics*, **52**, 1487–94.

Ehrlich, P. (1900). On immunity with special reference to cell life. Croonian Lecture. *Proceedings of the Royal Society, London*, **66**, 424–48.

Ehrlich, P. (1902). The relationship existing between chemical constitution, distribution and pharmacological action. In *The Collected Papers of Paul Ehrlich*, Vol. 1, ed. F. Himmelweit, pp. 596–618. London: Pergamon Press. Published in translation, 1956.

Ehrlich, P. (1909a). On partial functions of the cell. Nobel Lecture. In *The Collected Papers of Paul Ehrlich*, Vol. 3, ed. F. Himmelweit, pp. 183–94. London: Pergamon Press. Published in translation, 1956.

Ehrlich, P. (1909b). Chemotherapie von Infekkorskrankheiten. In *The Collected Papers of Paul Ehrlich*, Vol. 3, ed. F. Himmelweit, pp. 213–27. London: Pergamon Press. Published in translation, 1956.

Ehrlich, P. & Morgenroth, J. (1900). On haemolysis: third communication. In *The Collected Papers of Paul Ehrlich*, Vol. 2, ed. F. Himmelweit, pp. 205–12. London: Pergamon Press. Published in translation, 1956.

Ehrlich, P. & Morgenroth, J. (1901). On haemolysis: fifth communication. In *The Collected Papers of Paul Ehrlich*, Vol. 2, ed. F. Himmelweit, pp. 246–55. London: Pergamon Press. Published in translation, 1956.

Einstein, A. (1956). *Investigations on the Theory of Brownian Movement*. New York: Dover Publications.

Eisenberg, D. (1984). Three-dimensional structure of membrane and surface proteins. *Annual Reviews of Biochemistry*, **53**, 595–623.

Eisenberg, D., Schwarz, E., Komaromy, M. & Wall, R. (1984). Analysis of membrane and surface protein sequences with the hydrophobic moment plot. *Journal of Molecular Biology*, **179**, 125–42.

Eisenberg, D., Weiss, R. M. & Terwilliger, T. C. (1984). The hydrophobic moment detects periodicity in protein hydrophobicity. *Proceedings of the National Academy of Sciences*, **81**, 140–4.

Eisenberg, D., Weiss, R. M., Terwilliger, T. C. & Wilcox, W. (1982). Hydrophobic moments and protein structure. *Faraday Symposium Chemical Society*, **17**, 109–20.

Emmerson, J. & Mackay, D. (1981). A test of the null equation for functional antagonism. *British Journal of Pharmacology*, **73**, 135–41.

Fersht, A. R., Shi, J. P., Knill-Jones, J., Lowe, D. M., Wilkinson, A. J., Blow, D. M., Brick, P., Carter, P., Waye, M. M. Y. & Winter, G. (1985). Hydrogen bonding and biological specificity analysed by protein engineering. *Nature*, **314**, 235–8.

Filman, D. J., Bolin, J. T., Matthews, D. A. & Kraut, J. (1982). Crystal structures of *Escherichia coli* and *Lactobacillus casei* dihydrofolate reductase refined at 1.7 Å resolution. 2: Environment of bound NADPH and implications for catalysis. *Journal of Biological Chemistry*, **257**, 13663–72.

Finer-Moore, J. & Stroud, R. M. (1984). Amphipathic analysis and possible formation of the ion channel in an acetylcholine receptor. *Proceedings of the National Academy of Sciences*, **81**, 155–9.

Finney, J. L. (1984). Solvent effects in biomolecular systems and processes. *Journale de Physique*, **45**, Supplement 9, Communication 7, 197–210.

Franke, R. (1984). *Theoretical Drug Design Methods*. Amsterdam:Elsevier.

Free, S. M. & Wilson, J. N. (1964). A mathematical contribution to structure–activity studies. *Journal of Medicinal Chemistry*, **7**, 395–9.

Frohlich, H. (1958). *Theory of Dielectrics*. Oxford: Clarendon Press.

Fukui, K. (1975). *Theory of Orientation and Stereoselection*. New York: Springer-Verlag.

Furchgott, R. F. (1966). The use of β-haloalkylamines in the differentiation of receptors and in the determination of dissociation constants of receptor-agonist complexes. *Advances in Drug Research*, **3**, 21–55.

Gaddum, J. H. (1936). The quantitative effects of antagonistic drugs. *Journal of Physiology*, **89**, 7–9.

Gardner, P., Ogden, D. C. & Colquhoun, D. (1984). Conductances of single ion channels opened by nicotinic agonists are indistinguishable. *Nature*, **309**, 160–2.

Geiger, A. & Stanley, H. E. (1982). Low-density patches in the hydrogen-bond network of liquid water: evidence from molecular dynamics computer simulations. *Physical Review Letters*, **49**, 1749–52.

Geiger, A., Mausbach, P., Schnither, J., Blumberg, R. L. & Stanley, E. H. (1984). Structure and dynamics of the hydrogen bond network in water by computer simulations. *Journale de Physique*, **45**, Supplement 9, Communication 7, 13–30.

GENSTAT. (1983). *GENSTAT: A General Statistical Program*. Oxford: Numerical Algorithms Group.

Gilman, A. G. (1984). G-Proteins and dual control of adenylate cyclase. *Cell*, **36**, 577–9.

Golender, V. E. & Rosenblit, A. B. (1983). *Logical and Combinatorial Algorithms for Drug Design*. Chichester: Research Studies Press, John Wiley & Sons.

Goodfellow, J. M., Finney, J. L. & Barnes, P. (1982). Monte Carlo computer simulation of water-amino acid interactions. *Proceedings of the Royal Society, London B*, **214**, 213–28.

Gordon, M. S. & Pople, J. A. (1969). MBLD: standard geometric models and Cartesian coordinates of molecules. *QCPE*, **11**, 135.

Grant, E. H., Sheppard, R. J. & South, G. P. (1978). *Dielectric Behaviour of Biological Molecules in Solution*. Oxford: Clarendon Press.

Greer, J. & Bush, B. L. (1978). Macromolecular shape and surface maps by solvent exclusion. *Proceedings of the National Academy of Sciences*, **75**, 303–7.

Gresh, N., Claverie, P. & Pullman, A. (1984). Theoretical studies of molecular conformation. Derivative of an additive procedure for the computation of intramolecular interaction energies. Comparison with *ab initio* SCF computations. *Theoretica Chimica Acta, Berlin*, **66**, 1–20.

Gronenborn, A., Birdsall, B., Hyde, E. I., Roberts, G. C. K., Feeney, J. & Burgen, A. S. V. (1981). [1]H and ^{31}P nmr characterization of the conformations of the trimethoprim-NADP[1]-dihydrofolate reductase complex. *Molecular Pharmacology*, **20**, 145–53.

Hassall, C. H. (1985). Computer graphics as an aid to drug design. *Chemistry in Britain*, **21**, No. 1, 39–46.

Hassall, C. H., Krohn, A., Moody, C. J. & Thomas, W. A. (1984). The design and synthesis of new triazolo, pyrazolo and pyridazo-pyridazine derivatives as inhibitors of angiotensin converting enzyme. *Journal of the Chemical Society, Perkin Transactions*, **1**, 155–64.

Havere van, W., Lenstra, A. T. H., Geise, H. J., Berg van den, G. R. & Benschop, H. P. (1982). 2-((Hydroxyimino)methyl)-1-methylpyridinium chloride. *Acta Crystallographica*, **B38**, 2516–8.

Hayes, D. M. & Kollman, P. A. (1976a). Electrostatic potentials of proteins. 1, Carboxypeptidase A. *Journal of the American Chemical Society*, **98**, 3335–45.

Hayes, D. M. & Kollman, P. A. (1976b). Electrostatic potentials of proteins. 2, Role of electrostatics in a possible catalytic mechanism for carboxypeptidase A. *Journal of the American Chemical Society*, **98**, 7811–6.

Hayes-Roth, F., Waterman, D. & Lenat, D. (1983). *Building Expert Systems*. Reading, Massachusetts: Addison-Wesley.

Henry, N. F. M. & Lonsdale, K. (1969). *International Tables for X-ray Crystallography*. Birmingham: Kynoch Press.

Hermann, R. B. (1972). Theory of hydrophobic bonding. 2, The correlation of hydrocarbon solubility in water with solvent cavity surface area. *Journal of Physical Chemistry*, **76**, 2754–9.

Hill, A. V. (1909). The mode of action of nicotine and curari, determined by the form of the contraction curve and the method of temperature coefficients. *Journal of Physiology*, **39**, 361–73.

Hill, A. V. (1913). The combination of haemoglobin with oxygen and with carbon monoxide. *Biochemical Journal*, **7**, 471–80.

Hindin, H. J. (1985). Parallel processing promises faster program execution. *Computer Design*, **24**, No. 10, 57–66.

Hol, W. G. J. & Wierenga, R. K. (1984). The α-helix dipole and the binding of phosphate groups of coenzymes and substrates by proteins. In *X-ray Crystallography and Drug Action*, ed. A. S. Horn & C. J. De Ranter, pp. 151–68. Oxford: Clarendon Press.

Hopp, T. P. & Woods, K. R. (1981). Prediction of protein antigenic determinants from amino acid sequences. *Proceedings of the National Academy of Sciences*, **78**, 3824–8.

Hyde, E. I., Birdsall, B., Roberts, G. C. K., Feeney, J. & Burgen, A. S. V. (1980a). Proton nuclear magnetic resonance saturation transfer studies of coenzyme binding to *Lactobacillus casei* dihydrofolate reductase. *Biochemistry*, **19**, 3738–46.

Hyde, E. I., Birdsall, B., Roberts, G. C. K., Feeney, J. & Burgen, A. S. V. (1980b). Phosphorus-31 nuclear magnetic resonance studies of the binding of oxidised coenzymes to *Lactobacillus casei* dihydrofolate reductase. *Biochemistry*, **19**, 3746–54.

International Tables for X-ray Crystallography (1969). Birmingham: Kynoch Press.

Jeffrey, G. A. (1982). Hydrogen bonding in amino acids and carbohydrates. In *Molecular Structure and Biological Activity*, ed. J. F. Griffin & W. L. Duax, pp. 135–48. New York: Elsevier.

Jeffrey, G. A. & Maluszynska, H. (1982). A survey of hydrogen bond geometries in the crystal structure of amino acids. *International Journal of Biological Macromolecules*, **4**, 173–85.

Jeffrey, G. A. & Robinns, A. (1978). 2,3-Dimethyl-2,3-butanediol (pinacol). *Acta Crystallographica*, **B34**, 3817–20.

Johari, G. P. (1981). The dipolar correlation factor, the electrostatic field, the dipole moment and the Coulombic-interaction energy of water molecules in clathrate hydrates. *Journal of Chemical Physics*, **74**, 1326–36.

Johnson, A. P. (1985). Computer aids to synthesis planning. *Chemistry in Britain*, **21**, No. 1, 59.

Jolles, G. & Wooldridge, K. R. H. (1984). *Drug Design: Fact or Fantasy?* London: Academic Press.

Kabasch, W. & Sander, C. (1983). *A Program to Define Secondary Structure and Solvent Exposure of Proteins from Atomic Coordinates*. Brookhaven: Protein Data Bank.

Kao, P. N., Dwork, A. J., Kaldany, R. R. J., Silver, M. L., Wideman, J., Stein, S. & Karlin, A. (1984). Identification of the α subunit half-cystine specifically labelled by an affinity reagent for the acetylcholine receptor binding site. *Journal of Biological Chemistry*, **259**, 11662–5.

Katz, B. & Miledi, R. (1970). Membrane noise produced by acetylcholine. *Nature*, **226**, 962–3.

Katz, B. & Miledi, R. (1972). The statistical nature of the acetylcholine potential and its molecular components. *Journal of Physiology*, **224**, 665–99.

Kauzmann, W. (1959). Some factors in the interpretation of protein denaturation. *Advances in Protein Chemistry*, **14**, 1–63.

Klopman, G. (1984). Artificial intelligence approach to structure activity studies. Computer automa͂ted structure evaluation of biological activity of organic molecules. *Journal of the American Chemical Society*, **106**, 7315–21.

Klopman, G. & McGonigal, M. (1981). Computer simulation of physical-chemical properties of organic molecules: molecular system identification. *Journal of Chemical Information and Computing Science*, **21**, 48–52.

Komatsu, K., Nakamura, H., Nakagawa, S. & Umeyama, H. (1984). Electrostatic forces in the inhibition of dihydrofolate reductase by methotrexate. A field potential study. *Chemical and Pharmaceutical Bulletin*, **32**, 3313–6.

Kowalski, B. R. (1984). *Chemometrics: Mathematics and Statistics in Chemistry.* Dordrecht: D. Reidel Publishing Company.

Krugh, T. R., Wittlin, F. N. & Cramer, S. P. (1975). Ethidium bromide–dinucleotide complexes. Evidence for intercalation and sequence preferences in binding to double-stranded nucleic acids. *Biopolymers*, **14**, 197–210.

Kuntz, I. D., Blaney, J. M., Oatley, S. J., Langridge, R. & Ferrin, T. E. (1982). A geometric approach to macromolecule–ligand interactions. *Journal of Molecular Biology*, **161**, 269–88.

Kyte, J. & Doolittle, R. F. (1982). A simple method for displaying the hydropathic character of a protein. *Journal of Molecular Biology*, **157**, 105–32.

Lackey, S., Veres, J. & Ziegler, M. (1985). Supercomputer expands parallel processing options. *Computer Design*, **24**, No. 10, 76–81.

Langlet, J., Claverie, P. & Caron, F. (1981). Intermolecular interactions in an external electric field: application to the analysis of the evaluation of interaction energies from field mass spectrometry experiments. In *Intermolecular Forces*, ed. B. Pullman, pp. 397–429. Dordrecht: D. Reidel.

Langley, J. N. (1873). On the physiological action of jaborandi. *Proceedings of the Cambridge Philosophical Society*, April, 402.

Langley, J. N. (1878). On the physiology of salivary secretion II. On the mutual antagonism of atropin and pilocarpin, having special reference to their relations in the submaxillary gland of the cat. *Journal of Physiology*, **1**, 339–69.

Langley, J. N. (1880). On the antagonism of poisons. *Journal of Physiology*, **3**, 11–21.

Langley, J. N. (1906). On nerve endings and on special excitable substances in cells. Croonian Lecture. *Proceedings of the Royal Society, London* B, **78**, 170–94.

Langley, J. N. (1907). On the contraction of muscle, chiefly in relation to the presence of 'receptive' substances (Part 1). *Journal of Physiology*, **36**, 347–84.

Langley, J. N. (1908a). On the contraction of muscle, chiefly in relation to the presence of 'receptive' substances (Part 2). *Journal of Physiology*, **37**, 165–212.

Langley, J. N. (1908b). On the contraction of muscle, chiefly in relation to the presence of 'receptive' substances (Part 3). The reaction of frog's muscle to nicotine after denervation. *Journal of Physiology*, **37**, 285–300.

Langley, J. N. (1909). On the contraction of muscle chiefly in relation to the presence of 'receptive' substances (Part 4). The effect of curari and of some other substances on the nicotine response of the sartorius and gastrocnemius muscles of the frog. *Journal of Physiology*, **39**, 235–95.

Langley, J. N. & Dickinson, W. L. (1890a). Pituri and nicotin. *Journal of Physiology*, **11**, 265–306.

Langley, J. N. & Dickinson, W. L. (1890b). The action of various poisons upon nerve fibres and peripheral nerve cells. *Journal of Physiology*, **11**, 509–27.

Langridge, R., Ferrin, T. E., Kuntz, I. & Connolly, M. (1981). Real time colour graphics in studies of molecular interactions. *Science*, **211**, 661–6.

Lavery, R., Pullman, A. & Pullman, B. (1982). The electrostatic field of B-DNA. *Theoretica Chimica Acta, Berlin*, **62**, 93–106.

Lee, B. & Richards, F. M. (1971). The interpretation of protein structures: estimation of static accessibility. *Journal of Molecular Biology*, **55**, 379–400.

Lee, J. S. & Waring, M. J. (1978). Bifunctional intercalation and sequence specificity in the binding of quinomycin and triostin antibiotics to deoxyribonucleic acid. *Biochemical Journal*, **173**, 115–28.

Leo, A. & Weininger, D. (1984). CLOGP3: a database for calculating log *P*. *Pomona College Medicinal Chemistry Project*, Claremont, California, U.S.A.

Leslie, A. G. W., Arnott, S., Chandrasekaran, R. & Ratliff, R. L. (1980). Polymorphism of DNA double helices. *Journal of Molecular Biology*, **143**, 49–72.

Levine, R. D. & Bernstein, R. B. (1974). *Molecular Reaction Dynamics*. Oxford: Oxford University Press.

Levitt, M. (1983a). Computer simulation of DNA double-helix dynamics. *Cold Spring Harbor Symposia on Quantitative Biology*, **47**, 251–62.

Levitt, M. (1983b). Molecular dynamics of native protein. 1. *Journal of Molecular Biology*, **168**, 595–620.

Levitt, M. (1983c). Molecular dynamics of native protein. 2. *Journal of Molecular Biology*, **168**, 621–57.

Levitt, M. & Chothia, C. (1976). Structural patterns in globular proteins. *Nature*, **261**, 552–8.

Levitzki, A. (1981). Negative cooperativity at the insulin receptor. *Nature*, **289**, 442–3.

Levitzki, A. (1984). *Receptors: A Quantitative Approach*. California: Benjamin/Cummings.

Lewi, P. J. (1984). Multidimensional data representation in medicinal chemistry. In *Chemometrics: Mathematics and Statistics in Chemistry*, ed. B. R. Kowalski, pp. 351–76. Dordrecht: D. Reidel Publishing Company.

Lim, V. I. (1974a). Structural principles of the globular organization of protein chains. A stereochemical theory of globular protein secondary structure. *Journal of Molecular Biology*, **88**, 857–72.

Lim, V. I. (1974b). Algorithms for prediction of α-helical and β-structural regions in globular proteins. *Journal of Molecular Biology*, **88**, 873–94.

Long, A. K., Rubenstein, S. D. & Joncas, L. J. (1983). A computer program for organic synthesis. *Chemistry and Engineering News*, **61**, May, 22.

Mackay, D. (1981). An analysis of functional antagonism and synergism. *British Journal of Pharmacology*, **73**, 127–34.

Mager, P. P. (1985). *Multidimensional Pharmacochemistry: Design of Safer Drugs*. New York: Academic Press.

Maitland, G. C., Rigby, M., Smith, E. B. & Wakeham, W. A. (1981). *Intermolecular Forces: Their Origin and Determination*. Oxford: Clarendon Press.

Mardia, K. V., Kent, J. T. & Bibby, J. M. (1979). *Multivariate Analysis*. London: Academic Press.

Marx, J. L. (1983). Cloning of acetylcholine receptor genes. *Science*, **219**, 1055–6.

McLachlan, A. D. (1965). Effect of the medium on dispersion forces in liquids. *Discussions of the Faraday Society*, **40**, 239–45.

McLachlan, A. D. (1982). Rapid comparison of protein structures. *Acta Crystallographica*, **A38**, 871–3.

Metropolis, N., Rosenbluth, A. W., Rosenbluth, M. M., Teller, A. H. & Teller, E. (1953). Equation of state calculations by fast computing machines. *Journal of Chemical Physics*, **21**, 1087–92.

Mezei, M. & Beveridge, D. L. (1981). Monte Carlo studies of the structure of dilute aqueous solutions of Li^+, Na^+, K^+, F^- and Cl^-. *Journal of Chemical Physics*, **74**, 6902–10.

Mezei, M., Beveridge, D. L., Berman, H. M., Goodfellow, J. M., Finney, J. L. & Neidle, S. (1983). Monte Carlo studies on water in the dCpG/proflavine crystal hydrate. *Journal of Biomolecular Structure and Dynamics*, **1**, 287–97.

Mishina, M., Tobimatsu, T., Imoto, K., Tanaka, K., Fujita, Y., Fukuda, K., Kurasaki, M., Takahashi, H., Morimoto, Y., Hirose, T., Inayama, S., Takahashi, T., Kuno, M. & Numa, S. (1985). Location of functional regions of acetylcholine receptor α-subunit by site-directed mutagenesis. *Nature*, **313**, 364–9.

Mokhoff, N. (1984). Parallelism makes strong bid for next generation computers. *Computer Design*, **23**, No. 10, 104–31.

Morffew, A. J. (1983). The use of animated difference matrices for analysing protein molecular dynamics simulation data. *Journal of Molecular Graphics*, **1**, 43–7.

Morffew, A. J., & Todd, S. J. P. (1984). The use of Balasubramanian plots in the analysis of protein dynamics data. *Journal of Molecular Graphics*, **2**, 18–20.

Morrison, D. F. (1978). *Multivariate Statistical Methods*. London: McGraw Hill.

Mulliken, R. S. (1955). Electronic population analysis on LCAO-MO molecular wave functions 1. *Journal of Chemical Physics*, **23**, 1833–40.

Murray-Rust, P. & Bland, R. (1978). Computer retrieval and analysis of molecular geometry. 2, Variance and its interpretation. *Acta Crystallographica*, **B34**, 2527–33.

Murray-Rust, P. & Motherwell, W. D. S. (1978a). Computer retrieval and analysis of molecular geometry. 1, General principles and methods. *Acta Crystallographica*, **B34**, 2518–26.

Murray-Rust, P. & Motherwell, W. D. S. (1978b). Computer retrieval and analysis of molecular geometry. 3, Geometry of the β-1^1-aminofuranoside fragment. *Acta Crystallographica*, **B34**, 2534–46.

Nakamura, H., Komatsu, K., Nakagawa, S. & Umeyama, H. (1985). Visualization of electrostatic recognition by enzymes for their ligands and cofactors. *Journal of Molecular Graphics*, **3**, 2–11.

Namasivayam, S. & Dean, P. M. (1986). Statistical method for surface pattern-matching between dissimilar molecules: electrostatic potentials and accessible surfaces. *Journal of Molecular Graphics*, **4**, 46–50.

Neidle, S. & Berman, H. M. (1982). Drug intercalation in nucleic acids: the current state of knowledge. In *Molecular Structure and Biological Activity*, ed. J. F. Griffin & W. L. Duax, pp. 287–301. Amsterdam: Elsevier.

Neidle, S., Berman, H. M. & Shieh, H. S. (1980). Highly structured water network in crystals of a deoxydinucleoside–drug complex. *Nature*, **288**, 129–33.

Nickerson, M. (1956). Receptor occupancy and tissue response. *Nature*, **178**, 697–8.

Nishikawa, K. & Ooi, T. (1974). Comparison of homologous tertiary structures of proteins. *Journal of Theoretical Biology*, **43**, 351–74.

Noda, M., Shimiza, S., Tanabe, T., Takai, T., Kayano, T., Ikeda, T., Takahashi, H., Nakayama, H., Kanoaka, Y., Minamino, M., Kangawa, K., Matsuo, H., Rafferty, M. A., Hirose, T., Seuchi, I., Hayashida, H., Miyata, T. & Numa, S. (1984). Primary structure of *Electrophorus electricus* sodium channel deduced from c DNA sequence. *Nature*, **312**, 121–7.

Noda, M., Takahashi, H., Tanabe, T., Toyosato, M., Furutani, Y., Hirose, T., Asai, M., Inayama, S., Miyata, T. & Numa, S. (1982). Primary structure of α-subunit precursor of *Torpedo californica* acetylcholine receptor deduced from c DNA sequence. *Nature*, **299**, 793–7.

Noda, M., Takahashi, H., Tanabe, T., Toyosato, M., Kikyotani, S., Hirose, T., Asai, M., Takashima, H., Inayama, S., Miyata, T. & Numa, S. (1983). Primary structures of β- and α-subunit precursors of *Torpedo californica* acetylcholine receptor deduced from c DNA sequences. *Nature*, **301**, 251–5.

Nuss, M. E., Marsh, F. J. & Kollman, P. A. (1979). Theoretical studies of drug–dinucleotide interactions. Empirical energy function calculations on the interaction of ethidium, 9-aminoacridine, and proflavin cations with the base-paired dinucleotides GpC and CpG. *Journal of the American Chemical Society*, **101**, 825–33.

Olson, W. K. (1982). Theoretical studies of nucleic acid conformation. Potential energies, chain statistics and model building. In *Topics in Nucleic Acid Structure, Part 2*, ed. S. Neidle, pp. 1–79. London: Macmillan.

Osman, R., Weinstein, H. & Topiol, S. (1981). Models for active sites of metalloenzymes. 2, Interactions with a model substrate. *Annals of the New York Academy of Sciences*, **367**, 356–69.

Paddon, D. J. (1984). *Supercomputers and Parallel Computation*. Oxford: Clarendon Press.

Padlan, E. A. & Davies, D. R. (1975). Variability of three dimensional structure in immunoglobulins. *Proceedings of the National Academy of Sciences*, **72**, 819–23.

Papadimitriou, C. H. & Steiglitz, K. (1982). *Combinatorial Optimization: Algorithms and Complexity*. New Jersey: Prentice-Hall.

Parthasarathy, R., Srikrishnan, T., Ginell, S. L. & Guru Row, T. N. (1982). Intercalation of water molecules between nucleic acid base: characteristics of the intercalated water molecules and their possible biological implications. In *Molecular Structure and Biological Activity*, ed. J. F. Griffin & W. L. Duax, pp. 351–64. Amsterdam: Elsevier.

Paton, W. D. M. (1961). A theory of drug action based on the rate of drug-receptor combination. *Proceedings of the Royal Society, London* B, **154**, 21–69.

Pawley, G. S. & Dove, M. T. (1983). Molecular dynamics on a parallel computer. *Helvetica Physica Acta*, **56**, 583–92.

Pawley, G. S. & Thomas, G. W. (1982). Implementations of lattice calculations on the DAP. *Journal of Computational Physics*, **47**, 165–78.

Pedersen, B. F. (1974). The geometry of hydrogen bonds from donor water molecules. *Acta Crystallographica*, **B30**, 289–91.

Peinel, G., Frischleder, H. & Birnstock, S. (1980). The electrostatic molecular potential – a tool for the prediction of electrostatic molecular interaction properties. *Theoretica Chimica Acta, Berlin*, **57**, 245–53.

Perahia, D. & Pullman, B. (1971). The conformational energy map for the disulphate bridge in proteins. *Biochemical and Biophysical Research Communications*, **43**, 65–8.

Perutz, M. F. (1985). The birth of protein engineering. *New Scientist*, 13th June, 12–15.

Pfaffinger, P. J., Martin, J. M., Hunter, D. D., Nathanson, N. M. & Hille, B. (1985). GTP-binding proteins couple cardiac muscarinic channels to a K channel. *Nature*, **317**, 536–8.

Podelski, T. R. & Changeux, J. P. (1970). On the excitability and cooperativity of the electroplax membrane. In *Fundamental Concepts in Drug–Receptor Interactions*, ed. J. F. Danielli, J. F. Moran & D. J. Triggle, pp. 93–119. New York: Academic Press.

Politzer, P. & Daiker, K. C. (1981). Models for chemical reactivity. In *The Force Concept in Chemistry*, ed. B. M. Deb, pp. 294–387. New York: Van Nostrand Reinhold.

Port, G. N. J. & Pullman, A. (1974). Quantum-mechanical studies of environmental effects on biomolecules. *Ab initio* studies of the hydration of peptides and proteins. *International Journal of Quantum Chemistry, Quantum Biology Symposium*, 1, 21–32.

Pullman, A. & Pullman, B. (1974). New paths in molecular orbital approach to solvation of biological molecules. *Quarterly Reviews of Biophysics*, 7, 505–66.

Pullman, A., Pullman, B. & Berthod, H. (1978). An SCF *ab initio* investigation of the 'through-water' interaction of the phosphate anion with the Na^+ cation. *Theoretica Chimica Acta*, 47, 175–92.

Pullman, B. (1976). Proteins, nucleic acids and their constituents. In *Quantum Mechanics of Molecular Conformations*, ed. B. Pullman, pp. 295–383. London: John Wiley.

Pullman, B. & Courriere, P. (1972). On the conformation of acetylcholine and acetylthiocholine. *Molecular Pharmacology*, 8, 371–3.

Quarendon, P., Naylor, C. B. & Richards, W. G. (1984). Display of quantum mechanical properties on Van der Waals' surfaces. *Journal of Molecular Graphics*, 2, 4–7.

Ragazzi, M., Ferro, D. R. & Clementi, E. (1979). Analytical potentials from *ab initio* computations for the interaction between biomolecules. 5, Formyl-triglycylamide and water. *Journal of Chemical Physics*, 70, 1040–50.

Raventos, J. (1937). Pharmacological actions of quaternary ammonium salts. *Quarterly Journal of Experimental Physiology*, 26, 361–74.

Raventos, J. (1938). Synergisms and antagonisms of acetylcholine by quaternary ammonium salts. *Quarterly Journal of Experimental Physiology*, 27, 99–111.

Rekker, R. F. (1977). *The Hydrophobic Fragmental Constant: its Derivation and Application*. Oxford: Elsevier.

Remington, S. J. & Matthews, B. W. (1980). A systematic approach to the comparison of protein structure. *Journal of Molecular Biology*, 140, 77–99.

Richards, W. G. & Mangold, L. (1983). Computer aided molecular design. *Endeavour*, 7, 2–4.

Richards, W. G. (1983). *Quantum Pharmacology*, 2nd edn. London: Butterworths.

Richardson, J. S. (1981). Anatomy and taxonomy of protein structure. *Advances in Protein Chemistry*, 34, 167–339.

Richmond, T. J. & Richards, F. M. (1978). Packing of α-helices: geometrical constraints and contact areas. *Journal of Molecular Biology*, 119, 537–55.

Roberts, G. C. K. (1983). Flexible keys and deformable locks: ligand binding to dihydrofolate reductase. In *Quantitative Approaches to Drug Design*, ed. J. C. Dearden, pp. 91–8. Amsterdam: Elsevier.

Rogers, N. K. & Sternberg, M. J. E. (1984). Electrostatic interactions in globular proteins. Different dielectric models applied to the packing of α-helices. *Journal of Molecular Biology*, 174, 527–42.

Rollett, J. S. (1965). *Computing Methods in Crystallography*. Oxford: Pergamon Press.

Rossky, P. J. & Karplus, M. (1979). Solvation. A molecular-dynamics study of a dipeptide in water. *Journal of the American Chemical Society*, 101, 1913–37.

Rossky, P. J., Karplus, M. & Rahman, A. (1979). A model for the simulation of an aqueous dipeptide solution. *Biopolymers*, 18, 825–54.

Rupley, J. A., Gratton, E. & Careri, G. (1983). Water and globular proteins. *Trends in Biochemical Sciences*, 8, 18–22.

Santary, M. & Kypr, J. (1984). A fast computer algorithm for finding an optimum geometrical interaction of two macromolecules. *Journal of Molecular Graphics*, **2**, 47–9.

SAS. (1982). *SAS Users Guide: Statistics*. North Carolina: SAS Institute.

Schild, H. O. (1947). pA. A new scale for the measurement of drug antagonism. *British Journal of Pharmacology*, **2**, 189–206.

Schild, H. O. (1949). pA_x and competitive drug antagonism. *British Journal of Pharmacology*, **4**, 277–80.

Schuster, P. (1978). The fine structure of the hydrogen bond. In *Intermolecular Interactions: from Diatomics to Biopolymers*, ed. B. Pullman, pp. 363–432. New York: Wiley Interscience.

Schuster, P., Zundel, G. & Sandorfy, C. (1976). *The Hydrogen Bond: Recent Developments in Theory and Experiments*, Vol. 1. New York: North Holland.

Schwarzenbach, D. (1968). Structure of piperazine hexahydrate. *Journal of Chemical Physics*, **48**, 4134–40.

Scordamaglia, R., Cavalonne, F. & Clementi, E. (1977). Analytical potentials from *ab initio* computations for the interaction between biomolecules. 2, Water with the four bases of DNA. *Journal of the American Chemical Society*, **99**, 5545–50.

Shieh, H. S., Berman, H. M., Dabrow, M. & Neidle, S. (1980). The structure of drug–deoxydinucleoside phosphate complex: generalized conformational behaviour of intercalation complexes with RNA and DNA fragments. *Nucleic Acids Research*, **8**, 85–97.

Simon, Z., Chiriac, A., Holban, S., Ciubotaru, D. & Mihalas, G. I. (1984). *Minimum Steric Difference: The MTD Method for QSAR Studies*. Letchworth: Research Studies Press, John Wiley & Sons.

Sobell, H. M., Sakore, R. D., Jain, S. C., Banerjee, A., Bhandary, K. K., Reddy, B. S. & Lozansky, E. D. (1983). β-Kinked DNA – a structure that gives rise to drug intercalation and DNA breathing – and its wider significance in determining the premelting and melting behaviour of DNA. *Cold Spring Harbor Symposia on Quantitative Biology*, **47**, 293–314.

Sperelakis, N. (1984). Hormonal and neurotransmitter regulation of Ca^{++} influx through voltage dependent slow channels in cardiac muscle membrane. *Membrane Biochemistry*, **5**, 131–66.

Stanley, E. H., Blumberg, R. L. & Geiger, A. (1983). Gelation models of hydrogen bond networks in liquid water. *Physical Review B*, **28**, 1626–9.

Steiner, E. (1976). *The Determination and Interpretation of Molecular Wave Functions*. Cambridge: University Press.

Stephenson, R. P. (1956). A modification of receptor theory. *British Journal of Pharmacology*, **11**, 379–93.

Sternberg, M. J. E. (1983). The analysis and prediction of protein structure. In *Computing in Biological Science*, ed. M. J. Geisow & A. N. Barrett, pp. 143–77. Amsterdam: Elsevier Biomedical.

Sternberg, M. J. E. & Cohen, F. (1982). Prediction of the secondary and tertiary structures of interferon from four homologous amino acid sequences. *International Journal of Biological Macromolecules*, **4**, 137–44.

Stillinger, F. H. & Rahman, A. (1974). Improved simulation of liquid water by molecular dynamics. *Journal of Chemical Physics*, **60**, 1545–57.

Stuper, A. J., Brugger, W. E. & Jurs, P. C. (1979). *Computer Assisted Studies of Chemical Structure and Biological Function*. New York: John Wiley & Sons.

Sutton, L. E. (1958). *Tables of Interatomic Distances and Configuration in Molecules and Ions.* London: Chemical Society.

Tainer, J. A., Getzoff, E. D., Alexander, H., Houghton, R. A., Olson, A. J., Lerner, R. A. & Hendrickson, W. A. (1984). The reactivity of anti-peptide antibodies is a function of the atomic mobility of sites in a protein. *Nature,* **312,** 127–34.

Takeda, Y., Ohlendorf, D. H., Anderson, W. F. & Matthews, B. W. (1983). DNA-binding proteins. *Science,* **221,** 1020–6.

Taylor, R., Kennard, O. & Versichel, W. (1983). Geometry of the N—H . . . O=C hydrogen bond. 1, Lone-pair directionality. *Journal of the American Chemical Society,* **105,** 5761–6.

Taylor, R., Kennard, O. & Versichel, W. (1984a). Geometry of the N—H . . . O=C hydrogen bond. 2, Three-centre (bifurcated) and four-centre (trifurcated) bonds. *Journal of the American Chemical Society,* **106,** 244–8.

Taylor, R., Kennard, O. & Versichel, W. (1984b). The geometry of the N—H . . . O=C hydrogen bond. 3, Hydrogen bond distances and angles. *Acta Crystallographica,* **B40,** 280–8.

Teeter, M. M. (1984). Water structure of a hydrophobic protein at atomic resolution: pentagon rings of water molecules in crystals of crambin. *Proceedings of the National Academy of Sciences,* **81,** 6014–18.

Todd, S. & Gillett, J. (1983). Animation in the Winchester graphics system. *Journal of Molecular Graphics,* **1,** 39–41.

Vedani, A. & Dunitz, J. D. (1985). Lone-pair directionality in hydrogen bond potential functions for molecular mechanics computations: the inhibition of human carbonic anhydrase 2 by sulphonamides. *Journal of the American Chemical Society,* **107,** 7653–8.

Veillard, A. (1976). Small molecules and inorganic compounds. In *Quantum Mechanics of Molecular Conformations,* ed. B. Pullman, pp. 1–115. London: John Wiley.

Verloop, A., Hoogenstraaten, W. & Tipker, J. (1976). In *Drug Design,* Vol. 7, ed. E. J. Ariens, p. 165. New York: Academic Press.

Vinogradov, S. N. & Linell, R. H. (1971). *Hydrogen Bonding.* New York: Van Nostrand Reinhold.

Vinter, J. G. (1985a). *CONECT: A Program to Generate Full Connectivity.* Brookhaven: Protein Data Bank.

Vinter, J. G. (1985b). Molecular graphics for the medicinal chemist. *Chemistry in Britain,* **21,** No. 1, 32–8.

Volz, K. W., Matthews, D. A., Alden, R. A., Freer, S. T., Hansch, C., Kaufman, B. T. & Kraut, J. (1982). Crystal structure of avian dihydrofolate reductase containing phenyltriazine and NADPH. *Journal of Biological Chemistry,* **257,** 2528–36.

Wang, A. H. J., Ughetto, G., Quigley, G. J., Hakoshima, T., van der Marel, G. A., Van Boom, J. H. & Rich, A. (1984). The molecular structure of a DNA–triostin A complex. *Science,* **225,** 1115–21.

Waring, M. J. (1981). Inhibitors of nucleic acid synthesis. In *The Molecular Basis of Antibiotic Action,* by Gale, E. F., Cundliffe, E., Reynolds, P. E., Richmond, M. H. & Waring, M. J. London: John Wiley & Sons.

Weinstein, H., Osman, R., Topiol, S. & Green, J. P. (1981). Quantum chemical studies on molecular determinants for drug action. *Annals of the New York Academy of Sciences,* **367,** 434–48.

Weinstein, H., Osman, R., Topiol, S. & Venanzi, C. A. (1983). Molecular determinants for biological mechanisms: model studies of interactions in carboxypeptidase. In

Quantitative Approaches to Drug Design, ed. J. C. Dearden, pp. 81–90. Amsterdam: Elsevier.

Wells, T. N. C. & Fersht, A. R. (1985). Hydrogen bonding in enzymatic catalysis analysed by protein engineering. *Nature*, **316**, 656–7.

Wharton, R. P. & Ptashne, M. (1985). Changing the binding specificity of a repressor by redesigning an α-helix. *Nature*, **316**, 601–5.

Wheatley, P. J. & Jaffrey, P. (1981). *The Determination of Molecular Structure*. London: Constable.

Winter, G., Fersht, A. R., Wilkinson, A. J., Zoller, M. & Smith, M. (1982). Redesigning enzyme structure by site-directed mutagenesis: tyrosyl tRNA synthetase and ATP binding. *Nature*, **299**, 756–8.

Wodak, S. J. & Janin, J. (1980). Analytical approximation to the accessible surface area of proteins. *Proceedings of the National Academy of Sciences*, **77**, 1736–40.

Wold, S., Albano, C., Dunn, W. J., Edlund, U., Esbensen, K., Geladi, P., Hellberg, S., Johansson, E., Lindberg, W. & Sjostrom, M. (1984). Multivariate data analysis in chemistry. In *Chemometrics: Mathematics and Statistics in Chemistry*, ed. B. R. Kowalski, p. 17. Dordrecht: D. Reidel Publishing Company.

Wold, S., Dunn, W. J. & Hellberg, S. (1984). Pattern recognition as a tool for drug design. In *Drug Design: Fact or Fantasy?* ed. J. Jolles & K. R. H. Wooldridge, p. 95–115. London: Academic Press.

Woolfson, M. M. (1978). *An Introduction to X-ray Crystallography*. Cambridge: University Press.

Zingsheim, H. P., Neugebauer, D. C. L., Barrantes, F. J. & Frank, J. (1980). Structural details of membrane-bound acetylcholine receptor from *Torpedo-marmorata*. *Proceedings of the National Academy of Sciences*, **77**, 952–6.

Zoller, M. J. & Smith, M. (1983). Oligonucleotide-directed mutagenesis of DNA fragments cloned into M13 vectors. *Methods in Enzymology*, **100**, 468–500.

SUBJECT INDEX

A-DNA, 200–1, 238
accessible surface, 86, 107–8, 119–21, 145, 148–9, 196, 206, 209, 221, 286, 312, 318–9, 321
ACE, 324–7
acetylcholine, 9, 13, 26–30, 52, 125–6, 150–3, 226, 333, 346
acetylcholinesterase, 52
AchR, 153, 333–4
acridine, 276
actinomycin, 97, 254, 257
ADAPT, 323
adenylate–cyclase, 25–7
agonism, 15
AlaPro, 324
alloxan, 97
9-aminoacridine, 276, 278–9
aminoacylation, 335
p-aminobenzamide, 161
p-aminobenzoyl moiety, 179–80
aminofuranoside, 61
amphiphilicity, 223–6
angiotensin, 324
anion–sodium interaction, 235
antagonism, 2, 9, 27, 289
antibodies, 146
antifolate agent, 153
antigenic properties, 146, 222–3
antimalarial agent, 153
antitoxin, 4
arsenical, 7
artificial intelligence, 292, 338, 344–5
atom-pair potentials, 32, 271
atrial muscle, 22
atropine, 2–3, 9, 152
autoscaling, 302, 304

B-DNA, 200–1, 210, 214, 238–9, 241, 243, 252, 254, 267, 331
ball-and-stick models, 321

base-pair, 91, 331
Bayesian discriminants, 323
BIBSER, 58
bifunctional intercalators, 171
bis-arylhydroxysulphonic acid, 297
bond lengths, 60
Born–Oppenheimer approximation, 87
BPTI (Bovine Pancreatic Trypsin Inhibitor), 182, 220
Bragg condition, 48
branch-and-bound search, 116
Bravais lattice, 43, 45
Brookhaven Protein Data Bank, 58, 64, 116, 225, 348
Brownian motion, 32, 230
bungarotoxin, 333–4
3-butadiene, 342

caesium, 30
captopril, 324–6
carbachol, 28, 30–1
carbonium ion, 164
carboxypeptidase, 283
carboxyphenylethidium, 272
carcinogenicity, 347
Cartesian coordinates, 36–8, 41, 46, 53–7, 60, 96
CASE, 347
catalysis, 160, 337
CCDC, 58–60
centrosymmetric crystals, 50
charge–charge interactions, 286
CHEMGRAF, 322
CHEMMOVIE, 322
chemometrics, 300, 306, 310, 313–4
cholinesterase, 52
Clark's theory, 8, 9, 11–13
clathrate hydrate complexes, 196–8, 219, 230–1
cloning of genes, 31, 150–1, 153, 328–30

CNDO, 318
cocrystallization, 36, 67, 273, 319
CODEN, 59
coenzyme, 164, 177–80, 183, 199, 202, 288
cofactor, 158, 176–7, 202
coliphage, 331
combinatorial analysis, 140, 289–90, 307, 313–4, 319, 340–1
complementarity, 121, 149, 254, 321–2
COMPND, 58
computer-aided design, 321, 324, 327
computer-aided synthesis, 340
conductance across cell membranes, 28
conformations, 31, 107, 112, 125–8, 136, 138–9, 182–4, 195, 199, 236, 243, 286, 325–6
congeneric series, 6, 35, 55, 73, 123, 185–6, 290–1, 301, 306–7, 309, 313
CONN, 58–9
connectivities, 313
CONNSER, 58, 60–1
contact surface, 118–22, 148–9, 164, 169, 183, 194, 198, 228, 257
CORN, 66
counterions in solution, 269
covariance matrix, 302, 304
CpG, 241, 243, 243–4, 247
CPK, 147–8
Crambin, 197
CRAY-1, 351
critical points, 36, 101, 110, 115, 129, 264, 275
crystallography, 47–8, 50, 157, 160, 193
CSD, 58–9, 61

DAP, 351
DATA, 58–60
datacentres, 59
dC-dG, 241, 248
dCpdG, 172, 265, 278
dCpdGpdC, 267
DDM (Difference Distance Matrix), 112–6, 167
decamethonium, 30–1
decapeptide, 324
deoxyhaemoglobin, 293
depsipeptide, 175
descriptors, 291, 306, 308, 323, 345–8
dGpdC, 264–5, 267, 276, 278
DHFR, 153–9, 161, 165, 177, 179, 183–4, 198, 202, 258
dielectric effect, 33, 67, 79, 101–3, 191–2, 227–8
Diels–Alder reaction, 342
difference distance matrix (see DDM)
diffractometers, 49
DIHDRL, 318

dinucleotide, 172–3, 236–7, 241, 243, 251–2, 258, 260, 264, 266, 271, 275–6, 281
dipeptide, 126, 216, 324
diphosphoglycerate, 292–3, 334
diphtheria, 4
dipole moment, 72, 77–8, 160, 225, 229, 246, 271, 274
dipole–charge interactions, 286
dipole–dipole interactions, 76, 78, 81
direction field, 101, 264–6, 269, 273
dispersion energy, 80, 243
DM (Distance Matrix), 167
DNA-polymerase, 330
DOCKER, 286
docking manoeuvre, 91, 115, 140, 148, 165, 169, 236–8, 252, 263, 280, 283, 285, 311
dodecahedral water complexes, 194
dose–response relationship, 8, 14–5, 18, 20, 29–30
DPG, 293–5, 297
drug–dinucleotide interaction, 276
drug–receptor complex, 14, 22, 31, 33, 201, 204, 214, 237, 275, 278, 282
dyad axis, 238
dyestuffs, 3–4

ECEPP, 318
efficacy, 16–9, 21, 24, 35
Ehrlich's work, 1–7, 35
eigenvectors, 62
electric field, 79–80, 101, 226, 229, 243, 245–6, 263–4, 266–8, 271, 275, 280–1
electron density, 49, 50–1, 60, 68–9, 96, 109–10, 149, 322, 345
electrostatic potential, 76, 96–7, 101–3, 149, 251, 253–4, 257–8, 340
ELEMENT, 59
Enaplopril, 324
enthalpy, 186, 308–9
entropy, 72
enzyme–coenzyme interactions, 179
enzyme–substrate interactions, 35, 282
equipotential lines, 264
ESR, 192
ethidium, 171–2, 243, 250, 258, 260–1, 263, 271–4, 276, 278, 281–2
ethidium–dinucleotide interactions, 282
Euler angles, 41, 72, 78

FISIPL, 318
folding of proteins, 141, 148, 158
four-centre hydrogen bonds, 93
Fourier analysis, 48, 50, 226
furanose, 249, 260

Gaussian orbitals, 97
gene sequence, 142, 223, 329, 333

GEOM, 59
GEOMETRY, 35
glyceraldehyde-3-phosphate, 134
GpC, 241, 243–4, 247
GppNHp, 26–7
group-oriented strategy, 343
guanine-nucleotide-binding protein, 25
5'-guanylylimododiphosphate, 26

haeme, 169
haeme-myoglobin, 122
haemoglobin, 70, 121, 292, 297–8, 334
haemolysins, 5
halolactonization, 343
Hamiltonian operator, 87–8
Hammett equation, 310–11
Herman Mauguin groups, 45
hexagons, 196–7
Hill, 8, 30
histamine, 18
HOMO, 281–2, 311
homology in proteins, 150–1, 154, 156
Hoogsteen base pairing, 135, 172, 175,
    243, 246
HSITE, 348
hydrates, 230
hydration, 33, 191–4, 199–201, 205–9,
    210–5, 221, 231–5
hydrogen bond, 84, 86, 91, 93, 95, 143,
    168, 186, 198, 200, 219, 285
hydropathy, 221–3, 225
hydrophilic interactions, 66, 145, 196, 209,
    210, 221–2, 225, 310
hydrophilicity, 188, 221–3
hydrophobic interactions, 103, 108, 161,
    298, 309
hydrophobicity, 120, 140, 188, 222, 225
hydroxylamine, 52
hydroxymethylcyclohexane, 342

IC_{50}, 154
ileum, 18
induction energy, 79
insecticidal activity, 347
interaxial angles, 43, 45–6
intercalation, 171–3, 175, 201–2, 237–9,
    241, 243, 247, 249–51, 254–6, 276–9
intramolecular forces, 335
intrinsic activity, 13–4, 17
ion-channel, 226
ion–dipole interactions, 231–2
iontophoretic applications, 27, 333
islet activating protein, 26
isomorphous replacement, 49
isopotential lines, 99, 228
isoprenaline, 22, 27

ketoxamine, 347
key-hole analogy, 166
kinking of DNA, 240

KLN, 346–7
knowledge engineering, 338

Langmuir adsorption isotherm, 13
Laplacian operator, 88
LDH, 149
lead compound, 288, 319, 321, 340–1
Lennard–Jones potential function, 81–2,
    216
LHASA, 338, 341–2, 344
ligand design, 288, 290–2, 300, 306, 316,
    323
ligand-binding, 334
ligand-point, 183–4, 186, 298, 319–20
linear free enthalpy, 308, 313
lipid–water, 309
lipophilicity, 146, 348
LISP, 338
lock-and-key analogy, 254
log–log plots, 9
lone-pair electrons, 93, 207, 250, 281–2,
    284, 348
long-range interactions, 72–3, 79, 251, 311,
    343
LUMO, 281–2
lysozyme, 145, 192

MACCS, 323
mAchR, 26–7
mass-action law, 22
MBLD, 57
methanamide, 207, 283–5
methanoate, 207, 260
methotrexate, 153–4, 161–2, 176, 202, 258
methylacrylate, 342
metmyoglobin, 169
Metropolis condition, 211–2
Michaelis–Menten constant, 13–4, 186
MM2, 91
mobility of atoms, 145–6, 192–3, 205, 218,
    223
MOLAREA, 119
molecular dynamics, 83, 85, 91, 112, 128,
    141, 147, 190, 193, 216, 218–20, 279
molecular graphics, 147, 285, 321, 348
molecular shape, 109, 147
molecular surface, 119, 121–2, 149, 230,
    253, 255, 318
monoclinic crystals, 46
monodentate, 200
monopole, 77
Monte Carlo calculations, 83, 105, 193,
    211–2, 214, 216, 220, 228, 231
MOVS, 82
mRNA, 152–3, 329, 333
MTD, 313
Mulliken population analysis, 75–6
multi-dimensional space, 87
multipole, 77, 96, 101–2

multivariate analysis, 60, 87, 300–1, 306
muscarinic receptors, 22, 26–7, 150, 152
mustard, 22
mutagenesis, 329, 331–5
mutant DNA, 328–30, 334, 337
myasthenia gravis, 151
myoglobin, 348

N-body interactions, 81, 210
N-electron, 74
nAchR, 30–1
NADPH, 157–9, 163–4, 176–8, 202
neurotoxins, 151
neutron diffraction, 51
Newton–Raphson optimization, 90–1
nicotinamide, 160, 162–4, 177–8, 180,
    183–4, 202
nicotine, 7–8
nonadecapeptide, 223
nucleophile, 52
nucleophilic interactions, 96, 99, 285

octanol, 309
octapeptide, 324
octupoles, 77
oil/water partition coefficient, 30·1
oligonucleotide, 329–30
oocytes, 152, 328, 333–4
oxyhaemoglobin, 293

p-space, 304
pair-potential studies, 71–3, 80, 82–3, 108
PASCAL, 338
patch-clamp method, 28
pattern recognition, 291, 301
PCILO, 124, 137, 260, 280
penicillin, 58, 288
pentagonal water, 194, 198, 202, 205
pentapeptide, 254, 256
permittivity, 76, 230–1
phage, 330
pharmacokinetic parameters, 104
pharmacophore, 70, 87
phenanthridine, 276
phenanthridinium ring, 172, 250, 260–1
phenoxazone ring, 254
phi/psi plots, 318
phonon waves, 238–9
pilocarpin, 3
pinacol, 195–6, 205
piperazine, 194
pitch rotations, 260–2
PLUTO, 59
polarizability, 79
polyhedral arrangements, 196
polynucleotide, 137–8
polypeptide, 69, 106, 111–2, 116, 120–1,
    123, 129–31, 133, 139, 150–1, 220, 334

polystyrene, 68
polytope simplex, 314
porphyrins, 239
pralidoxime, 52, 55
prealbumin, 122
predictors for structure, 223
prepeptide, 151
principal component analysis, 303–5
prochirality, 160
proflavine, 202, 204–5, 276, 278
propeller twist, 136
propranolol, 27
n-propylamine, 196
propylbenzilylcholine, 22
protein structure, 67–8, 112, 134, 224
protein–DNA interactions, 332
pteridine ring, 153, 160, 162–4, 176, 179,
    202
pucker of sugar rings, 136–7, 175, 204,
    238, 241
purine bases, 134, 136, 171, 201, 237, 278
pyrimidine bases, 134, 171, 179, 201

QSAR, 96, 101, 104, 108, 123, 165, 184–5,
    287, 291, 300, 312, 323, 340, 345–6, 348
quadrupoles, 77
quantum-mechanical methods, 60
quinoxaline antibiotics, 172–3

radio-labelled ligands, 150
receptor–effector interactions, 22
receptor-induced ligand orientation, 258
receptors, 2–3, 5, 7, 11, 13–21, 25, 27,
    31–2, 33, 35, 152–3, 215
rectus abdominus muscle, 8–9, 13, 160
reentrant surface, 120, 148
REFCODE, 59
repressor–operator interactions, 331–2
repulsion energy, 80–2, 168, 241, 266, 273
retrosynthetic analysis, 341, 343
RNA, 134–5, 152, 254
roll rotation, 260, 262–3
Ro31–2201, 327
Ro31–2848, 327

salvarsan, 7
saxitoxin, 117, 307
Schiff base, 294
Schild's plot, 12
Schrödinger equation, 109
search-tree, 343
self-assembly systems, 331
self-complementary DNA, 204–5
separatrix, 110
short-range interactions, 73, 80–1, 126, 237
single-channel recordings, 28
site-directed mutagenesis, 328–35

site-points, 159, 161, 165, 176–7, 180, 184–5, 187, 204, 294, 298, 300, 312, 319–20, 322, 324, 340, 348
sodium ion, 152
software, 316, 321, 323, 345
space-filling models, 59
spacers for drug design, 201
Stephenson's efficacy, 17, 31
sterimol, 312
stimulus-chain, 22
STRUCTURAL, 35
sub-goal precursors, 342, 344
supercomputers, 350
supermolecule, 206–8
superpositioning, 326

tandem, 80
tertiary structure of proteins, 65, 116, 123, 139–42, 151, 223
tetrapeptides, 140
tetrodotoxin, 111–2
TGC, 333
THF, 153
thionicotinamide, 177
three-centre hydrogen bonds, 91
through-water interactions, 235
thymidylate synthetase, 154
thyroxine, 120, 169
TMA, 110
topology, 119, 133, 156, 224, 318
torsion angles, 37–8, 42, 60–2, 65, 69, 84, 86, 88–9, 91, 111, 123–31, 137–8, 143, 172, 175, 216, 237–9, 241, 243, 276, 278, 318–25
toxin, 4–5, 333
toxophile, 4–5
toxophore, 4, 6
trajectories for protein conformation, 236
transducer, 22, 24–5, 27
transfer-RNA, 329
transition state, 283, 285, 337

triclinic crystals, 46, 52
trimethoprim, 153–4, 178, 183, 184
trimethylammonium, 17
triostin, 136, 172–5, 239
tripeptide model, 126–7
tRNA, 329, 335
tubocurarine, 28
two-centre hydrogen bonds, 91
Tyr-tRNA, 335, 337

U-turns in protein, 127
unstacking of DNA, 241, 243, 247–9, 276, 278
unwinding of DNA, 171–2, 175, 237, 241, 251

van der Waals contact, 159, 161, 183, 258, 319
van der Waals interaction, 175, 239, 286, 298, 322, 337
van der Waals radius, 91, 103, 166, 312
van der Waals surface, 118–9, 147–8
Venn diagram, 184
voltage–current relationship, 30

water–ion interaction, 231
water–peptide interaction, 216
water–solute interaction, 108, 211
water–water interaction, 105, 211, 214, 216
Watson–Crick base pairing, 172, 175, 204, 243, 246
Weber–Fechner law, 20
Wiswesser line notation, 346

x-ray, 49–51, 58, 67–8, 83, 106, 130, 138–9, 144, 157, 193, 197, 201, 231, 239, 292, 332

yaw rotation, 258, 260–3

Z-DNA, 139, 199

AUTHOR INDEX

Abola, 318
Albano (see Wold)
Alden (see Volz)
Alexander (see Tainer)
Allen, 59, 320
Allinger, 91
Altona, 63
Alvey, 304
Amidon, 108
Amos, 79
Anderson, J. E., 331
Anderson, S., 323
Anderson, S. G. (see Bader)
Anderson, W. F. (see Takeda)
Ariens, 13, 14, 18
Arik, 108
Arnone, 293
Arnott (see Leslie)
Artymiuk, 145
Arunlakshana, 13
Asai (see Noda)

Bader, 110
Baker, 184
Ballivet (see Claudio)
Banerjee (see Sobell)
Barnard, 152
Barnes, 215
Barrantes (see Zingsheim)
Barrett, 110
Barry, 68
Beddell, 292, 294, 298
Beeman, 85
Bellard (see Allen)
Ben-Naim, 108
Benschop, 52
Berg, 52
Berman, 204, 214 (see Mezei)
Bernstein, F. C., 64, 318
Bernstein, R. B., 259

Berthod, 231, 235
Bevan (see Birdsall)
Beveridge, 231
Bhandary (see Sobell)
Bibby, 87, 304
Birdsall, 177, 179, 182, 184 (see
    Gronenborn, see Hyde)
Birnstock, 264
Black, 22, 25, 26
Blake (see Artymiuk, see Blaney)
Bland, 59
Blaney, 120 (see Kuntz)
Blow, 335 (see Fersht)
Blumberg, 191 (see Geiger)
Blundell, 47, 286
Bolin, 154, 161, 162, 184, 198, 202
Bolis, 81
Bonaccorsi, 99
Bonchev, 291
Boom (see Wang)
Breitwieser, 26, 27
Brice (see Allen, see Bernstein, F. C.)
Brick, 335 (see Fersht)
Brickenkamp, 196
Brickman, 148
Browman, 91, 318
Brown, 297
Brugger, 301
Buchanan, 339
Buckingham, 80
Burgen, 32, 177 (see Gronenborn, see
    Hyde)
Burkert, 91
Burridge (see Blaney)
Busetta, 286
Bush, 121

Careri (see Rupley)
Carhart, 346
Caron, 245

Carozzo, 81
Carruthers (see Browman)
Carter (see Fersht)
Cartwright (see Allen)
Cavalonne (see Clementi, E., see
  Scordamaglia)
Champness (see Baker)
Chandrasekaran (see Leslie)
Changeux, 30
Chiriac, 312
Chothia, 120, 133, 134, 221
Chou, 140, 223
Ciubotaru, 312
Clark, 8, 9, 11, 12, 13, 15, 20
Claudio, 151
Claverie, 124, 125, 245
Clementi, E., 81, 208, 210, 212, 214
Clementi, S., 301
Clore (see Birdsall)
Cohen, 140, 141
Colquhoun, 28 (see Gardner)
Conner (see Dickerson)
Connolly, 119, 149, 319 (see Langridge, see
  Blaney)
Cook, 74
Corongiu, 81, 210, 212, 214
Courriere, 125
Cramer (see Krugh)
Crandell, 346
Crippen, 185, 187, 346

Dabrow (see Shieh)
Daiker, 97
Dale, 1
Danziger, 116, 307, 348
Dashevsky, 105
Daudey (see Claverie)
Davies, 112
Dean, 97, 98, 101, 116, 247, 252, 254, 258,
  260, 264, 271, 280, 307, 321, 322
Deb, 77
Deisenhofer, 220
Del Bene, 95
Deming, 316
Dickerson, 136, 139, 199, 200
Dickinson, 3
Diner (see Claverie)
Dionne, 28
Doolittle, 221
Doubleday (see Allen)
Dove, 351, 352
Drew, 136, 199 (see Dickerson)
Duke (see Bader)
Dunitz, 95
Dunn, 291, 301
Dwork (see Kao)

Edlund (see Wold)
Edmonds, 102

Ehrlich, 1, 2, 3, 4, 5, 6, 7, 35
Einstein, 32
Eisenberg, 225, 226
Emmerson, 21, 27
Esbensen (see Wold)

Fasman, 140, 223
Feeney (see Birdsall, see Gronenborn, see
  Hyde)
Fero (see Ragazzi)
Ferrin (see Blaney, see Kuntz, see
  Langridge)
Fersht, 335, 336, 337
Filman, 154, 162, 178, 184, 202 (see Bolin)
Finer-Moore, 226
Finney, 191, 215 (see Mezei)
Frank (see Zingsheim)
Franke, 108, 301, 309, 310
Fratini (see Dickerson)
Free, 307
Freer (see Volz)
Frischleder, 264
Frohlich, 230
Fujita (see Mishina)
Fukuda (see Mishina)
Fukui, 311
Furchgott, 19, 20, 21
Furutani (see Noda)

Gaddum, 11, 12, 13
Galway, 304
Gardner, 28
Geiger, 190, 191, 230
Geise, 52
Geladi (see Wold)
Getzoff (see Tainer)
Giessner-Prettre (see Claverie)
Gilbert (see Claverie)
Gillett, 114
Gilman, 25
Ginell (see Parthasarathy)
Golender, 291, 345
Goodfellow, 215 (see Mezei)
Goodford, 292, 297, 298
Gordon, 57
Grace (see Artymiuk)
Grant, 228, 230
Gratton (see Rupley)
Green (see Weinstein)
Greer, 121
Gresh, 125
Gronenborn, 182 (see Birdsall)
Groot, 14
Guru Row (see Parthasarathy)

Hakoshima (see Wang)
Hamlin (see Bolin)
Hansch, 104 (see Volz)
Harrison, 331

Hassall, 324
Havere, 52
Hawkes, 28
Hayashida (see Noda)
Hayes, 97
Hayes-Roth, 339
Heinemann (see Claudio)
Hellberg, 291, 301
Hendrickson (see Tainer)
Henry, 45
Hermann, 108, 119
Higgs (see Allen)
Hill, 7, 8, 29, 30
Hille (see Pfaffinger)
Hindin, 349
Hirose (see Mishina, see Noda)
Hol, 160
Holban (see Simon)
Hoogenstraaten, 312
Hopp, 221, 222
Houghton (see Tainer)
Hummelink (see Allen)
Hummelink-Peters (see Allen)
Hunter (see Pfaffinger)
Hyde, 177 (see Gronenborn)

Ikeda (see Noda)
Imoto (see Mishina)
Inayama (see Mishina, see Noda)

Jaffrey, 47
Jain (see Sobell)
Janin, 119
Jeffrey, 94, 195
Johansson (see Wold)
Johari, 230
Johnson, A. P., 341
Johnson, L. N., 47
Jolles, 287
Joncas, 341, 342
Jorgensen (see Blaney)
Jurs, 301

Kabasch, 318
Kaldany (see Kao)
Kanaoka (see Noda)
Kangawa (see Noda)
Kao, 333
Karlin (see Kao)
Karplus, 216, 219, 220
Kashuba (see Browman)
Katz, 27
Kaufman (see Volz)
Kauzmann, 103
Kayano (see Noda)
Kennard, 51, 92, 320, 348 (see Allen, see
    Bernstein, F. C.)
Kent, 87, 304
Kikyotani (see Noda)

Klopman, 345, 346
Knill-Jones (see Fersht)
Koetzle (see Bernstein, F. C.)
Kollmann, 97, 241, 247, 276, 278
Komaromy, 226
Komatsu, 257, 321
Koopman, 18
Kopka (see Dickerson)
Kowalski, 301
Kraut (see Bolin, see Filman, see Volz)
Krohn (see Hassall)
Krugh, 171
Kuno (see Mishina)
Kuntz, 121, 122, 165, 166, 169 (see
    Langridge)
Kurusaki (see Mishina)
Kypr, 122
Kyte, 221

Lackey, 349
Lane, 304
Langlet, 245 (see Claverie)
Langley, 2, 3, 4, 7, 35
Langridge, 148 (see Blaney, see Kuntz)
Lavery, 101, 253, 264, 268, 270
Lee, B., 119, 120
Lee, J. S., 172
Leff, 22, 25, 26
Lenat, 339
Lenstra, 52
Leo, 310
Lerner (see Tainer)
Leslie, 201
Levine, 259
Levitt, 84, 86, 87, 133, 142, 145, 220, 239,
    240
Levitzki, 27, 29, 30, 31
Lewi, 301
Lim, 140
Lindberg (see Wold)
Linell, 95
Long, 341, 342
Lonsdale, 45
Lowe (see Fersht)
Lozansky (see Sobell)

Mackay, 20, 21, 27
Mager, 300
Maitland, 81
Malrieu (see Claverie)
Maluszynska, 94
Mangold, 322
Mardia, 87, 304
Marel (see Wang)
Marsh, 241, 247, 276, 278
Martin (see Pfaffinger)
Marx, 151
Matsuo (see Noda)
Matthews, B. W., 134 (see Takeda)

Matthews, D. A. (see Bolin, see Filman, see Volz)
Mausbach (see Geiger)
McGonigal, 346
McLachlan, 116, 117, 227
Metropolis, 211, 212
Meyer (see Bernstein, F. C.)
Mezei, 214, 231
Mihalas (see Simon)
Miledi, 27, 152
Minamino (see Noda)
Mishina, 333
Miyata (see Noda)
Mokhoff, 349
Monamy (see Browman)
Moody (see Hassall)
Morffew, 114, 128
Morgenroth, 5
Morimoto (see Mishina)
Morrison, 304
Motherwell, 59, 61, 62 (see Allen)
Mulliken, 75, 76
Murray-Rust, 59, 61, 62

Nakagawa, 321 (see Komatsu)
Nakamura, 321 (see Komatsu)
Nakayama (see Noda)
Namasivayam, 321
Nathanson (see Pfaffinger)
Naylor (see Quarendon)
Neidle, 204, 214 (see Mezei, see Shieh)
Neugebauer (see Zingsheim)
Nguyen-Dang (see Bader)
Nickerson, 18
Nishikawa, 112
Noda, 151, 223, 224
Norrington (see Baker)
North, 68
Numa (see Mishina, see Noda)
Nuss, 241, 247, 276, 278

Oatley (see Artymiuk, see Blaney, see Kuntz)
Ogden, 28
Ohlendorf (see Takeda)
Olson, 135 (see Tainer)
Ooi, 112
Osman, 283, 285

Paddon, 351
Padlan, 112
Panke, 196
Papadimitriou, 314
Parthasarathy, 201
Pascaul (see Birdsall)
Paton, 6, 31, 32
Patrick (see Claudio)
Pawley, 351, 352
Pedersen, 94
Peinel, 264

Perahia, 127
Perutz, 335
Petrongolo, 81
Pfaffinger, 26
Phillips (see Artymiuk)
Pincelli (see Claverie)
Pjura (see Dickerson)
Podelski, 30
Politzer, 97
Pople, 57, 95
Port, 207
Pottle (see Browman)
Ptashne, 331
Pullman, A., 101, 125, 207, 231, 235, 253, 264, 268, 270 (see Bonaccorsi)
Pullman, B., 101, 125, 126, 127, 207, 231, 235, 253, 264, 268, 270 (see Claverie)

Quarendon, 149
Quigley (see Wang)

Rafferty (see Noda)
Ragazzi, 81
Rahman, 190, 216
Ratliff (see Leslie)
Raventos, 12, 13
Reddy (see Sobell)
Rekker, 310
Remington, 134
Rich (see Wang)
Richards, F. M., 119, 120, 140
Richards, W. G., 88, 322 (see Quarendon)
Richardson, 134
Richmond, 140
Rigby, 81
Robbinns, 195
Roberts, 177, 184 (see Gronenborn, see Hyde)
Rogers, J. R. (see Allen, see Bernstein, F. C.)
Rogers, N. K., 102
Rollett, 46
Rosen (see Browman)
Rosenblit, 291, 345
Rosenbluth, A. W. (see Metropolis)
Rosenbluth, M. M. (see Metropolis)
Rossky, 216, 219, 220
Rossum, 18
Rubenstein, 341, 342
Rumsey (see Browman)
Rupley, 191, 193

Sakore (see Sobell)
Sander, 318
Sandorfy, 95
Santary, 122
Sarkisov, 105
Schild, 12, 13
Schnither (see Geiger)

Schuster, 95
Schwarz (see Eisenberg)
Schwarzenbach, 194
Scordamaglia, 81
Scrocco (see Bonaccorsi)
Seuchi (see Noda)
Sheppard, 228, 230
Shi (see Fersht)
Shieh, 204, 214
Shimanouchi (see Bernstein, F. C.)
Shimiza (see Noda)
Shortliffe, 339
Silver (see Kao)
Simon, 312, 313
Simonis, 14
Sjostrom (see Wold)
Smith, D. H., 346
Smith, D. R. (see Baker)
Smith, E. B., 81
Smith, M., 329
Sobell, 238, 240
South, 228, 230
Sperelakis, 27
Srikrishnan (see Parthasarathy)
Stammers, 298
Stanley, 190, 191 (see Geiger)
Steiglitz, 314
Stein (see Kao)
Steinbach, 28
Steiner, 75, 88
Stephenson, 15, 17, 19, 31
Sternberg, 102, 116, 141 (see Artymiuk)
Stevens, 28
Stillinger, 190
Stroud, 226
Stuper, 301
Sumikawa, 152
Sundaralingham, 63
Sutton, 57
Szabo, 26, 27

Tainer, 145, 146, 223
Takahashi (see Mishina, see Noda)
Takai (see Noda)
Takashima (see Noda)
Takeda, 121
Tal (see Bader)
Tanabe (see Noda)
Tanaka (see Mishina)
Tasumi (see Bernstein, F. C.)
Taylor, 51, 92, 348
Teeter, 197
Teller, A. H. (see Metropolis)
Teller, E. (see Metropolis)
Terwilliger, 226
Thomas, G. W., 351
Thomas, W. A. (see Hassall)
Tickle, 286
Tipker, 312

Tobimatsu (see Mishina)
Todd, 114, 128
Tomasi (see Bonaccorsi)
Topiol, 283, 285
Toyosato (see Noda)

Ughetto (see Wang)
Umeyama, 321 (see Komatsu)

Vedani, 95
Veillard, 125
Venanzi, 285
Venkataraghavan, 346
Veres, 349
Verloop, 312
Versichel, 51, 92, 348
Vinogradov, 95
Vinter, 316, 318
Volvani (see Amidon)
Volz, 154

Wakeham, 81
Wakelin, 97, 98, 247, 253, 254, 258, 260, 280
Wall, 226
Wang, 136, 172, 175
Waring, 171, 172
Waterman, 339
Watson (see Allen)
Waye (see Fersht)
Weininger, 310
Weinstein, 101, 283, 285
Weiss, 225
Wells, 337
Wharton, 331
Wheatley, 47
Wideman (see Kao)
Wierenga, 160
Wilcox (see Eisenberg)
Wilkinson, A. J. (see Fersht, see Winter)
Wilkinson, S. (see Beddell)
Williams (see Bernstein, F. C.)
Wilson, 307
Wing (see Dickerson)
Winter, 329 (see Fersht)
Wittlin (see Krugh)
Wodak, 119
Wold, 291, 301
Woods, 221, 222
Wooldridge, 287
Woolfson, 47
Wootton, 298

Yalkowsky, 108

Ziegler, 349
Zingsheim, 151
Zoller, 329
Zundel, 95